Lecture Notes in Mathematics

Volume 2311

This series reports on new developments in all areas of mathematics and their applications - quickly, informally and at a high level. Mathematical texts analysing new developments in modelling and numerical simulation are welcome. The type of material considered for publication includes:

1. Research monographs 2. Lectures on a new field or presentations of a new angle in a classical field 3. Summer schools and intensive courses on topics of current research.

Texts which are out of print but still in demand may also be considered if they fall within these categories. The timeliness of a manuscript is sometimes more important than its form, which may be preliminary or tentative.

Titles from this series are indexed by Scopus, Web of Science, Mathematical Reviews, and zbMATH.

Nima Moshayedi

Kontsevich's Deformation Quantization and Quantum Field Theory

 Springer

Nima Moshayedi
Department of Mathematics
University of California, Berkeley
CA, USA

Institut für Mathematik
Universität Zürich
Zürich, Switzerland

This work was supported by Schweizerischer Nationalfonds zur Förderung der Wissenschaftlichen Forschung (200020 192080).
Schweizerischer Nationalfonds zur Förderung der Wissenschaftlichen Forschung (P2ZHP2_199401).
National Centres of Competence in Research SwissMAP.

ISSN 0075-8434   ISSN 1617-9692   (electronic)
Lecture Notes in Mathematics
ISBN 978-3-031-05121-0   ISBN 978-3-031-05122-7   (eBook)
https://doi.org/10.1007/978-3-031-05122-7

Mathematics Subject Classification: 53D55, 81T20, 57R56, 81T45, 81T70, 81Q30, 81T18, 53D05, 53D20, 53D17, 58A05, 53Z05, 81Q60

This Springer imprint is published by the registered company Springer Nature Switzerland AG
The registered company address is: Gewerbestrasse 11, 6330 Cham, Switzerland

Preface

Deformation quantization originated in the field of theoretical physics, mainly from the ideas of Dirac and Weyl, in order to understand the mathematical structure when passing from a commutative classical algebra of observables to a non-commutative quantum algebra of observables. The rigorous formulation of this physical concept does actually lead to interesting mathematical notions and insights. In order to describe such a quantization procedure, one considers the theory of infinitesimally deforming algebras by using formal power series and methods of non-commutative geometry. It turned out that the mathematical concepts needed there are a combination of classical analytical methods of operator algebras together with modern methods of homotopy theory and algebraic and differential geometry. The connection to quantum field theory leads to the study of important new field theories which have been developed in this context and turn out to be also useful in order to get insights in other fields of mathematics such as, for example, in the field of symplectic or Poisson geometry and integrable systems.

This book began as lecture notes for the course "Poisson geometry and deformation quantization" given by me during the fall semester 2020 at the University of Zurich. The aim is mainly to get the reader familiar with the concepts of deformation quantization as in the construction of Kontsevich and understand the relation to quantum field theory as in the construction of Cattaneo and Felder. The main difference with other textbooks on deformation quantization is that we also explain intensively the relation of the algebraic methods with the field-theoretic ones, where we develop the deformation quantization as a semi-classical limit of the expectation value for a certain observable with respect to a special sigma model. Moreover, we try to provide a reference where most of the prerequisites, in order to understand the previously mentioned construction, are actually covered. The structure of the book does in fact try to allow the reader to already delve into the first chapter after a basic course in analysis, linear algebra, and topology.

The book is divided into five chapters. The first chapter is fairly standard and is devoted to the study of the main notions of differential geometry, in particular, manifolds, vector fields, differential forms, integration of manifolds, Stokes' theorem, de Rham cohomology and de Rham's theorem, Lie groups, Lie algebras, connections, real Hodge theory, and basics of category theory. The second chapter introduces the concept of symplectic structures, Lagrangian submanifolds, Weinstein's theorems, classical mechanics, moment maps, symplectic reduction, Kähler manifolds, and complex Hodge theory. The third chapter builds directly upon the second chapter by introducing Poisson structures, its relation to symplectic structures, their local behavior, Poisson cohomology and their interpretation, symplectic groupoids, and integrability conditions, Dirac manifolds and Morita equivalence for Poisson manifolds. These first three chapters are mainly covered to understand the last two chapters. In chapter four, we start with deformation quantization by introducing the concept of star products. We first look at important examples in the local symplectic setting and then move on to the global setting of Fedosov. Afterwards, we start to discuss Kontsevich's celebrated formality theorem and study the proof of it by using the notion of strong homotopy Lie algebras (also called L_∞-algebras) introduced by Stasheff. We are then ready to introduce Kontsevich's star product as a special case of the formality theorem for bivector fields and bidifferential operators. This is going to use the concept of graphs and configuration spaces. At the end of the chapter, we briefly discuss the approach of operads to deformation quantization (formality) constructed by Tamarkin. The fifth chapter is devoted to the understanding of Kontsevich's construction from the point of view of quantum field theory. We start by considering the functorial approach to quantum field theory provided by Atiyah and Segal and then move to the more important perturbative definition in terms of Feynman path integrals. At first, we will show how one can derive the Moyal product through such a path integral by using the field theory formulation of quantum mechanics. Afterwards, we consider the gauge formalisms of Faddeev–Popov and BRST and discuss special cases of the Poisson sigma model in these formalisms. Then we move on to the general Poisson sigma model and discuss several properties, such as the phase space reduction in its Hamiltonian formulation. In order to deal with the general form of the Poisson sigma model, we will introduce the gauge formalism of Batalin–Vilkovisky and show its relation to the Faddeev–Popov and BRST construction. Finally, we state the Cattaneo–Felder theorem and sketch the proof by using the aforementioned Batalin–Vilkovisky formalism. At the very end, we briefly consider the notion of AKSZ sigma models for which the Poisson sigma model is a special case.

This book mainly serves as a textbook for graduate students in mathematics or theoretical physics who want to learn about deformation quantization and its relation to quantum field theory. Each chapter contains examples and exercises for the reader in order to strengthen their understanding of the topic and the (sometimes quite abstract) objects. We want to mention that the first three chapters do not replace a full textbook on the corresponding subject and are rather meant to provide the reader with the necessary (plus some more) notions for the fourth and fifth chapters. Similarly, the fifth chapter should not replace a full textbook on quantum field theory.

Berkeley, CA, USA   Nima Moshayedi
Zurich, Switzerland

Acknowledgments

At first, I would like to thank my advisor A. S. Cattaneo for comments on a first version of these notes and his constant support during my time in Zurich and beyond. I am incredibly thankful to the two referees for providing valuable comments which helped me a lot to improve a first version of the book. I also want to thank U. McCrory for helping me with all the correspondences and her support. I want to thank my parents and Manuela for always cheering me up on times I did not feel very motivated. Many thanks also to N. Reshetikhin for his great support and mentoring in research as well as in everyday life. Finally, I want to thank the University of Zurich for giving me the opportunity to teach such a course and the University of California, Berkeley for providing a wonderful atmosphere and work environment where a substantial part of this book was written.

This research was supported by the NCCR SwissMAP, funded by the Swiss National Science Foundation, and by the SNF grants No. 200020_192080 and No. P2ZHP2_199401.

Contents

Chapter 1
Introduction

The concepts of symplectic and Poisson geometry appear naturally in theoretical physics in the context of the dynamics of a classical mechanical system. Mathematically, it lies in the intersection of differential and non-commutative geometry. Theoretical physicists are often interested in objects that they can measure which are called *observables*. Mathematically, observables can be described as elements of the algebra of smooth functions on the Euclidean space $M = \mathbb{R}^3$ (or more general, some *manifold M*) endowed with an additional algebraic structure rather than just point-wise multiplication. In fact, the algebra of smooth functions on M, denoted by $C^\infty(M)$, with some bilinear map $\{ \ , \ \}\colon C^\infty(M) \times C^\infty(M) \to C^\infty(M)$, satisfying certain properties. Such a bracket $\{ \ , \ \}$ is called a *Poisson bracket*. The dynamics of an observable $\mathcal{O} \in C^\infty(M)$, i.e. its time evolution, is directly encoded in the geometric structure induced by the Poisson bracket and some special function $H \in C^\infty(M)$, called the *Hamiltonian* of the classical system. It particular, it is described by the differential equation

$$\frac{\mathrm{d}\mathcal{O}}{\mathrm{d}t} = \{H, \mathcal{O}\}.$$

Classical states of some particle moving in 3-dimensional Euclidean space are described uniquely by its *position* and corresponding *momentum* coordinates in the configuration space given by $\mathbb{R}^6$. This configuration space is usually called the *phase space* in the physics literature. In general, if we consider the Euclidean space $\mathbb{R}^n$ instead of $\mathbb{R}^3$, we get the phase space $\mathbb{R}^{2n}$, i.e. to n position coordinates $q_1, \ldots, q_n$ we have associated n momentum coordinates $p_1, \ldots, p_n$. Since the phase space is always even-dimensional, it induces a geometric structure called *symplectic structure*. One can actually show that the geometric data of a symplectic structure on the phase space induces the algebraic data given by a Poisson bracket on the algebra of observables described by smooth functions. If we go one more step towards generality and consider a manifold M instead of Euclidean space $\mathbb{R}^n$, we can describe the phase space by its *cotangent bundle T^*M*. The algebraic data given

© The Author(s), under exclusive license to Springer Nature Switzerland AG 2022
N. Moshayedi, *Kontsevich's Deformation Quantization and Quantum Field Theory*, Lecture Notes in Mathematics 2311, https://doi.org/10.1007/978-3-031-05122-7_1

by the Poisson bracket can be directly transferred into geometric data by expressing
the Poisson bracket in terms of a *bivector field* on M. The geometric data given by
the symplectic structure is expressed as a *differential form* of degree 2 on M and
both data are directly related. However, mathematically it is possible to consider
a manifold which is endowed with a Poisson structure that is not induced from
a symplectic structure, which is of course interesting to study on its own. Let us
now see what are the corresponding geometric and algebraic structures in *quantum
mechanics*.

In theoretical physics, a *quantum system* is described by a complex Hilbert space
$\mathcal{H}$ together with an operator $\widehat{H} \colon \mathcal{H} \to \mathcal{H}$, called the *Hamiltonian*. The Hamiltonian
takes the place of the Hamiltonian function H in the classical setting. A *quantum
state* of such a system is represented by an element in $\mathcal{H}$, whereas the *quantum
observables* are given by *self-adjoint operators* on $\mathcal{H}$. The dynamics of the quantum
observables, i.e. its time evolution, is measured through the algebraic structure given
by the commutator $[\ ,\ ]$ of operators and the distinguished Hamiltonian $\widehat{H}$. In
particular, it is given by the differential equation

$$\frac{\mathrm{d}\widehat{\mathcal{O}}}{\mathrm{d}t} = \frac{\mathrm{i}}{\hbar}\left[\widehat{H}, \widehat{\mathcal{O}}\right].$$

This is usually called the *Heisenberg equation*. Here $\mathrm{i} := \sqrt{-1}$ denotes the
imaginary unit. The constant $\hbar$ is called the *reduced Planck constant* which is a
small physical constant naturally appearing at the quantum level. Numerically, it
is given by $\hbar := 1.054 \times 10^{-34}\mathrm{m}^2\mathrm{kg/s}$, but it should serve as a parameter for
the underlying mathematical theory. The connection of this setting to the setting
of classical mechanics is usually given by the introduction of position $\widehat{q}_i$ and
momentum $\widehat{p}_j$ operators on $\mathcal{H}$ which satisfy the following commutation relation:

$$\left[\widehat{p}_i, \widehat{q}_j\right] = \frac{\mathrm{i}}{\hbar}\delta_{ij}.$$

In this way, one can obtain classical mechanics from the quantum theory in the limit
where $\hbar \to 0$. This is a general concept that has to be satisfied whenever a quantum
theory, depending on the parameter $\hbar$, is constructed. If one has a quantum theory
for which the limit $\hbar \to 0$ does not produce the classical theory, it is physically not
correct.

An important question is whether there is a precise mathematical formulation
of such a *quantization procedure* in terms of a well-defined map between classical objects and their quantum counterparts. There are different mathematical
approaches to quantization of a given classical system. The method of *geometric
quantization* (see e.g. [Kir85, Woo97, BW12]) focuses on the geometric structure
on the phase space. There, the idea is to quantize the classical phase space $\mathbb{R}^{2n}$ to
the corresponding Hilbert space $\mathcal{H} = L^2(\mathbb{R}^n)$ on which a dynamical equation for
quantum states, the *Schrödinger equation*, is defined. The classical counterpart to
this equation is given by the *Hamilton–Jacobi equation. Deformation quantization*

is another approach which focuses on the classical algebra of observables. There, one wants to capture the non-commutativity structure of the space of operators out of the commutative structure given by point-wise multiplication on $C^\infty(\mathbb{R}^{2n})$. An important well-known result of *Groenewold* [Gro46] states that it is actually impossible to quantize the Poisson algebra $C^\infty(\mathbb{R}^{2n})$ in a way where the Poisson bracket of two functions is sent onto the commutator bracket of the corresponding operators on $L^2(\mathbb{R}^n)$. To overcome this issue, one can instead consider a small *deformation* of the point-wise product on $C^\infty(\mathbb{R}^{2n})$ to an associative but non-commutative product, usually called a *star product*.

Deformation quantization originated from the work of *Weyl* [Wey31], who gave an explicit formula for the operator $\mathcal{O}_f$ on $L^2(\mathbb{R}^n)$ associated to a function $f \in C^\infty(\mathbb{R}^{2n})$:

$$\mathcal{O}_f := \int_{\mathbb{R}^{2n}} \check{f}(\xi, \eta) \exp\left(\frac{\mathrm{i}}{\hbar}(P\xi + Q\eta)\right) \mathrm{d}^n \xi \mathrm{d}^n \eta,$$

where $\check{f}$ denotes the *inverse Fourier transform* of f. Here, $P = (P_i)$ and $Q = (Q_j)$ denote operators satisfying the commutation relation $[P_i, Q_j] = \frac{\mathrm{i}}{\hbar}\delta_{ij}$. Moreover, the integral is considered in the *weak* sense. An inverse map was later found by *Wigner* [Wig32], who gave a way to recover the corresponding classical observable by taking the *symbol* of the operator $\mathcal{O}_f$. It was then *Moyal* [Moy49] who interpreted the symbol of the commutator of two operators $\mathcal{O}_f$ and $\mathcal{O}_g$ corresponding to the functions f and g as what is today called the *Moyal bracket*:

$$\mathcal{M}(f, g) := \frac{\sinh(\varepsilon P)}{\varepsilon}(f, g) = \sum_{k=0}^{\infty} \frac{\varepsilon^{2k}}{(2k+1)!} P^{2k+1}(f, g),$$

where $\varepsilon := \frac{\mathrm{i}\hbar}{2}$ and P^k is the k-th power of the Poisson bracket on $C^\infty(\mathbb{R}^{2n})$. Already Groenewold had a similar formula for the symbol of the product $\mathcal{O}_f\mathcal{O}_g$ which today can be interpreted as the first appearance of the Moyal star product $\star$. The Moyal bracket can then be rewritten in terms of this star product as

$$\mathcal{M}(f, g) = \frac{1}{2\varepsilon}(f \star g - g \star f).$$

It was *Flato* who recognized that this star product is a deformation of the commutative point-wise product on $C^\infty(\mathbb{R}^{2n})$. This is marked as the beginning of the mathematical program of *deformation quantization*. He conjectured the problem of giving a general recipe to deform the point-wise product on $C^\infty(M)$ in such a way that $\frac{1}{2\varepsilon}(f \star g - g \star f)$ still remains a deformation of the Poisson structure on M. Following this conjecture, a first way of formulating quantum mechanics as such a deformation of classical mechanics had been discovered. The work of *Bayen, Flato, Fronsdal, Lichnerowicz* and *Sternheimer* was essential for the formulation of the deformation problem for symplectic spaces and the physical applications

[Bay+78a, Bay+78b]. Later, *de Wilde* and *Lecomte* [DL83] have proven the existence of a star product on a generic symplectic manifold by using *Darboux's theorem* which tells that locally any symplectic manifold of dimension $2n$ can be identified with $\mathbb{R}^{2n}$ by a choice of *Darboux charts*. Using cohomological arguments, one can construct such a star product by a correct gluing of the locally defined Moyal product. Independently of the previous result, *Fedosov* [Fed94, Fed96] gave an explicit construction of a star product on a generic symplectic manifold. The generalization to the Poisson manifold $\mathbb{R}^d$ endowed with a general Poisson structure was given through the *formality theorem* of *Kontsevich* [Kon03]. He derived an explicit formula for the product on $\mathbb{R}^d$ by using special graphs and configuration space integrals. The globalization procedure, in order to obtain Kontsevich's star product on any Poisson manifold, was described by *Cattaneo, Felder* and *Tomassini* [CFT02b, CFT02a] by extending Fedosov's construction to the Poisson case where they used notions of *formal geometry* developed by *Gelfand–Fuks* [GF69, GF70], *Gelfand–Kazhdan* [GK71] and *Bott* [Bot10]. This completed the program of Flato proposed thirty years before. Another approach to Kontsevich's formality result was given by *Tamarkin* [Tam98, Tam03] who used the notion of *operads* in order to prove the formality theorem. This has opened more insights towards deformation quantization in a higher setting which turned out to be quite useful when consider the concept of stratified manifolds.

However, from the physical point of view, it was not really clear how Kontsevich's construction can be interpreted. *Cattaneo* and *Felder* [CF00] have shown that Kontsevich's formula can actually be formulated in terms of a perturbative *quantum field theory*, i.e. a perturbative expression of the functional integral of a certain 2-dimensional topological field theory. This theory is called the *Poisson sigma model* which was discovered independently by *Ikeda* [Ike94] and *Schaller–Strobl* [SS94, SS95] by an attempt to combine 2-dimensional gravity with Yang–Mills theories. In fact, it can be regraded as a 2-dimensional bosonic string theory. Using this field theory, Cattaneo and Felder have shown that the graphs, which have been constructed by Kontsevich, arise naturally in the perturbative expansion as the *Feynman diagrams* appearing as the coefficients in the formal power series in $\hbar$, subject to the expectation of certain observables placed at the boundary of the 2-dimensional disk. Since the Poisson sigma model is a theory with *symmetry*, i.e. a *gauge theory*, one has to understand how the perturbative expansion can be done rigorously. This leads to several gauge formalisms such as the *Faddeev–Popov ghost method* [FP67] or the *BRST method* [BRS74, BRS75, BRS76, Tyu76]. Unfortunately, it turns out that for the general Poisson sigma model these formalisms are not enough and thus we need to pass to more sophisticated methods. The gauge formalism which is needed is called the *Batalin–Vilkovisky formalism* [BV77, BV81, BV83]. In particular, the Poisson sigma model turns out to be an example of a more general type of field theories that can be treated nicely in this gauge formalism. These theories are called *AKSZ theories* after *Alexandrov, Kontsevich, Schwarz* and *Zaboronsky* [Ale+97].

Another approach, using methods of *derived algebraic geometry*, regarding a higher version is encoded in the setting of *shifted symplectic and Poisson structures*.

A way of dealing with these concepts and constructing a higher shifted deformation quantization was more recently formulated by *Calaque, Pantev, Toën, Vaquié* and *Vezzosi* [Pan+13, Cal+17]. This point of view leads to insights towards *extended topological field theories* in the sense of *Baez–Dolan* [BD95] and *Lurie* [Lur09], and also a shifted version of the AKSZ construction.

Chapter 2
Foundations of Differential Geometry

This chapter introduces the main concepts and notions of differential geometry which are needed in order to be able to understand the discussions in the following chapters. The modern methods of differential geometry started with the work of Gauss and Riemann through the development of curved objects and the notion of a manifold in the nineteenth century. As it turned out in the beginning of the twentieth century, the methods of differential geometry where extremely useful to understand various aspects of the physical world, especially for Einstein's theory of general relativity or modern aspects of Maxwell's theory of electrodynamics as a classical field theory. These methods have gained more and more importance when theoretical physicists where studying quantum phenomena, in particular, the quantum theory of fields on curved space-times, starting in the middle of the twentieth century. We will start this chapter by defining the most basic notion in differential geometry, that of a *manifold*, and discuss various properties such as their local behaviour, the concept of vector fields and differential forms. One of the main notions will be integration on manifolds through densities and differential forms. A key result is given by *Stokes' theorem* which gives a statement regarding the situation when the manifold has non-vanishing *boundary*. We will also discuss the notion of singular homology and de Rham cohomology, (real) Hodge Theory, Lie groups and Lie algebras and (affine) connections. At the end of this chapter we will introduce an important notion of a subject called *category theory* which is not in particular part of differential geometry but will be important for further discussions. We also point out the excellent references [War83, KN63, KN69, Tu17, BT82, Cat18, Lee02, DK00] for extended and more detailed discussions of the field of differential geometry.

© The Author(s), under exclusive license to Springer Nature Switzerland AG 2022
N. Moshayedi, *Kontsevich's Deformation Quantization and Quantum Field Theory*,
Lecture Notes in Mathematics 2311, https://doi.org/10.1007/978-3-031-05122-7_2

2.1 Smooth Manifolds

2.1.1 Charts and Atlases

Definition 2.1.1 (Chart) A *chart* on a set M is a pair (U, ϕ) where $U \subset M$ is a subset and ϕ is an injective map $U \to \mathbb{R}^n$ for some n.

We call ϕ a *chart map* or a *coordinate map*. Sometime we also refer only to ϕ as a chart since U is already contained inside the definition of ϕ. Let (U, ϕ_U) and (V, ϕ_V) be charts on M. Then we can compose the bijections $(\phi_U)|_{U \cap V} : U \cap V \to \phi_U(U \cap V)$ and $(\phi_V)|_{U \cap V} : U \cap V \to \phi_V(U \cap V)$ to the bijection

$$\phi_{U,V} : (\phi_V)|_{U \cap V} \circ ((\phi_U)|_{U \cap V})^{-1} : \phi_U(U \cap V) \to \phi_V(U \cap V).$$

We call this the *transition map* between the charts (U, ϕ_U) and (V, ϕ_V). Moreover, we refer to n as being the *dimension* of M (we will give another definition for the dimension later on). See Fig. 2.1 for a visualization.

Definition 2.1.2 (Atlas) An *atlas* on a set M is a collection of charts $\{(U_\alpha, \phi_\alpha)\}_{\alpha \in I}$, where I is an index set such that $\bigcup_{\alpha \in I} U_\alpha = M$.

Remark 2.1.3 We will denote the transition maps between two charts (U_α, ϕ_α) and (U_β, ϕ_β) simply by $\phi_{\alpha\beta}$.

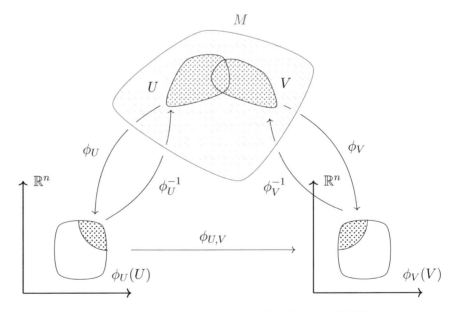

Fig. 2.1 Example of charts and transition map on an n-dimensional manifold M

If $\phi_\alpha(U_\alpha)$ is open for all $\alpha \in I$, then the atlas $\mathcal{A} = \{(U_\alpha, \phi_\alpha)\}_{\alpha \in I}$ induces a topology on M. This topology is given by

$$\mathcal{O}_\mathcal{A}(M) := \{V \subset M \mid \phi_\alpha(V \cap U_\alpha) \text{ is open } \forall \alpha \in I\}.$$

Definition 2.1.4 (Open Atlas) An atlas is said to be *open* if $\phi_\alpha(U_\alpha \cap U_\beta)$ is open for all $\alpha, \beta \in I$.

Definition 2.1.5 (Differentiable Atlas) An atlas is said to be *differentiable* if it is open and all transition functions are C^k-maps for $k = 0, 1, \ldots, \infty$.

Definition 2.1.6 (Smooth Atlas) An atlas is said to be *smooth* if it is open and all transition functions are C^∞-maps.

Definition 2.1.7 (C^k-equivalence) Two C^k-atlases on the same set are said to be C^k-*equivalent* if their union is a C^k-atlas for $k = 0, 1, \ldots, \infty$.

Definition 2.1.8 (C^k-manifold) A C^k-*manifold* is an equivalence class of C^k-atlases for $k = 0, 1, \ldots, \infty$.

Definition 2.1.9 (Smooth Manifold) A *smooth manifold* is an equivalence class of C^∞-atlases.

Remark 2.1.10 A manifold is usually called *topological*, if it is an equivalence class of C^0-atlases and it its usually called *differentiable*, if it is an equivalence class of C^k-atlases for $k = 1, 2, \ldots, \infty$.

Exercise 2.1.11 Let $\mathbb{R}$ be endowed with the atlas $\{(\mathbb{R}, \mathrm{id}), (\mathbb{R}, x \mapsto \mathrm{sgn}(x)\sqrt{x})\}$ where $\mathrm{sgn}(x)$ is the sign function $x \mapsto \begin{cases} 1, & x \geq 0 \\ -1, & x < 0 \end{cases}$. Show that $\mathbb{R}$ endowed with this atlas is a C^0-manifold but not a C^k-manifold for $k \geq 1$.

Example 2.1.12 (n-Sphere) Define

$$S^n := \{q \in \mathbb{R}^{n+1} \mid \|q\| = 1\}, \quad n \geq 1.$$

We claim that this is an n-dimensional C^0-manifold. We can construct a smooth atlas $\{(U_1, \phi_1), (U_2, \phi_2)\}$ by using the *stereographic projection*. Define $U_1 := S^n \setminus \{(0, \ldots, 0, 1)\}$ and $U_2 := S^n \setminus \{(0, \ldots, 0, -1)\}$. We call the point $N = (0, \ldots, 0, 1)$ the *north pole* and the point $S = (0, \ldots, 0, -1)$ the *south pole*. It is easy to see that $U_1 \cup U_2 = S^n$. Moreover, we define the map $\phi_1 : U_1 \to \mathbb{R}^n$ by

$$\phi_1(q_1, \ldots, q_{n+1}) = \left(\frac{q_1}{1 - q_{n+1}}, \ldots, \frac{q_n}{1 - q_{n+1}} \right).$$

This map is called *stereographic projection*. Its inverse $\phi^{-1}\colon \mathbb{R}^n \to U_1$ is given by

$$\phi^{-1}(p_1,\ldots,p_n) = \left(\frac{2p_1}{\sum_{i=1}^n p_i^2 + 1}, \ldots, \frac{2p_n}{\sum_{i=1}^n p_i^2 + 1}, 1 - \frac{2}{\sum_{i=1}^n p_i^2 + 1}\right).$$

It is not hard to see that ϕ and ϕ^{-1} are both continuous and hence ϕ^{-1} is a homeomorphism. For the second coordinate chart (U_2, ϕ_2) we consider the stereographic projection from the south pole which is given by $\phi_2 := -\phi_1 \circ \alpha$ where $\alpha\colon S^n \to S^n$ is the antipode map defined as $q \mapsto -q$. Also α is a homeomorphism and hence the map $\phi_2\colon U_2 \to \mathbb{R}^n$ is a homeomorphism. Finally, we have

$$\phi_2 \circ \phi_1^{-1}(p_1,\ldots,p_n) = \frac{1}{\sum_{i=1}^n p_i^2}(p_1,\ldots,p_n).$$

This shows that S^n is indeed an n-dimensional C^0-manifold.

Exercise 2.1.13 Show that S^n is a smooth n-manifold for all $n \geq 1$.

Definition 2.1.14 (Diffeomorphism) We will call a map of manifolds a *diffeomorphism*, if it is invertible with smooth inverse. If there exists a diffeomorphism between two manifolds M and N, we say that they are *diffeomorphic* and write $M \cong N$.

Exercise 2.1.15 (Torus) Define the set

$$T^2 := \left\{(x, y, z) \in \mathbb{R}^3 \;\middle|\; \left(\sqrt{y^2 + z^2} - b\right)^2 + z^2 = a\right\}$$

with $a, b \in \mathbb{R}_{>0}$ fixed. Show that T^2 defines a smooth 2-dimensional manifold. Moreover, show that it is diffeomorphic to the product of two circles, i.e. $T^2 \cong S^1 \times S^1$. *Hint: Use the parametrization*

$$\phi\colon \mathbb{R}^2 \to \mathbb{R}^3,$$

$$(\theta, \varphi) \mapsto \begin{pmatrix} \cos(2\pi\theta) & 0 & -\sin(2\pi\theta) \\ 0 & 1 & 0 \\ \sin(2\pi\theta) & 0 & \cos(2\pi\theta) \end{pmatrix} \begin{pmatrix} 0 \\ \sqrt{a}\cos(2\pi\varphi) \\ b + \sqrt{a}\sin(2\pi\varphi) \end{pmatrix}.$$

and show that it is smooth, $(\mathbb{Z} \times \mathbb{Z})$-periodic, thus it induces a map $(\mathbb{R}/\mathbb{Z})^2 \to \mathbb{R}^3$, and that it gives the desired diffeomorphism (See Fig. 2.2).

Exercise 2.1.16 (Real Projective Space) Define

$$\mathbb{R}P^n := (\mathbb{R}^{n+1} \setminus \{0\})/\sim,$$

Fig. 2.2 Two-dimensional torus

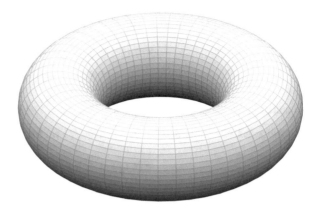

where $x \sim tx$ for all $t \in \mathbb{R} \setminus \{0\}$ with $x \in \mathbb{R}^{n+1}$, to be the *n-dimensional real projective space*. Show that $\mathbb{R}P^n$ is a smooth manifold and that $\mathbb{R}P^1$, the *real projective line*, is diffeomorphic to S^1.

Remark 2.1.17 From now on, if we call a set M a manifold, we will always mean a smooth manifold when not stated otherwise. Moreover, all maps between manifolds will be regarded as smooth maps.

Definition 2.1.18 (Submanifold) Let N be an n-dimensional manifold. A k-dimensional *submanifold*, with $k \leq n$, is a subset M of N such that there is an atlas $\{(U_\alpha, \phi_\alpha)\}_{\alpha \in I}$ of N with the property that for all α with $U_\alpha \cap M \neq \emptyset$ we have $\phi_\alpha(U_\alpha \cap M) = W_\alpha \times \{x\}$ with $W_\alpha \subset \mathbb{R}^k$ an open subset and $x \in \mathbb{R}^{n-k}$.

Remark 2.1.19 Any chart with this property is called an *adapted chart* and an atlas consisting of adapted charts is called an *adapted atlas*. Moreover, by a diffeomorphism of $\mathbb{R}^n$ we can always assume that $x = 0$.

Example 2.1.20 (Graphs) Let F be a smooth map from an open subset $V \subset \mathbb{R}^k$ to $\mathbb{R}^{n-k}$ and consider its graph

$$M = \{(x, y) \in V \times \mathbb{R}^{n-k} \mid y = F(x)\}.$$

Then M is a submanifold of $N := V \times \mathbb{R}^{n-k}$. As an adapted atlas we may take the one consisting of a single chart (N, ι), where $\iota \colon N \hookrightarrow \mathbb{R}^n$ denotes the inclusion map.

2.1.2 Pullback and Push-forward

Let M and N be two manifolds and consider a map $F \colon M \to N$.

Definition 2.1.21 (Pullback) The $\mathbb{R}$-linear map

$$F^*: C^\infty(N) \to C^\infty(M),$$

$$f \mapsto f \circ F.$$

is called *pullback* by F.

Exercise 2.1.22 Show that for $f, g \in C^\infty(N)$ we have

$$F^*(fg) = F^*(f)F^*(g).$$

Moreover, if $G: N \to Z$ is a map between manifolds N and Z, show that

$$(G \circ F)^* = F^* \circ G^*.$$

Definition 2.1.23 (Push-forward) Using F as before with the additional condition that it is a diffeomorphism, we define the *push-forward* to be the inverse of the pullback F^* which we denote by F_*. In fact, we get

$$F_*: C^\infty(M) \to C^\infty(N),$$

$$f \mapsto f \circ F^{-1}.$$

Exercise 2.1.24 Show that

$$F_*(fg) = F_*(f)F_*(g),$$

and

$$(G \circ F)_* = G_* \circ F_*.$$

2.1.3   Tangent Space

Let M be a smooth manifold.

Definition 2.1.25 (Coordinatized Tangent Vector) A *coordinatized tangent vector* at $q \in M$ is a triple (U, ϕ_U, v) where (U, ϕ_U) is a chart with $U \ni q$ and v is an element of $\mathbb{R}^n$.

We say that two coordinatized tangent vectors (U, ϕ_U, v) and (V, ϕ_V, w) are equivalent if

$$w = \mathrm{d}_{\phi_U(q)}\phi_{U,V}\, v.$$

Fig. 2.3 The tangent space
of the sphere S^2 at a point
$q \in S^2$

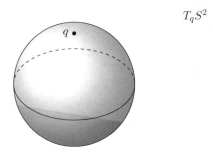

$T_q S^2$

Definition 2.1.26 (Tangent Vector) A *tangent vector* at $q \in M$ is an equivalence class of coordinatized tangent vectors at q.

Definition 2.1.27 (Tangent Space) The *tangent space* of M at $q \in M$ is given by the set of all tangent vectors at q.

Note that each chart (U, ϕ_U) at $q \in M$ defines a bijection of sets

$$\Phi_{q,U} \colon T_q M \to \mathbb{R}^n,$$

$$[(U, \phi_U, v)] \mapsto v.$$

We will also just write Φ_U when the base point $q \in M$ is understood. This bijection allows us to transfer the vector space structure of $\mathbb{R}^n$ to $T_q M$ and gives $\Phi_{q,U}$ the structure of a linear isomorphism (Fig. 2.3).

Lemma 2.1.28 $T_q M$ *has a canonical vector space structure for which* $\Phi_{q,U}$ *is a linear isomorphism for every chart* (U, ϕ_U) *containing* q.

Proof Given a chart (U, ϕ_U), the bijection Φ_U defines the linear structure

$$\lambda \cdot_U [(U, \phi_U, v)] = [(U, \phi_U, \lambda v)], \qquad \forall \lambda \in \mathbb{R},$$

$$[(U, \phi_U, v)] +_U [(U, \phi_U, v')] = [(U, \phi_U, v + v')], \qquad \forall v, v' \in \mathbb{R}^n$$

If (V, ϕ_V) is another chart, we have

$$\begin{aligned}
\lambda \cdot_U [(U, \phi_U, v)] &= [(U, \phi_U, \lambda v)] \\
&= [(V, \phi_V, d_{\phi_U(q)} \phi_{U,V} \lambda v)] \\
&= [(V, \phi_V, \lambda d_{\phi_U(q)} \phi_{U,V} v)] \\
&= \lambda \cdot_V [(V, \phi_V, d_{\phi_U(q)} \phi_{U,V} v)] \\
&= \lambda \cdot_V [(U, \phi_U, v)],
\end{aligned}$$

Fig. 2.4 Hyperboloid

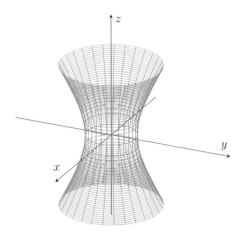

and so $\cdot_U = \cdot_V$. Similarly, we have

$$
\begin{aligned}
[(U, \phi_U, v)] +_U [(U, \phi_U, v')] &= [(U, \phi_U, v + v')] \\
&= [(V, \phi_V, d_{\phi_U(q)} \phi_{U,V}(v + v'))] \\
&= [(V, \phi_V, d_{\phi_U(q)} \phi_{U,V} v + d_{\phi_U(q)} \phi_{U,V} v')] \\
&= [(V, \phi_V, d_{\phi_U(q)} \phi_{U,V} v)] +_V [(V, \phi_V, d_{\phi_U(q)} \phi_{U,V} v')] \\
&= [(U, \phi_U, v)] +_V [(U, \phi_U, v')],
\end{aligned}
$$

and so $+_U = +_V$. $\square$

Example 2.1.29 (Tangent Space of Hyperboloid) Let $f : \mathbb{R}^3 \to \mathbb{R}$ be given by $f(x, y, z) = x^2 + y^2 - z^2 - a$ for $a \in \mathbb{R}$. Then we get $df = (2x, 2y, -2z)$. This is isomorphic to a basis vector of $T_{f(x)}\mathbb{R} \cong \mathbb{R}$ whenever any of x, y, z are non-zero. This has to hold for $f(x, y, z) = 0$ and thus $H := f^{-1}(0)$ is indeed a manifold, the *hyperboloid* (see Fig. 2.4). Note that since $df_x^{-1}(0) = T_x f^{-1}(0)$, we can compute the tangent space $T_x H$ by considering all vectors v with $df_x(v) = 0$. Note that on $f^{-1}(0)$, we get that $z = \sqrt{x^2 + y^2 - a}$, thus

$$
df_x(v) = \begin{pmatrix} 2x \\ 2y \\ -2\sqrt{x^2 + y^2 - a} \end{pmatrix} \begin{pmatrix} v_1 \\ v_2 \\ v_3 \end{pmatrix} = -2\left(xv_1 + yv_2 - \sqrt{x^2 + y^2 - a}\, v_3\right).
$$

Clearly, this expression vanishes if $v_1 = \frac{1}{x}\left(\sqrt{x^2 + y^2 - a}\, v_3 - yv_2\right)$, so the tangent space is indeed a 2-dimensional space which depends on x, y, z and (except for

$x = 0$) is given by

$$T_{(x,y,z)}H = \left\{(v_1, v_2, v_3) \in \mathbb{R}^3 \,\middle|\, (v_1, v_3) \in \mathbb{R}^2, \; v_1 = \frac{1}{x}\left(\sqrt{x^2 + y^2 - a}\,v_3 - yv_2\right)\right\}.$$

Note that this is only defined for $x^2 + y^2 \geq a$. It is easy to see that at the point $(\sqrt{a}, 0, 0)$ we get

$$T_{(\sqrt{a},0,0)}H = \{(v_1, v_2, v_3) \in \mathbb{R}^3 \mid (v_2, v_3) \in \mathbb{R}^2, \; v_1 = 0\} \cong \mathbb{R}^2.$$

Definition 2.1.30 (Dimension of a Manifold) We define the *dimension* of a manifold M by

$$\dim M := \dim T_q M, \quad q \in M.$$

Let $F: M \to N$ be a differentiable map between two manifolds M and N. Given a chart (U, ϕ_U) of M containing $q \in M$ and a chart (V, ψ_V) of N containing $F(q) \in N$, we have a linear map

$$d_q^{U,V} F := \Phi_{F(q),V}^{-1} d_{\phi_U(q)} F_{U,V} \Phi_{q,U} : T_q M \to T_{F(q)} N.$$

Lemma 2.1.31 (Differential/Tangent Map) *The map* $d_q^{U,V} F$ *does not depend on the choice of charts, so we get a canonically defined linear map*

$$d_q F : T_q M \to T_{F(q)} N,$$

called the differential *(or* tangent map*) of F at $q \in M$.*

Proof Let $(U', \phi_{U'})$ be another chart of M containing $q \in M$ and $(V', \psi_{V'})$ another chart of N containing $F(q) \in N$. Then

$$
\begin{aligned}
d_q^{U,V} F[(U, \phi_U, v)] &= [(V, \psi_V, d_{\phi_U(q)} F_{U,V} v)] \\
&= [(V', \psi_{V'}, d_{\psi_{V'}(F(q))} \psi_{V,V'} d_{\phi_U(q)} F_{U,V} v)] \\
&= [(V', \psi_{V'}, d_{\phi_{U'}(q)} F_{U',V'} (d_{\phi_U(q)} \phi_{U,U'})^{-1} v)] \\
&= d_q^{U',V'} F[(U', \phi_{U'}, (d_{\phi_U(q)} \phi_{U,U'})^{-1} v)] \\
&= d_q^{U',V'} F[(U, \phi_U, v)],
\end{aligned}
$$

so we get $d_q^{U,V} F = d_q^{U',V'} F$. $\square$

We immediately get the following lemma.

Lemma 2.1.32 *Let $F: M \to N$ and $G: N \to Z$ be maps between manifolds M, N, Z. Then*

$$\mathrm{d}_q(G \circ F) = \mathrm{d}_{F(q)}G \circ \mathrm{d}_q F, \quad \forall q \in M.$$

Example 2.1.33 (Tangent Space of n-Sphere) Consider the n-sphere $S^n := \{q \in \mathbb{R}^{n+1} \mid \|q\| = 1\}$ Note that 0 is a regular point for the function $f(q) = \|q\|^2 - 1$. Moreover, note that $\mathrm{d}_q f(v) = 2\langle q, v \rangle =: q^\perp$. Thus, we can write

$$T_q S^n = \ker \mathrm{d}_q f = \ker q^\perp.$$

2.2 Vector Fields and Differential 1-Forms

2.2.1 Tangent Bundle

We can glue all the tangent spaces of a manifold M together and obtain the following definition.

Definition 2.2.1 (Tangent Bundle) The *tangent bundle* of a manifold M is given by

$$TM := \bigsqcup_{q \in M} T_q M.$$

An element of TM is of the form (q, v), where $q \in M$ and $v \in T_q M$. Consider the surjective map $\pi : TM \to M$, $(q, v) \mapsto q$. Then the *fiber* $T_q M$ can be denoted by $\pi^{-1}(q)$.

Proposition 2.2.2 *There exists an atlas which gives TM the structure of a manifold.*

Proof Let $\{(U_\alpha, \phi_\alpha)\}_{\alpha \in I}$ be an atlas in the equivalence class defining M. We set $\hat{U}_\alpha := \pi^{-1}(U_\alpha)$ and

$$\begin{aligned} \hat{\phi}_\alpha : \hat{U}_\alpha &\to \mathbb{R}^n \times \mathbb{R}^n, \\ (q, v) &\mapsto (\phi_\alpha(q), \Phi_{q, U_\alpha} v). \end{aligned} \tag{2.1}$$

Note that the chart maps are linear in the fibers. The transition maps are then given by

$$\hat{\phi}_{\alpha\beta}(x, w) = (\phi_{\alpha\beta}(x), \mathrm{d}_x \phi_{\alpha\beta} w).$$

We can then define the tangent bundle of M to be the equivalence class of the atlas $\{(\hat{U}_\alpha, \hat{\phi}_\alpha)\}_{\alpha \in I}$.   □

Remark 2.2.3 Note that $\dim TM = 2 \dim M$.

2.2.2   Vector Bundles

Definition 2.2.4 (Vector Bundle) A *vector bundle* of rank r over a manifold M of dimension n is a manifold E together with a surjection $\pi : E \to M$ such that

(1) $E_q := \pi^{-1}(q)$ is an r-dimensional vector space for all $q \in M$.
(2) E possesses an atlas of the form $\{(\tilde{U}_\alpha, \tilde{\phi}_\alpha)\}_{\alpha \in I}$ with $\tilde{U}_\alpha = \pi^{-1}(U_\alpha)$ for an atlas $\{(U_\alpha, \phi_\alpha)\}_{\alpha \in I}$ of M and

$$\tilde{\phi}_\alpha : \tilde{U}_\alpha \to \mathbb{R}^n \times \mathbb{R}^r$$
$$(q, v \in E_q) \mapsto (\phi_\alpha(q), A_\alpha(q)v), \tag{2.2}$$

where $A_\alpha(q)$ is a linear isomorphism for all $q \in U_\alpha$.

Remark 2.2.5 We usually call E the *total space* and M the *base space*. Moreover, we usually call $E_q := \pi^{-1}(q)$ the *fiber* at $q \in M$.

The maps

$$A_{\alpha\beta} : U_\alpha \cap U_\beta \to \mathrm{End}(\mathbb{R}^r),$$
$$q \mapsto A_{\alpha\beta}(q) := A_\beta(q)A_\alpha(q)^{-1} : \mathbb{R}^r \to \mathbb{R}^r. \tag{2.3}$$

are smooth for all $\alpha, \beta \in I$. The transition maps

$$\tilde{\phi}_{\alpha\beta}(x, u) = (\phi_{\alpha\beta}(x), A_{\alpha\beta}(\phi_\alpha^{-1}(x))u)$$

are linear in the second factor $\mathbb{R}^r$.

Example 2.2.6 (Tangent Bundle) The *tangent bundle* TM of a manifold M is an example of a vector bundle.

Example 2.2.7 (Trivial Bundle) Let M be a manifold. Then we can define a vector bundle of rank n with total space $E = M \times \mathbb{R}^n$ and base space M. Note that $\pi : E \to M$ is the projection onto the first factor. We call this a *trivial bundle* over M.

Example 2.2.8 (Möbius Band) We can view the Möbius band E as a vector bundle of rank 1 over the circle S^1 (see Fig. 2.5). Each fiber at a point $x \in S^1$ is given in the form $U \times \mathbb{R}$, where $U \subset S^1$ is an open arc on the circle including x. Note that the total space E is not given by the trivial bundle $S^1 \times \mathbb{R}$ which would be the cylinder.

Fig. 2.5 The Möbius band

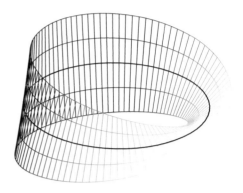

Definition 2.2.9 (Line Bundle) We call a vector bundle a *line bundle* if it has rank 1.

Remark 2.2.10 Example 2.2.8 is an example of a line bundle.

Definition 2.2.11 (Morphism of Vector Bundles) Let (E_1, M_1, π_1) and (E_2, M_2, π_2) be two vector bundles. A *morphism* between (E_1, M_1, π_1) and (E_2, M_2, π_2) is given by a pair of continuous maps $f: E_1 \to E_2$ and $g: M_1 \to M_2$ such that $g \circ \pi_1 = \pi_2 \circ f$ and for all $q \in M_1$, the map $\pi_1^{-1}(q) \to \pi_2^{-1}(g(q))$ induced by f is a homomorphism of vector spaces.

Definition 2.2.12 (Section) Let $\pi: E \to M$ be a vector bundle. A *section* on an open subset $U \subset M$ is a continuous map $\sigma: U \to E$ such that $\pi \circ \sigma = \mathrm{id}_U$.

Remark 2.2.13 We denote the space of sections of a vector bundle $E \to M$ by $\Gamma(E)$.

Exercise 2.2.14 For any two arbitrary vector bundles E, F over M, one can take the *tensor product bundle* $E \otimes F$ of E and F. We will consider just the tensor product of line bundles. Recall that a vector bundle is completely specified by an atlas $\{(U_\alpha, \phi_\alpha)\}_{\alpha \in I}$, an integer r (the rank of the vector bundle) and transition functions $A_{\alpha\beta}: U_{\alpha\beta} \to \mathrm{End}(\mathbb{R}^r)$ for all $\alpha, \beta \in I$ satisfying the conditions

$$A_{\alpha\beta}(q)A_{\beta\alpha}(q) = \mathrm{id}_{\mathbb{R}^r}, \qquad\qquad \forall q \in U_{\alpha\beta}, \qquad (2.4)$$

$$A_{\beta\gamma}(q)A_{\alpha\beta}(q) = A_{\alpha\gamma}(q) \qquad\qquad \forall q \in U_{\alpha\beta\gamma}. \qquad (2.5)$$

For a line bundle, $r = 1$. Suppose we have two line bundles E and F with the same trivializing atlas (this can always be achieved by passing to suitable refinements of trivialising atlases, and simplifies the discussion considerably) $\{U_\alpha, \phi_\alpha\}_{\alpha \in I}$ and transition functions $A_{\alpha\beta}: U_\alpha \cap U_\beta \to \mathrm{End}(\mathbb{R}) \cong \mathbb{R}$ and $B_{\alpha\beta}: U_{\alpha\beta} \to \mathrm{End}(\mathbb{R}) \cong \mathbb{R}$ respectively. We now define the tensor product $E \otimes F$ by transition functions

$$C_{\alpha\beta}: U_{\alpha\beta} \to \mathbb{R},$$

$$q \mapsto A_{\alpha\beta}(q) \cdot B_{\alpha\beta}(q).$$

(1) Show that this defines a line bundle by checking conditions (2.4) and (2.5) explicitly.
(2) Show that for any line bundle L on M, the tensor product $L^* \otimes L$ is isomorphic to the trivial bundle $M \times \mathbb{R}$.
(3) Show that isomorphism classes of line bundles over a manifold M form a group. This is called the *Picard group* of M.

2.2.3   Vector Fields

Definition 2.2.15 (Vector Field) Let M be a manifold. A section $X \in \Gamma(TM)$ of the tangent bundle TM is called a *vector field*.

In an atlas $\{(U_\alpha, \phi_\alpha)\}_{\alpha \in I}$ of M and the corresponding atlas $\{(\tilde{U}_\alpha, \tilde{\phi}_\alpha)\}_{\alpha \in I}$ of TM, a vector field X is represented by a collection of smooth maps $X_\alpha \colon \phi_\alpha(U_\alpha) \to \mathbb{R}^n$. All these maps are related by

$$X_\beta(\phi_{\alpha\beta}(x)) = d_x \phi_{\alpha\beta} X_\alpha(x), \quad \forall \alpha, \beta \in I, \forall x \in \phi_\alpha(U_\alpha \cap U_\beta).$$

Remark 2.2.16 The vector at $q \in M$ defined by the vector field X is usually denoted by X_q as well as by $X(q)$. The latter notation is often avoided as one may apply a vector field to a function f, and in this case the standard notation is $X(f)$. One also uses X_α to denote the representation of X in the chart with index α, but this should not create confusion with the notation X_q for X at the point q.

We can add and multiply vector fields. Let $X, Y \in \Gamma(TM)$ be two vector fields on M, $\lambda \in \mathbb{R}$, and $f \in C^\infty(M)$. Then

$$(X + Y)_q := X_q + Y_q, \quad \forall q \in M, \tag{2.6}$$

$$(\lambda X)_q := \lambda X_q, \quad \forall q \in M, \tag{2.7}$$

$$(fX)_q := f(q) X_q, \quad \forall q \in M. \tag{2.8}$$

Remark 2.2.17 We denote the space of vector fields on a manifold M by $\mathfrak{X}(M) := \Gamma(TM)$.

Let $F \colon M \to N$ be a diffeomorphism between two manifolds. If X is a vector field on M, then $d_q F X_q$ is a vector in $T_{F(q)} N$ for each $q \in M$. If F is a diffeomorphism, we can perform this construction for each $y \in N$, by setting $q = F^{-1}(y)$, and define a vector field, denoted by $F_* X$ on N:

$$(F_* X)_{F(q)} := d_q F X_q, \quad \forall q \in M,$$

or, equivalently,

$$(F_*X)_y = d_{F^{-1}(y)}FX_{F^{-1}(y)}, \quad \forall y \in N.$$

Definition 2.2.18 (Push-forward of a Vector Field) For a map of manifolds $F: M \to N$, the $\mathbb{R}$-linear map $F_*: \mathfrak{X}(M) \to \mathfrak{X}(N)$ is called the *push-forward* of vector fields.

Remark 2.2.19 Note that if $G: N \to Z$ is another diffeomorphism between manifolds, we immediately have

$$(G \circ F)_* = G_*F_*.$$

Moreover, we have $(F_*)^{-1} = (F^{-1})_*$.

If $U \subset \mathbb{R}^n$ is an open subset, X a smooth vector field and f a smooth map, then we can define

$$X(f) = \sum_{i=1}^{n} X^i \frac{\partial f}{\partial x^i},$$

where on the right-hand-side we regard X as a map $U \to \mathbb{R}^n$. Each X^i is given by a map in $C^\infty(M)$. Note that the map $C^\infty(M) \to C^\infty(M)$, $f \mapsto X(f)$, is $\mathbb{R}$-linear and satisfies the Leibniz rule

$$X(fg) = X(f)g + fX(g), \quad \forall f, g \in C^\infty(M).$$

This means, choosing local coordinates (x^i) on M, we can write a vector field as

$$X = \sum_{i=1}^{n} X^i \frac{\partial}{\partial x^i} \in \mathfrak{X}(M).$$

Definition 2.2.20 (Commutator of Vector Fields) One can define an anti-symmetric $\mathbb{R}$-bilinear map on $\mathfrak{X}(M)$ by

$$[\ , \ ]: \mathfrak{X}(M) \times \mathfrak{X}(M) \to \mathfrak{X}(M),$$
$$(X, Y) \mapsto [X, Y] := XY - YX, \tag{2.9}$$

called the *commutator* of vector fields.

Exercise 2.2.21 (Jacobi Identity) Show that for all $X, Y, Z \in \mathfrak{X}(M)$

$$[X, [Y, Z]] + [Y, [Z, X]] + [Z, [X, Y]] = 0.$$

Exercise 2.2.22 (Derivation) Show that $[\ ,\ ]$ is a *derivation*, i.e. for $X, Y \in \mathfrak{X}(M)$ and $f \in C^\infty(M)$ we have

$$[X, fY] = X(f)Y + f[X, Y].$$

Definition 2.2.23 (Immersion) A smooth map $F: M \to N$ between two manifolds is called an *immersion* if for all $q \in M$ the differential $d_q F: T_q M \to T_{F(q)} N$ is injective.

Definition 2.2.24 (Submersion) A smooth map $F: M \to N$ between two manifolds is called a *submersion* if for all $q \in M$ the differential $d_q F: T_q M \to T_{F(q)} N$ is surjective.

Definition 2.2.25 (Embedding) An injective immersion $F: N \to M$ is called an *embedding* if it is a homeomorphism onto its image $F(N) \subseteq M$.

Exercise 2.2.26 Let $I \subseteq \mathbb{R}$ be an open interval and $\gamma: I \to \mathbb{R}^2$ be given by

$$\gamma(t) = (\cos(3t)\cos(t), \cos(3t)\sin(t))$$

Check whether (I, γ) is an immersion, a submanifold, or an embedding, for the following choices of I:

(1) $I = (0, \pi)$,
(2) $I = (-\frac{\pi}{6}, \frac{\pi}{2})$,
(3) $I = (0, \frac{\pi}{3})$.

Exercise 2.2.27 Let $T^2 = S^1 \times S^1$ be the 2-dimensional torus and let γ_k be the parametrized curve given by

$$\gamma: \mathbb{R} \to T^2,$$

$$t \mapsto (\exp(2\pi i t), \exp(2\pi i k t)),$$

where $k \in \mathbb{R}$.

(1) Prove that γ_k is an immersion for any $k \in \mathbb{R}$.
(2) Let $k = \frac{m}{n} \in \mathbb{Q}$, where $m, n \in \mathbb{Z}$ are supposed to be coprime. Show that γ_k descends to a continuous map $\bar{\gamma}_k$ on the quotient $\mathbb{R}/n\mathbb{Z}$.
(3) Check that we can put a smooth structure on $\mathbb{R}/n\mathbb{Z}$. Show that γ_k defines an *embedding* of this manifold into T^2.
(4) Let $k \in \mathbb{R} \setminus \mathbb{Q}$. Show that $\gamma_k(\mathbb{R}) \subseteq T^2$ is dense. Conclude that it cannot be an embedding.

Exercise 2.2.28 (Hairy Ball Theorem) Let X be a vector field on S^2. We want to show that there is a point $p \in S^2$ with $X_p = 0$.

(1) First, show that the statement implies that $T S^2$ is a non-trivial vector bundle.

(2) Define the *stereographic projections* $\phi_N \colon S^1 \setminus \{N = (0, 1)\} \to \mathbb{R}$, $(x, y) \mapsto \frac{x}{1-y}$, $\phi_S \colon S^1 \setminus \{S = (0, -1)\}$, $(x, y) \mapsto \frac{x}{1+y}$, similarly as in Example 2.1.12. Define $X_N := \phi_N^{-1} \circ X \circ \phi_N \colon \mathbb{R}^2 \to \mathbb{R}^2$, and similarly X_S. Show that X is everywhere non-zero if and only if X_N, X_S are non-zero when restricted to the unit disk D.

(3) Show that for $u \in S^1$, $\phi_N \circ \phi_S(u) = u$ and $X_S(u) = X_N(u) - 2\langle X_N(u), u \rangle u$, where $\langle\ ,\ \rangle$ denotes the standard inner product on $\mathbb{R}^2$.

(4) Recall the following facts from topology:

- Let $f \colon D \to \mathbb{R}^2$ be continuous and define $\gamma \colon S^1 = \partial D \to \mathbb{R}^2$ as the restriction of f to ∂D. Then f is non-zero on D precisely if the winding number of γ around 0 vanishes.
- The winding number is invariant under homotopy.

Now, assume that X_N is non-zero when restricted to D and define γ_S as the restriction of X_S to S^1. Show that γ_S is homotopic to a path with winding number 2 around 0. Conclude the statement.

2.2.4 Flow of a Vector Field

To a vector field $X \in \mathfrak{X}(M)$ we associate the ODE

$$\dot{q} = X(q). \tag{2.10}$$

Definition 2.2.29 (Integral Curve) A solution of (2.10) is called an *integral curve*. It is given by a path $q \colon I \to M$ such that $\dot{q}(t) = X(q(t)) \in T_{q(t)}M$ for all $t \in I$. Here, $I := [a, b] \subset \mathbb{R}$ denotes some interval.

Definition 2.2.30 (Maximal Integral Curve) An integral curve is called *maximal* if it is not the restriction of a solution to a proper subset of its domain.

Definition 2.2.31 (Flow) Let M be a manifold and let $X \in \mathfrak{X}(M)$ be a vector field. The *flow* Φ_t^X of X is given as follows: For $q \in M$ and t in a neighborhood of 0, $\Phi_t^X(q)$ is the unique solution at time t to the Cauchy problem with initial condition at $q \in M$. Explicitly,

$$\frac{\partial}{\partial t} \Phi_t^X(q) = X(\Phi_t^X(q)),$$

$$\Phi_0^X(q) = q.$$

Remark 2.2.32 One can then show that

$$\Phi^X_{t+s}(q) = \Phi^X_t(\Phi^X_s(q)), \quad \forall q \in M \tag{2.12}$$

and for all $t, s \in \mathbb{R}$ such that the flow is defined.

Definition 2.2.33 (Complete Vector Field) A vector field is called *complete* if all its integral curves exist for all $t \in \mathbb{R}$.

Example 2.2.34 Let $M = \mathbb{R}$ and consider $X = x^2 \sin^2(x) \frac{\partial}{\partial x}$. We can see that X is complete. Indeed, let $q : I \to \mathbb{R}$ be some integral curve of X such that $q(0) = t$ for some $t \in \mathbb{R}$. Then either $q = 0$ (if $t = \pi k$ for some $k \in \mathbb{Z}$) or $q(I) \subset (\pi k, \pi(k+1))$ (if $\pi k < t < \pi(k+1)$ for some $k \in \mathbb{Z}$). It is easy to see that both situations imply that $I = \mathbb{R}$.

Exercise 2.2.35 Show that the sum of two complete vector fields X and Y is not complete. *Hint: use the vector fields $X = y^2 \frac{\partial}{\partial x}$ and $Y = x^2 \frac{\partial}{\partial y}$ on $\mathbb{R}^2$.*

Definition 2.2.36 (Global Flow) The flow of a complete vector field $X \in \mathfrak{X}(M)$ is a diffeomorphism $\Phi^X_t : M \to M$ for all $t \in \mathbb{R}$. We then call Φ^X_t the *global flow* of X.

Remark 2.2.37 The condition (2.12) for a global flow can then be written more compactly as

$$\Phi^X_{t+s} = \Phi^X_t \circ \Phi^X_s$$

2.2.5 Cotangent Bundle

If E is a vector bundle over M, then the union of the dual spaces E^*_q is also a vector bundle, called the *dual bundle* of E. Namely, let $E^* = \bigsqcup_{q \in M} E^*_q$. We denote an element of E^* as a pair (q, ω) with $\omega \in E^*_q$. We let $\tilde{\pi} : E^* \to M$ to be such that $\tilde{\pi}(q, \omega) = q$ (the projection onto the first factor). To an atlas $\{(\tilde{U}_\alpha, \tilde{\phi}_\alpha)\}_{\alpha \in I}$ of E we associate the atlas $\{(\hat{U}_\alpha, \hat{\phi}_\alpha)\}_{\alpha \in I}$ of E^* with $\hat{U}_\alpha = \tilde{\pi}^{-1}(U_\alpha) = \bigsqcup_{q \in M} E^*_q$ and

$$\hat{\phi}_\alpha : \hat{U}_\alpha \to \mathbb{R}^n \times (\mathbb{R}^r)^*,$$

$$(q, \omega \in E^*_q) \mapsto (\phi_\alpha(q), (A_\alpha(q)^*)^{-1}\omega), \tag{2.13}$$

where we regard $(\mathbb{R}^r)^*$ as the manifold $\mathbb{R}^r$ with its standard structure.

Definition 2.2.38 (Cotangent Bundle) The dual bundle of the tangent bundle TM of a manifold M, denoted by T^*M, is called the *cotangent bundle* of M.

2.2.6 Differential 1-Forms

Similarly as we have defined vector fields as sections of the tangent bundle, we can construct a *differential 1-form* to be a section of the cotangent bundle. We denote the space of 1-forms on a manifold M by

$$\Omega^1(M) := \Gamma(T^*M).$$

Definition 2.2.39 (de Rham Differential) The *de Rham differential* is defined as the map

$$d: C^\infty(M) \to \Omega^1(M), \qquad f \mapsto df, \tag{2.14}$$

which is $\mathbb{R}$-linear and satisfies the Leibniz rule

$$d(fg) = df\,g + f\,dg, \quad \forall f, g \in C^\infty(M).$$

Note that a 1-form $\omega \in \Gamma(T^*M)$ is dual to a vector field $X \in \Gamma(TM)$, so there is a pairing between them. The pairing is often denoted by $\iota_X \omega$. The operation ι is often called the *contraction* (sometimes also *interior multiplication*). If $V \subset \mathbb{R}^n$ is open, we can consider the differentials dx^i of the coordinate functions x^i. Note that we have

$$\iota_{\partial_j} dx^i = \frac{\partial x^i}{\partial x^j} = \delta^i_j,$$

where $\partial_j := \frac{\partial}{\partial x^j}$. Thus, we can write a 1-form as

$$\omega = \sum_{i=1}^n \omega_i\, dx^i,$$

where $\omega_i \in C^\infty(M)$ are uniquely determined functions for $1 \le i \le n$. If $f \in C^\infty(M)$, then

$$df = \sum_{i=1}^n \partial_i f\, dx^i.$$

One can define a new notion of derivative by using the notion of a flow of a vector field. We define the *Lie derivative* of a 1-form by

$$L_X\omega := \frac{\partial}{\partial t}\Big|_{t=0} (\Phi^X_{-t})_*\omega = \frac{\partial}{\partial t}\Big|_{t=0} (\Phi^X_t)^*\omega, \quad \forall X \in \mathfrak{X}(M), \forall \omega \in \Omega^1(M).$$

Exercise 2.2.40 (1-Forms)

(1) Let $f : U \to \mathbb{R}$, $(x, y) \mapsto \arctan\left(\frac{y}{x}\right)$ be a function on $U = \mathbb{R}^2 \setminus \{x = 0\}$ and $X = -y\frac{\partial}{\partial x} + x\frac{\partial}{\partial y}$ be a vector field on $\mathbb{R}^2$. Also, let $\omega = \exp(xy)dx + \sin(y)dy$ be a 1-form on $\mathbb{R}^2$.

    (a) Compute the differential df and $df(X)$.
    (b) Compute $L_X\omega$.
    (c) Compute $\omega(X)$.

(2) Let $\omega = xy\,dx + 2z\,dy - y\,dz$ be a 1-form on $\mathbb{R}^3$ and let $\alpha : \mathbb{R}^2 \to \mathbb{R}^3$, $(u, v) = (uv, u^2, 3u + v)$. Compute $\alpha^*\omega$.
(3) Let $\omega = \sum_{i=1}^{n} \omega_i dx^i$ be a 1-form on $\mathbb{R}^n$, $v \in \mathbb{R}^n \setminus \{0\}$ and consider the embeddings $\iota_v : \mathbb{R} \to \mathbb{R}^n$, $x \mapsto x \cdot v$. Compute $(\iota_v^* w)_x$ for all $x \in \mathbb{R}$.
(4) For a 1-form ω on a manifold M and a curve $\gamma : [a, b] \to M$, one can define the integral of ω along γ to be

$$\int_\gamma \omega = \int_a^b \omega(\dot\gamma(t))dt.$$

Compute the integral of $\frac{x}{x^2+y^2}dy - \frac{y}{x^2+y^2}dx$ along a circle of radius R around 0 in $\mathbb{R}^2$.

2.3 Tensor Fields

2.3.1 Tensor Bundle

Let V be a vector space over a field K. We define the k-th *tensor power* by

$$V^{\otimes k} := \underbrace{V \otimes \cdots \otimes V}_{k \text{ times}}.$$

By convention we have $V^{\otimes 0} := K$. Moreover, note that $\dim V^{\otimes k} = (\dim V)^k$.

Definition 2.3.1 (Tensor) We call an element of $V^{\otimes k}$ a tensor of order k.

Let $(e_i)_{i \in I}$ be a basis of V. Then $(e_{i_1} \otimes \cdots \otimes e_{i_k})_{i_1,\dots,i_k \in I}$ is a basis of $V^{\otimes k}$ and a tensor T of order k can be uniquely written by

$$T = \sum_{i_1,\dots,i_k \in I} T^{i_1 \cdots i_k} e_{i_1} \otimes \cdots \otimes e_{i_k}, \quad T^{i_1 \cdots i_k} \in K, \forall i_1, \dots, i_k \in I.$$

Definition 2.3.2 (Tensor Algebra) The *tensor algebra* of a vector space V is defined as

$$T(V) := \bigoplus_{k=0}^{\infty} V^{\otimes k}.$$

Definition 2.3.3 (Tensor of Type (k, s)**)** We define a *tensor of type* (k, s) for a vector space V as an element of

$$T_s^k(V) := V^{\otimes k} \otimes (V^*)^{\otimes s}.$$

Remark 2.3.4 Tensors of type $(0, s)$ are called *covariant* tensors of order s and tensors of type $(k, 0)$ are called *contravariant* of order k.

If we pick a basis $(e_i)_{i \in I}$ of V and consider the dual basis $(e^j)_{j \in I}$ of V^*. Then we get a basis of $T_s^k(V)$ as

$$(e_{i_1} \otimes \cdots \otimes e_{i_k} \otimes e^{j_1} \otimes \cdots \otimes e^{j_s})_{i_1, \dots, i_k, j_1, \dots, j_s \in I}.$$

A tensor of type (k, s) can then be uniquely written as

$$T = \sum_{i_1, \dots, i_k, j_1, \dots, j_s \in I} T_{j_1 \dots j_s}^{i_1 \dots i_k} e_{i_1} \otimes \cdots \otimes e_{i_k} \otimes e^{j_1} \otimes \cdots \otimes e^{j_s}.$$

Definition 2.3.5 (Tensor Bundle) If E is a vector bundle over M, we define $T_s^k(E)$ as the vector bundle whose fiber at q is $T_s^k(E_q)$. Namely, to an adapted atlas $\{(\tilde{U}_\alpha, \tilde{\phi}_\alpha)\}_{\alpha \in I}$ of E over the trivializing atlas $\{(U_\alpha, \phi_\alpha)\}_{\alpha \in I}$ of M, we associate the atlas $\{(\hat{U}_\alpha, \hat{\phi}_\alpha)\}_{\alpha \in I}$ of $T_s^k(E)$ with $\hat{U}_\alpha = \tilde{\pi}^{-1}(U_\alpha) = \bigsqcup_{q \in U_\alpha} T_s^k(E_q)$ and

$$\hat{\phi}_\alpha : \hat{U}_\alpha \to \mathbb{R}^n \times T_s^k(\mathbb{R}^r),$$
$$(q, \omega \in T_s^k(E_q)) \mapsto (\phi_\alpha(q), (A_\alpha(q))_s^k \omega), \tag{2.15}$$

where we identify $T_s^k(\mathbb{R}^r)$ with $\mathbb{R}^{r(k+s)}$.

We have the transition maps

$$\hat{\phi}_{\alpha\beta}(x, u) = (\phi_{\alpha\beta}(x), (A_{\alpha\beta}(\phi_\alpha^{-1}(x)))_s^k u).$$

2.3.2 Multivector Fields and Differential s-Forms

The important case is when $E = TM$ is the tangent bundle of a n-dimensional manifold M. Denote by Alt_s^k the tensor bundle induced by the *alternating tensor product* (wedge product) denoted by $\wedge$. Then a contravariant tensor field of order k is also called a *multivector field* and a covariant tensor field of order s is called a *differential s-form*. In particular, the space of multivector fields of order k on an n-dimensional manifold M is given by

$$\mathfrak{X}^k(M) := \Gamma(\mathrm{Alt}_0^k(TM)) = \Gamma\left(\bigwedge^k TM\right).$$

Choosing local coordinates (x^i) on M we can represent an element $X \in \mathfrak{X}^k(M)$ as

$$X = \sum_{1 \leq i_1 < \cdots < i_k \leq n} X^{i_1 \cdots i_k} \partial_{i_1} \wedge \cdots \wedge \partial_{i_k}.$$

The space of differential s-forms is then given by

$$\Omega^s(M) := \Gamma(\mathrm{Alt}_s^0(TM)) = \Gamma\left(\bigwedge^s T^*M\right).$$

Choosing local coordinates (x^i) on M we can represent an element $\omega \in \Omega^s(M)$ as

$$\omega = \sum_{1 \leq i_1 < \cdots < i_s \leq n} \omega_{i_1 \cdots i_s} \, dx^{i_1} \wedge \cdots \wedge dx^{i_s}.$$

We can define a product $\wedge$ between differential forms. For $\omega \in \Omega^s(M)$ and $\eta \in \Omega^\ell(M)$ it is given by

$$\omega \wedge \eta = \sum_{1 \leq i_1 < \cdots < i_s \leq n} \sum_{1 \leq j_1 < \cdots < j_\ell \leq n} \omega_{i_1 \cdots i_s} \eta_{j_1 \cdots j_\ell} dx^{i_1} \wedge \cdots \wedge dx^{i_s} \wedge dx^{j_1} \wedge \cdots \wedge dx^{j_\ell}.$$

Note that if $s + \ell > 0$, then $\omega \wedge \eta = 0$.

Exercise 2.3.6 Show that if $\omega \in \Omega^s(M)$ and $\eta \in \Omega^\ell(M)$, then

$$\omega \wedge \eta = (-1)^{s+\ell} \eta \wedge \omega.$$

One can extend the de Rham differential to general s-forms

$$d \colon \Omega^s(M) \to \Omega^{s+1}(M),$$
$$\omega \mapsto d\omega, \tag{2.16}$$

and obtain a sequence

$$0 \to C^\infty(M) \xrightarrow{d} \Omega^1(M) \xrightarrow{d} \Omega^2(M) \xrightarrow{d} \cdots \xrightarrow{d} \Omega^s(M) \xrightarrow{d} \Omega^{s+1}(M) \xrightarrow{d} \cdots \xrightarrow{d} \Omega^n(M) \to 0.$$
$$(2.17)$$

This sequence is called *de Rham complex*. Note that by construction $\Omega^0(M) = C^\infty(M)$. In particular, for a differential form $\omega = \sum_{1 \le i_1 < \cdots < i_s \le n} \omega_{i_1 \cdots i_s} \, dx^{i_1} \wedge \cdots \wedge dx^{i_s} \in \Omega^s(M)$ we get

$$d\omega = \sum_{1 \le i_1 < \cdots < i_s \le n} \sum_{j=1}^{n} \partial_j \omega_{i_1 \cdots i_s} \, dx^j \wedge dx^{i_1} \wedge \cdots \wedge dx^{i_s}.$$

Exercise 2.3.7 Show that $d^2 = 0$. *Hint: use that derivatives commute, i.e.* $\frac{\partial}{\partial x^i} \frac{\partial}{\partial x^j} = \frac{\partial}{\partial x^j} \frac{\partial}{\partial x^i}$.

Exercise 2.3.8 Show that if $\omega \in \Omega^s(M)$ and $\eta \in \Omega^\ell(M)$ then

$$d(\omega \wedge \eta) = d\omega \wedge \eta + (-1)^s \omega \wedge d\eta.$$

We can extend the Lie derivative and the contraction to any differential s-forms.

Definition 2.3.9 (Lie Derivative) We define the *Lie derivative* of an s-form ω with respect to a vector field X as

$$L_X \omega := \lim_{t \to 0} \frac{(\Phi_t^X)^* \omega - \omega}{t},$$

where Φ_t^X denotes the flow of X at time t. Explicitly, we have

$$L_X \omega = \sum_{1 \le i_1 < \ldots < i_s \le n} X(\omega_{i_1 \cdots i_s}) dx^{i_1} \wedge \cdots \wedge dx^{i_s} +$$
$$+ \sum_{1 \le i_1 < \ldots < i_s \le n} \sum_{1 \le k, r \le n} (-1)^{k-1} \omega_{i_1 \cdots i_s} \partial_r X^{i_k} dx^r \wedge dx^{i_1} \wedge \cdots \wedge \widehat{dx^{i_k}} \wedge \cdots \wedge dx^{i_s},$$
$$(2.18)$$

where $\widehat{}$ means that this element is omitted.

Exercise 2.3.10 (Properties for Lie Derivative) Let $X, Y \in \mathfrak{X}(M)$. Show that

- $L_X f = X(f), \forall f \in C^\infty(M)$,
- $L_X(\omega \wedge \eta) = L_X \omega \wedge \eta + \omega \wedge L_X \eta, \forall \omega \in \Omega^s(M), \forall \eta \in \Omega^\ell(M)$,
- $L_X d\omega = dL_X \omega, \forall \omega \in \Omega^s(M)$,
- $L_X L_Y \omega - L_Y L_X \omega = L_{[X,Y]} \omega, \forall \omega \in \Omega^s(M)$.

Definition 2.3.11 (Contraction) The *contraction* of a vector field X with a differential s-form ω is the differential $(s-1)$-form given by

$$\iota_X \omega = \sum_{1 \le i_1 < \ldots < i_s \le n} \sum_{k=1}^{n} (-1)^k \omega_{i_1 \ldots i_s} X^{i_k} dx^{i_1} \wedge \cdots \wedge \widehat{dx^{i_k}} \wedge \cdots \wedge dx^{i_s}.$$

Remark 2.3.12 If $\omega \in \Omega^0(M)$, i.e. $s = 0$, then for any $X \in \mathfrak{X}(M)$ we get that $\iota_X \omega$ is automatically zero.

Exercise 2.3.13 Let $X \in \mathfrak{X}(M)$ and let $\omega \in \Omega^s(M)$, $\eta \in \Omega^\ell(M)$. Show that

$$\iota_X(\omega \wedge \eta) = \iota_X \omega \wedge \eta + (-1)^s \omega \wedge \iota_X \eta.$$

Moreover, show that if $Y \in \mathfrak{X}(M)$ is another vector field, we have

$$\iota_X \iota_Y \omega = -\iota_Y \iota_X \omega,$$

and

$$\iota_X L_Y \omega - L_Y \iota_X \omega = \iota_{[X,Y]} \omega.$$

Theorem 2.3.14 (Cartan's Magic Formula) *Let M be a manifold. Let $\omega \in \Omega^s(M)$ and let $X \in \mathfrak{X}(M)$. Then*

$$L_X \omega = \iota_X d\omega + d\iota_X \omega. \tag{2.19}$$

Exercise 2.3.15 Prove Theorem 2.3.14.

Exercise 2.3.16 Let $\omega_1 = \exp(xy)dx + \sin(x^2 y)dy \in \Omega^1(\mathbb{R}^2)$ and $\omega_2 = \log(y)dx + \cos(x)dy \in \Omega^1(\mathbb{R}^2)$ be 1-forms on $\mathbb{R}^2$. Compute $d\omega_1$, $d\omega_2$ and $\omega_1 \wedge \omega_2$.

Exercise 2.3.17 Consider the 1-forms $\tau_1 = \frac{-y dx}{x^2 + y^2} + \frac{x dy}{x^2 + y^2} \in \Omega^1(\mathbb{R}^2)$ and $\tau_2 = \cot\left(\frac{xy}{x^2+y^2}\right)(dx + dy) \in \Omega^1(\mathbb{R}^2)$. Compute $d\tau_1$, $d\tau_2$ and $\tau_1 \wedge \tau_2$.

Exercise 2.3.18 Let $\beta = -x dx \wedge dy + y dy \wedge dz \in \Omega^2(\mathbb{R}^3)$ and $X = y\frac{\partial}{\partial x} + z\frac{\partial}{\partial z} \in \mathfrak{X}(\mathbb{R}^3)$. Compute $d\beta$ and $\iota_X \beta$.

2.4 Integration on Manifolds and Stokes' Theorem

2.4.1 Integration of Densities

Let M be a manifold endowed with an atlas $\{(U_\alpha, \phi_\alpha)\}_{\alpha \in I}$. The differential $d_x \phi_{\alpha\beta}$ is a linear map

$$T_x \phi_\alpha (U_\alpha \cap U_\beta) \rightarrow T_{\phi_{\alpha\beta}(x)} \phi_\beta (U_\alpha \cap U_\beta).$$

Since the image of the charts are open subsets of $\mathbb{R}^n$ we can identify the tangent spaces with $\mathbb{R}^n$. Hence, the linear map $d_x \phi_{\alpha\beta}$ is canonically given by an $n \times n$ matrix. Let $s \in \mathbb{R}$ and

$$A_{\alpha\beta}(q) = |\det d_{\phi_\alpha(q)} \phi_{\alpha\beta}|^{-s}.$$

Exercise 2.4.1 Show that this defines a line bundle $|\Lambda M|^s$ over M.

Definition 2.4.2 (s-Density) A section of $|\Lambda M|^s$ is called an s-*density*.

Let $\sigma \in \Gamma(|\Lambda M|^s)$ be an s-density. We represent it in the chart (U_α, ϕ_α) by the smooth function σ_α. It satisfies the transition rule

$$\sigma_\beta(\phi_{\alpha\beta}(x)) = |\det d_q \phi_{\alpha\beta}|^{-s} \sigma_\alpha(x), \quad \forall \alpha, \beta \in I, \forall x \in \phi_\alpha(U_\alpha \cap U_\beta). \tag{2.20}$$

For $s = 1$ one simply speaks of a *density*. Densities are the natural objects to integrate on a manifold. Let σ be a density on M and let $\{(U_\alpha, \phi_\alpha)\}_{\alpha \in I}$ be an atlas on M. One can show that, if M is a paracompact Hausdorff space, there always exists a finite partition of unity $\{\rho_j\}_{j \in J}$ subordinate to $\{U_\alpha\}_{\alpha \in I}$, i.e. for each $j \in J$ we have an α_j with $\text{supp}\, \rho_j \subset U_{\alpha_j}$. In fact, $\text{supp}\, \rho_j$ is compact. Since ϕ_{α_j} is a homeomorphism, also $\phi_{\alpha_j}(\text{supp}\, \rho_j)$ is compact. The representation $(\rho_j \sigma)_{\alpha_j}$ of the density $\rho_j \sigma$ in the chart $(U_{\alpha_j}, \phi_{\alpha_j})$ is smooth in $\phi_{\alpha_j}(U_{\alpha_j})$, so it is integrable on $\phi_{\alpha_j}(\text{supp}\, \rho_j)$. We define

$$\int_{M; \{(U_\alpha, \phi_\alpha)\}; \{\rho_j\}} \sigma := \sum_{j \in J} \int_{\phi_{\alpha_j}(\text{supp}\, \rho_j)} (\rho_j \sigma)_{\alpha_j} d^n x, \tag{2.21}$$

where $d^n x$ denotes the Lebesgue measure on $\mathbb{R}^n$ (also denoted by $dx^1 \cdots dx^n$).

Lemma 2.4.3 *The integral defined in (2.21) does not depend on the choice of atlas and partition of unity.*

Proof Consider an atlas $\{(\bar{U}_{\bar{\alpha}}, \bar{\phi}_{\bar{\alpha}})\}_{\bar{\alpha} \in \bar{I}}$ and a finite partition of unity $\{\bar{\rho}_{\bar{j}}\}_{\bar{j} \in \bar{J}}$ subordinate to it. From $\sum_{\bar{j} \in \bar{J}} \bar{\rho}_{\bar{j}} = 1$, it follows that $\sigma = \sum_{\bar{j} \in \bar{J}} \bar{\rho}_{\bar{j}} \sigma$, so we have

$$\int_{M; \{(U_\alpha, \phi_\alpha)\}; \{\rho_j\}} \sigma = \sum_{j \in J} \int_{\phi_{\alpha_j}(\operatorname{supp} \rho_j)} (\rho_j \sigma)_{\alpha_j} d^n x = \sum_{\bar{j} \in \bar{J}} \sum_{j \in J} \int_{\phi_{\alpha_j}(\operatorname{supp} \rho_j)} (\rho_j \bar{\rho}_{\bar{j}} \sigma)_{\alpha_j} d^n x,$$

(2.22)

where we have taken out the finite sum $\sum_{\bar{j} \in \bar{J}}$. Observe that

$$S_{j\bar{j}} := \int_{\phi_{\alpha_j}(\operatorname{supp} \rho_j)} (\rho_j \bar{\rho}_{\bar{j}} \sigma)_{\alpha_j} d^n x$$

$$= \int_{\phi_{\alpha_j}(\operatorname{supp} \rho_j \cap \operatorname{supp} \bar{\rho}_{\bar{j}})} (\rho_j \bar{\rho}_{\bar{j}} \sigma)_{\alpha_j} d^n x$$

$$= \int_{\phi_{\bar{\alpha}_{\bar{j}} \alpha_j}^{-1} (\bar{\phi}_{\bar{\alpha}_{\bar{j}}}(\operatorname{supp} \rho_j \cap \operatorname{supp} \bar{\rho}_{\bar{j}}))} (\rho_j \bar{\rho}_{\bar{j}} \sigma)_{\alpha_j} d^n x,$$

(2.23)

where $\phi_{\bar{\alpha}_{\bar{j}} \alpha_j}$ denotes the transition map $\phi_{\alpha_j}(U_{\alpha_j}) \rightarrow \bar{\phi}_{\bar{\alpha}_{\bar{j}}}(\bar{U}_{\bar{\alpha}_{\bar{j}}})$. Since $\rho_j \bar{\rho}_{\bar{j}} \sigma$ is a density, we have

$$(\rho_j \bar{\rho}_{\bar{j}} \sigma)_{\alpha_j}(x) = |\det d_x \phi_{\bar{\alpha}_{\bar{j}} \alpha_j}| (\rho_j \bar{\rho}_{\bar{j}} \sigma)_{\bar{\alpha}_{\bar{j}}}(\bar{x}),$$

with $\bar{x} := \phi_{\bar{\alpha}_{\bar{j}} \alpha_j}(x)$ and $x \in \phi_{\alpha_j}(\operatorname{supp} \rho_j \cap \operatorname{supp} \bar{\rho}_{\bar{j}})$. By change-of-variables we get

$$S_{j\bar{j}} = \int_{\bar{\phi}_{\bar{\alpha}_{\bar{j}}}(\operatorname{supp} \rho_j \cap \operatorname{supp} \bar{\rho}_{\bar{j}})} (\rho_j \bar{\rho}_{\bar{j}} \sigma)_{\bar{\alpha}_{\bar{j}}} d^n \bar{x} = \int_{\bar{\phi}_{\bar{\alpha}_{\bar{j}}}(\operatorname{supp} \bar{\rho}_{\bar{j}})} (\rho_j \bar{\rho}_{\bar{j}} \sigma)_{\bar{\alpha}_{\bar{j}}} d^n \bar{x}.$$

Hence, we have

$$\sum_{j \in J} S_{j\bar{j}} = \int_{\bar{\phi}_{\bar{\alpha}_{\bar{j}}}(\operatorname{supp} \bar{\rho}_{\bar{j}})} (\bar{\rho}_{\bar{j}} \sigma)_{\bar{\alpha}_{\bar{j}}} d^n \bar{x}$$

and

$$\int_{M; \{(U_\alpha, \phi_\alpha)\}; \{\rho_j\}} \sigma = \sum_{\bar{j} \in \bar{J}} \sum_{j \in J} S_{j\bar{j}} = \sum_{\bar{j} \in \bar{J}} \int_{\bar{\phi}_{\bar{\alpha}_{\bar{j}}}(\operatorname{supp} \bar{\rho}_{\bar{j}})} (\bar{\rho}_{\bar{j}} \sigma)_{\bar{\alpha}_{\bar{j}}} d^n \bar{x} = \int_{M; \{(\bar{U}_{\bar{\alpha}}, \bar{\phi}_{\bar{\alpha}})\}; \{\bar{\rho}_{\bar{j}}\}} \sigma.$$

We can drop the choice of atlas and partition of unity and simply define

$$\boxed{\int_M \sigma := \sum_{j \in J} \int_{\phi_{\alpha_j}(\operatorname{supp} \rho_j)} (\rho_j \sigma)_{\alpha_j} d^n x}$$

(2.24)

Exercise 2.4.4 Let M_1 and M_2 be disjoint open subsets of a manifold M such that $M \setminus (M_1 \cup M_2)$ has measure zero. Show that

$$\int_M \sigma = \int_{M_1} \sigma + \int_{M_2} \sigma.$$

We can also pullback densities by diffeomorphisms $F : M \to N$. We define the pullback by

$$(F^*\sigma)_q := (d_q F)^* \sigma_{F(q)}, \quad \forall q \in M.$$

We can also formulate it in terms of coordinates. Let $\{(U_\alpha, \phi_\alpha)\}_{\alpha \in I}$ be an atlas of M and let $\{(V_j, \psi_j)\}_{j \in J}$ be an atlas of N. Let $\{\sigma_j\}$ denote the representation of the density σ in the atlas of N and $\{F_{\alpha j}\}$ the representation of F with respect to the two atlases. Then we have

$$(F^*\sigma)_\alpha(x) = |\det d_x F_{\alpha j}|^s \sigma_j(F_{\alpha j}(x)), \quad \forall x \in \phi_\alpha(U_\alpha).$$

Exercise 2.4.5 Show that F^* is linear and that

$$F^*(\sigma_1 \sigma_2) = F^* \sigma_1 F^* \sigma_2.$$

Exercise 2.4.6 Consider $S^1 \subset \mathbb{R}^2$ with the atlas consisting of the four charts where S^1 is a graph, namely the four charts

$$U_1 = S^1 \cap \{(x, y) \in \mathbb{R}^2 \mid x > 0\}, \quad \phi_1(x, y) = y,$$

$$U_2 = S^1 \cap \{(x, y) \in \mathbb{R}^2 \mid x < 0\}, \quad \phi_2(x, y) = y,$$

$$U_3 = S^1 \cap \{(x, y) \in \mathbb{R}^2 \mid y > 0\}, \quad \phi_3(x, y) = x,$$

$$U_4 = S^1 \cap \{(x, y) \in \mathbb{R}^2 \mid y < 0\}, \quad \phi_4(x, y) = x.$$

Now, consider the functions σ_α defined on $\phi_\alpha(U_\alpha)$ by

$$\sigma_1 : y \mapsto \frac{1}{\sqrt{1 - y^2}},$$

$$\sigma_2 : y \mapsto \frac{1}{\sqrt{1 - y^2}},$$

$$\sigma_3 : x \mapsto \frac{1}{\sqrt{1 - x^2}},$$

$$\sigma_4 : x \mapsto \frac{1}{\sqrt{1 - x^2}}.$$

(1) Show that these functions define a *density* on S^1, namely, that for any two α, β, we have that

$$\sigma_\beta(\phi_{\alpha\beta}(x)) = |\det d_x \phi_{\alpha\beta}(x)|^{-1} \sigma_\alpha(x).$$

(2) Compute the integral of this density over S^1.

Exercise 2.4.7 Show that if $F: M \to N$ is a diffeomorphism and σ is a density on M, then

$$\int_M \sigma = \int_N F_* \sigma.$$

2.4.2 Integration of Differential Forms

Similarly as for densities we can define the integral of a differential form for some manifold M as in (2.24). The difference is that we want to consider the notion of orientation on M. This corresponds to the notion of an *oriented atlas*.

Definition 2.4.8 A diffeomorphism $F: U \to V$ of open subsets of $\mathbb{R}^n$ is *orientation preserving* (*orientation reversing*) with respect to the standard orientation if and only if $\det F > 0$ ($\det F < 0$).

Definition 2.4.9 We call an atlas *oriented* if all its transition maps are orientation preserving.

At first, one defines the notion of an *orientable manifold*.

Definition 2.4.10 (Top Form) A differential n-form on an n-dimensional manifold is called a *top form*. We will denote the space of top forms on a manifold M by $\Omega^{\text{top}}(M)$ without mentioning the dimension of M.

Definition 2.4.11 (Volume Form) A *volume form* on a manifold M is a nowhere vanishing top form.

Definition 2.4.12 (Orientable Manifold) A manifold M is called *orientable* if it admits a volume form.

Let $\{(U_\alpha, \phi_\alpha)\}_{\alpha \in I}$ be an atlas of a manifold M. Consider a volume form $v \in \Omega^{\text{top}}(M)$ whose representation in the given atlas is denoted by v_α. Let $v_\alpha = \underline{v_\alpha} dx^1 \wedge \cdots \wedge dx^n$. Then the functions $\underline{v_\alpha}$ transform as

$$\underline{v_\alpha}(x) = \det d_x \phi_{\alpha\beta} \, \underline{v_\beta}(\phi_{\alpha\beta}(x)), \quad \forall \alpha, \beta \in I, \forall x \in \phi_\alpha(U_\alpha \cap U_\beta).$$

This is almost the same as the transformation rule (2.20) for densities. To fix this, we can take absolute values of the representations v_α. We define the absolute value $|v|$ as the density with representations $|v_\alpha|$. Moreover, we want to restrict everything to top forms that do not change sign (at least locally). Choosing a volume form v on M, we can define a $C^\infty(M)$-linear isomorphism

$$\phi_v \colon \Omega^{\mathrm{top}}(M) \to \mathrm{Dens}(M)$$

as follows: since v is nowhere vanishing, for every top form ω there is a uniquely defined function f such that $\omega = fv$; the corresponding density is then defined to be $f|v|$. Formally, we may write

$$\phi_v \omega = \omega \frac{|v|}{v}.$$

Two volume forms v_1 and v_2 yield the same isomorphism, i.e. $\phi_{v_1} = \phi_{v_2}$, if and only if there is a positive function g such that $v_1 = gv_2$.

Exercise 2.4.13 Show that this defines an equivalence relation on the set of volume forms on M.

Definition 2.4.14 (Orientation) An equivalence class of volume forms on an orientable manifold M is called an *orientation*. An orientable manifold with a choice of orientation is called *oriented*.

We denote by $[v]$ an orientation, by $(M, [v])$ the corresponding oriented manifold and by $\phi_{[v]}$ the isomorphism given by ϕ_v for any $v \in [v]$. If M admits a partition of unity, we can define the integral of a top form $\omega \in \Omega^{\mathrm{top}}(M)$ by

$$\int_{(M,[v])} \omega := \int_M \phi_{[v]}\omega,$$

where we use the already defined integration of densities. More explicitly, for an atlas $\{(U_\alpha, \phi_\alpha)\}_{\alpha \in I}$ we have the representations $\omega_\alpha = \underline{\omega_\alpha}\mathrm{d}^n x$, for uniquely defined maps $\underline{\omega_\alpha}$, and we get

$$(\phi_{[v]}\omega)_\alpha = \underline{\omega_\alpha}|\mathrm{d}^n x|.$$

Hence, the integral of ω on an oriented manifold M is given by

$$\int_{(M,[v])} \omega = \sum_{j \in J} \int_{\phi_{\alpha_j}(\mathrm{supp}\,\rho_j)} (\rho_j)_{\alpha_j}\underline{\omega_{\alpha_j}}\mathrm{d}^n x, \qquad (2.25)$$

where $\{(U_\alpha, \phi_\alpha)\}_{\alpha \in I}$ is an oriented atlas corresponding[1] to $[v]$ and $\{\rho_j\}_{j \in J}$ is a partition of unity subordinate to $\{U_\alpha\}_{\alpha \in I}$. If the identification of differential forms

---

[1] This means that it is an atlas in which any $v \in [v]$ is represented by a positive volume form.

and densities on $\phi_{\alpha_j}(U_{\alpha_j})$ is understood, then we can write

$$\int_{(M,[v])} \omega = \sum_{j \in J} \int_{\phi_{\alpha_j}(\operatorname{supp}\rho_j)} (\rho_j\omega)_{\alpha_j}.$$

Lemma 2.4.15 *Let* $\{(U_\alpha, \phi_\alpha)\}_{\alpha \in I}$ *be an oriented atlas of* $(M, [v])$ *and* $\{\rho_j\}_{j \in J}$ *a partition of unity subordinate to it. Let* $\omega \in \Omega^{\text{top}}(M)$ *with* $\operatorname{supp}\omega \subset U_{\alpha_k}$ *for some* $k \in J$. *Then*

$$\int_{(M,[v])} \omega = \int_{\phi_{\alpha_k}(U_{\alpha_k})} \omega_{\alpha_k}.$$

Proof We have

$$\int_{(M,[v])} \omega = \sum_{j \in J} \int_{\phi_{\alpha_j}(\operatorname{supp}\rho_j)} (\rho_j\omega)_{\alpha_j} = \sum_{j \in J} \int_{\phi_{\alpha_j}(\operatorname{supp}\rho_j \cap \operatorname{supp}\omega)} (\rho_j\omega)_{\alpha_j}$$

$$= \sum_{j \in J} \int_{\phi_{\alpha_j}(U_{\alpha_j} \cap U_{\alpha_k})} (\rho_j\omega)_{\alpha_j} = \sum_{j \in J} \int_{\phi_{\alpha_j}(U_{\alpha_j} \cap U_{\alpha_k})} (\rho_j\omega)_{\alpha_k}$$

$$= \sum_{j \in J} \int_{\phi_{\alpha_k}(U_{\alpha_k})} (\rho_j\omega)_{\alpha_k} = \int_{\phi_{\alpha_k}(U_{\alpha_k})} \sum_{j \in J} (\rho_j\omega)_{\alpha_k} = \int_{\phi_{\alpha_k}(U_{\alpha_k})} \omega_{\alpha_k}.$$

$\square$

Lemma 2.4.16 *A connected oriented manifold admits two orientations.*

Exercise 2.4.17 Prove Lemma 2.4.16.

Definition 2.4.18 (Orientation Preserving/Reversing) A diffeomorphism F of connected oriented manifolds $(M, [v_M])$ and $(N, [v_N])$ is called *orientation preserving* if $F^*[v_N] = [v_M]$ and *orientation reversing* if $F^*[v_N] = -[v_M]$.

Proposition 2.4.19 (Change of Variables) *Let* $(M, [v_M])$ *and* $(N, [v_N])$ *be connected oriented manifolds,* $F: M \to N$ *a diffeomorphism and* ω *a top form on* N. *Then*

$$\int_{(M,[v_M])} F^*\omega = \pm \int_{(N,[v_N])} \omega,$$

with plus sign if F *is orientation preserving and the minus sign if* F *is orientation reversing.*

Exercise 2.4.20 Prove Proposition 2.4.19.

Typically, the chosen orientation is understood, so one simply writes

$$\int_M \omega.$$

Exercise 2.4.21 Let (θ, φ) spherical coordinates on S^2 and consider the 2-form $\omega = \cos(\theta)\mathrm{d}\theta \wedge \mathrm{d}\varphi$. Compute the integral

$$\int_{S^2} \omega.$$

Exercise 2.4.22 Let T^2 be the 2-dimensional torus defined by rotating the circle $(x - 3)^2 + y^2 = 1$ around the z-axis in $\mathbb{R}^3$. Compute the integral

$$\int_{T^2} xz\mathrm{d}x \wedge \mathrm{d}z + y\mathrm{d}x \wedge \mathrm{d}y.$$

2.4.3 Stokes' Theorem

We will denote the space of compactly supported differential s-forms on a manifold M by $\Omega_c^s(M)$. Moreover, we define the *n-dimensional upper half-space* by

$$\mathbb{H}^n := \{(x^1, \dots, x^n) \in \mathbb{R}^n \mid x^n \geq 0\}. \tag{2.26}$$

The boundary of the upper half-space is given by

$$\partial\mathbb{H}^n = \{(x^1, \dots, x^n) \in \mathbb{R}^n \mid x^n = 0\}.$$

On the boundary, we can take the orientation induced by the *outward pointing vector field* $-\partial_n$, i.e.

$$[i^* \iota_{-\partial_n}\mathrm{d}^n x] = (-1)^n[\mathrm{d}x^1 \wedge \cdots \wedge \mathrm{d}x^{n-1}],$$

where $i : \partial\mathbb{H}^n \hookrightarrow \mathbb{H}^n$ denotes the inclusion.

Lemma 2.4.23 *Let $\omega \in \Omega_c^{n-1}(\mathbb{H}^n)$. Then, using the orientations defined as before, we get*

$$\int_{\mathbb{H}^n} \mathrm{d}\omega = \int_{\partial\mathbb{H}^n} \omega.$$

Proof We write

$$\omega = \sum_{j=1}^{n}(-1)^{j-1}\omega^j dx^1 \wedge \cdots \wedge \widehat{dx^j} \wedge \cdots \wedge dx^n.$$

The components ω^j are then related to the components $\omega_{i_1 \cdots i_n}$ by a sign. Then we have

$$d\omega = \sum_{j=1}^{n} \partial_j \omega^j d^n x.$$

Using the standard orientation, we get

$$\int_{\mathbb{H}^n} d\omega = \sum_{j=1}^{n} \int_{\mathbb{H}^n} \partial_j \omega^j d^n x.$$

We use *Fubini's theorem* to integrate the j-th term along the j-th axis. Then, since ω has compact support, we get

$$\int_{-\infty}^{+\infty} \partial_j \omega^j dx^j = 0, \quad \forall j < n,$$

but for $j = n$ we have

$$\int_0^{+\infty} \partial_n \omega^n dx^n = -\omega^n|_{x^n=0}.$$

Hence, we get

$$\int_{\mathbb{H}^n} d\omega = - \int_{\partial \mathbb{H}^n} \omega^n d^{n-1} x.$$

On the other hand, we have

$$i^*\omega = (-1)^{n-1}\omega^n|_{x^n=0} dx^1 \wedge \cdots \wedge dx^{n-1}.$$

Thus, using the orientation of $\partial \mathbb{H}^n$ as before, we get

$$\int_{\partial \mathbb{H}^n} \omega = - \int_{\partial \mathbb{H}^n} \omega^n d^{n-1} x,$$

which concludes the proof.   □

  Similarly, we get the following lemma.

Lemma 2.4.24 *Let $\omega \in \Omega_c^{n-1}(\mathbb{R}^n)$. Then*

$$\int_{\mathbb{R}^n} d\omega = 0.$$

Definition 2.4.25 (Manifold with Boundary) An n-dimensional *manifold with boundary* is an equivalence class of atlases whose charts take values in $\mathbb{H}^n$.

Remark 2.4.26 One considers $\mathbb{H}^n$ as a topological space with topology induced from $\mathbb{R}^n$.

Remark 2.4.27 Let M be a manifold with boundary. For any $q \in M$ one can show that if there is a chart map sending q to an interior point of $\mathbb{H}^n$, then any chart map will send it to an interior point of $\mathbb{H}^n$. On the other hand, if there is a chart map sending q to a boundary point of $\mathbb{H}^n$, then any chart map will send q to a boundary point of $\mathbb{H}^n$. Hence, one can induce a manifold structure on the interior points $\overset{\circ}{M}$ and boundary points ∂M of M out of the manifold structure of M, such that $\dim \overset{\circ}{M} = \dim M = \dim \partial M + 1$, by restricting atlases of M. The manifold ∂M is called the *boundary* of M.

Theorem 2.4.28 (Stokes) *Let M be an n-dimensional oriented manifold with boundary and let $\omega \in \Omega_c^{n-1}(M)$. Then*

$$\boxed{\int_M d\omega = \int_{\partial M} \omega} \tag{2.27}$$

where we use the induced orientation on ∂M.

Remark 2.4.29 If M has no boundary, we get $\int_M d\omega = 0$.

Remark 2.4.30 Theorem 2.4.28 was never officially proved by Stokes but appeared (in some version) in Maxwell's book on electrodynamics from 1873 [Max73] where he mentions in a footnote that the idea comes from Stokes who used this theorem in the Smith's Prize Examination of 1854. This is the reason why we call it Stokes' theorem today. A first proof of this theorem was given by Hermann in 1861 [Her61] who does not mention Stokes at all. See also [Kat79] for some more historical facts on Stokes' theorem.

Proof of Theorem 2.4.28 Let $\{(U_\alpha, \phi_\alpha)\}_{\alpha \in I}$ be an orientable atlas of M corresponding to the given orientation and let $\{\rho_j\}_{j \in J}$ be a partition of unity subordinate to it. First, observe that

$$d\omega = d\left(\sum_{j \in J} \rho_j \omega\right) = \sum_{j \in J} d(\rho_j \omega).$$

Note that supp $(\rho_j\omega) \subset U_{\alpha_j}$ and, by Lemma 2.4.15, we have

$$\int_M d(\rho_j\omega) = \int_{\phi_{\alpha_j}(U_{\alpha_j})} d(\rho_j\omega)_{\alpha_j}.$$

If $\phi_{\alpha_j}(U_{\alpha_j})$ is contained in the interior of $\mathbb{H}^n$, then we regard $d(\rho_j\omega)_{\alpha_j}$ as a compactly supported top form on $\mathbb{R}^n$ by extending it by zero outside of its support. Hence, by Lemma 2.4.24, we get

$$\int_{\phi_{\alpha_j}(U_{\alpha_j})} d(\rho_j\omega)_{\alpha_j} = 0.$$

Otherwise, we regard $d(\rho_j\omega)_{\alpha_j}$ as a compactly supported top form on $\mathbb{H}^n$ by again extending by zero outside of its support. Hence, by Lemma 2.4.23, we get

$$\int_{\phi_{\alpha_j}(U_{\alpha_j})} d(\rho_j\omega)_{\alpha_j} = \int_{\partial(\phi_{\alpha_j}(U_{\alpha_j}))} (\rho_j\omega)_{\alpha_j}.$$

Note that $\partial(\phi_{\alpha_j}(U_{\alpha_j})) = \phi_{\alpha_j}(\partial U_{\alpha_j})$ by definition and both are oriented by outward pointing vectors. Thus, again by Lemma 2.4.15, we get

$$\int_M d(\rho_j\omega) = \int_{\partial M} \rho_j\omega.$$

Summing over j yields the result. □

Example 2.4.31 (Curl and Divergence Theorem) When we restrict Theorem 2.4.28 to surfaces and volumes in $\mathbb{R}^3$, we get special forms of the formula which are important in the study of electrodynamics. Define the *curl* of a vector field $K \in \mathfrak{X}(\mathbb{R}^3)$, given in components by $\vec{K} = (K_1, K_2, K_3)^T$, by

$$\text{curl}\, \vec{K} = \vec{\nabla} \times \vec{K} = \begin{pmatrix} \frac{\partial K_3}{\partial y} - \frac{\partial K_2}{\partial z} \\ \frac{\partial K_1}{\partial z} - \frac{\partial K_3}{\partial x} \\ \frac{\partial K_2}{\partial x} - \frac{\partial K_1}{\partial y} \end{pmatrix},$$

where $\vec{\nabla} = (\partial/\partial x, \partial/\partial y, \partial/\partial z)^T$ and $\times$ denotes the cross product in $\mathbb{R}^3$ (Fig. 2.6).

Then we get the following corollary:

Corollary 2.4.32 (Stokes) *Let $S \subset \mathbb{R}^3$ be a compact and oriented surface with positive oriented boundary ∂S. Then we get*

$$\iint_S \langle \text{curl}\, \vec{K}, \vec{n} \rangle \, d\sigma = \oint_{\partial S} \langle K, \vec{ds} \rangle,$$

Fig. 2.6 Example of the
vector field
$$\vec{K} = \left(\frac{-2y}{\sqrt{x^2+y^2}}, \frac{2x}{\sqrt{x^2+y^2}}, 0 \right)$$
in the plane with vanishing
curl

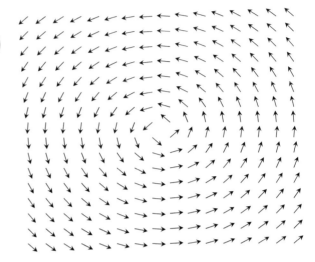

Fig. 2.7 Example of an
oriented surface S in $\mathbb{R}^3$ with
oriented boundary ∂S. The
light gray region visualizes
the mapping of the surface S
to $\mathbb{R}^2$ via a chart map ϕ

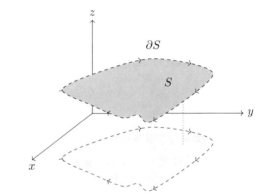

where $\langle \, , \, \rangle$ denotes the standard inner product on $\mathbb{R}^3$, $d\sigma$ is the induced Lebesgue measure on S, $\vec{n}$ denotes the normalized normal vector field to S and $\vec{ds}$ is the oriented line element describing the infinitesimal tangent vector field of the closed curve ∂S (Fig. 2.7).

An integral of the form $\int_S \langle \vec{K}, \vec{n} \rangle d\sigma$ is usually called the *flux* of the vector field $\vec{K}$ through the surface S. This is an important notion in the physics literature, especially in classical electrodynamics, where $\vec{K}$ is often either the electric field $\vec{E}$ or the magnetic field $\vec{B}$.

Moreover, define the *divergence* of $\vec{K}$ as

$$\operatorname{div} \vec{K} = \langle \vec{\nabla}, \vec{K} \rangle = \frac{\partial K_1}{\partial x} + \frac{\partial K_2}{\partial y} + \frac{\partial K_3}{\partial z}.$$

Fig. 2.8 A 3-dimensional region in $\mathbb{R}^3$ with surface element $d\sigma$ on its boundary and volume element dVol inside. The flux is measured by the angle between the normal $\vec{n}$ and the vector field $\vec{K}$ at each surface element

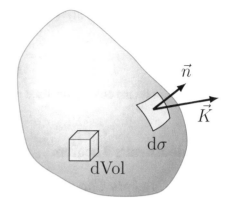

Corollary 2.4.33 (Gauss) *Let $V \subset \mathbb{R}^3$ be a 3-dimensional bounded region in $\mathbb{R}^3$ with closed boundary surface ∂V. Then we get*

$$\iiint_V \operatorname{div} \vec{K} \, d\text{Vol} = \oiint_{\partial V} \langle \vec{K}, \vec{n} \rangle \, d\sigma,$$

where $\vec{n}$ denotes the outer normalized normal vector field of ∂V and dVol $= dx\,dy\,dz$ denotes the volume form in $\mathbb{R}^3$ on ∂V (Fig. 2.8).

These special forms of Theorem 2.4.28 are especially used in order to derive the integral form of Maxwell's equations for electrodynamics.

Exercise 2.4.34 Prove Corollaries 2.4.32 and 2.4.33 by using Theorem 2.4.28.

Exercise 2.4.35 Let $\omega = dx \wedge dy \wedge dz \in \Omega^3(\mathbb{R}^3)$. Compute the integral

$$\int_B \omega,$$

where $B \subset \mathbb{R}^3$ is a ball of radius r.

Exercise 2.4.36 Let $S = \{(x, y, z) \in \mathbb{R}^3 \mid x^2 + y^2 + z^2 = 1, z \geq 0\}$ be the upper hemisphere, $\iota \colon S \hookrightarrow \mathbb{R}^3$ be the inclusion and $\omega = dx \wedge dy - dx \wedge dz + dy \wedge dz \in \Omega^2(\mathbb{R}^3)$. Compute the integral

$$\int_S \iota^* \omega$$

Exercise 2.4.37 (Green's Formula) Let $U \subseteq \mathbb{R}^2$ be a bounded open set with smooth boundary ∂U, and let $\vec{F} \in \mathfrak{X}(U)$ be a smooth vector field with a smooth extension to the boundary ∂U. Use Theorem 2.4.28 to prove Green's formula

$$\oint_{\partial U} \langle \vec{F}, \vec{ds} \rangle = \iint_U \langle \operatorname{curl} \vec{F}, \vec{n} \rangle \, dx \, dy,$$

where $\vec{n}$ denotes the outer normalized normal vector field of ∂U.

2.5 de Rham's Theorem

2.5.1 (Co)chain Complexes

An important notion for the study of differential forms and the area of *homological algebra* is the one of a chain or cochain complex.

Definition 2.5.1 (Chain Complex) Let k be a ring. A *chain complex* of k-modules is a sequence

$$\cdots \to C_{n+1} \xrightarrow{d_{n+1}} C_n \xrightarrow{d_n} C_{n-1} \to \cdots$$

where C_n is a k-module and $d_n \in \operatorname{Hom}_k(C_n, C_{n-1})$ satisfies $d_n \circ d_{n+1} = 0$ for all $n \in \mathbb{Z}$. It is often denoted by $(C_\bullet, d)$ or simply by $C_\bullet$.

Remark 2.5.2 The integer n is usually called the *degree* of the k-module C_n and the maps d_n are k-linear maps called *differential*. Moreover, a chain complex $C_\bullet$ is called *positive* if $C_n = 0$ for all $n \in \mathbb{Z}_{<0}$.

Definition 2.5.3 (Chain Map) A *chain map* between two chain complexes $(C_\bullet, d)$ and $(D_\bullet, d')$ is denoted by $f_\bullet : C_\bullet \to D_\bullet$ and is given by a collection of k-linear maps $f_n : C_n \to D_n$ such that $f_n \circ d_{n+1} = d'_{n+1} \circ f_{n+1}$ for all $n \in \mathbb{Z}$ such that the following diagram commutes:

$$
\begin{array}{ccccccccc}
\cdots & \xrightarrow{d_{n+2}} & C_{n+1} & \xrightarrow{d_{n+1}} & C_n & \xrightarrow{d_n} & C_{n-1} & \xrightarrow{d_{n-1}} & \cdots \\
& & \downarrow{f_{n+1}} & & \downarrow{f_n} & & \downarrow{f_{n-1}} & & \\
\cdots & \xrightarrow{d'_{n+2}} & D_{n+1} & \xrightarrow{d'_{n+1}} & D_n & \xrightarrow{d'_n} & D_{n-1} & \xrightarrow{d'_{n-1}} & \cdots
\end{array}
$$

Definition 2.5.4 (Subcomplex) A *subcomplex* $C'_\bullet$ of a chain complex $(C_\bullet, d)$ is a collection of k-modules $C'_n \subseteq C_n$ such that $d_n(C'_n) \subset C'_{n-1}$ for all $n \in \mathbb{Z}$.

Remark 2.5.5 Note that $(C'_\bullet, d)$ becomes then a chain complex itself and the inclusion $C'_\bullet \hookrightarrow C_\bullet$, induced by the canonical inclusion of $C'_n \hookrightarrow C_n$ for all $n \in \mathbb{Z}$, is a chain map.

Definition 2.5.6 (Quotient Complex) If $C'_\bullet$ is a subcomplex of a chain complex $C_\bullet$, then the *quotient complex* $C_\bullet/C'_\bullet$ is given by the collection of k-modules C_n/C'_n with the differentials $\bar{d}_n : C_n/C'_n \to C_{n-1}/C'_{n-1}$ which are uniquely determined by the universal property of the quotient.

Remark 2.5.7 The quotient map $\pi_\bullet : C_\bullet \to C_\bullet/C'_\bullet$ defined by the canonical projection $\pi_n : C_n \to C_n/C'_n$ for all $n \in \mathbb{Z}$ is a chain map.

Definition 2.5.8 (Kernel/Image/Cokernel) Let $f_\bullet : C_\bullet \to D_\bullet$ b a chain map between two chain complexes $(C_\bullet, d)$ and $(D_\bullet, d')$. Then the *kernel* of $f_\bullet$ is the subcomplex of $C_\bullet$ defined by $\ker(f_\bullet) := (\{\ker(f_n)\}_{n \in \mathbb{Z}}, d)$, the *image* of $f_\bullet$ is the subcomplex of $D_\bullet$ defined by $\mathrm{im}(f_\bullet) := (\{\mathrm{im}(f_n)\}_{n \in \mathbb{Z}}, d')$ and the *cokernel* of $f_\bullet$ is the quotient complex $\mathrm{coker}(f_\bullet) := D_\bullet/\mathrm{im}(f_\bullet)$.

Definition 2.5.9 (Cycle) Let $(C_\bullet, d)$ be a chain complex of k-modules. An n-cycle is an element of $\ker(d_n) =: Z_n(C_\bullet) =: Z_n$.

Definition 2.5.10 (Boundary) Let $(C_\bullet, d)$ be a chain complex of k-modules. An n-boundary is an element of $\mathrm{im}(d_{n+1}) =: B_n(C_\bullet) =: B_n$.

Definition 2.5.11 (Homology) Let $(C_\bullet, d)$ be a chain complex of k-modules. The n-th homology group of $C_\bullet$ is given by $H_n(C_\bullet) := Z_n/B_n$.

The notion of a *cochain complex* is exactly the same as for a chain complex whereas there all the arrows are reversed. The homology of a cochain complex is called *cohomology*.

2.5.2 Singular Homology

Definition 2.5.12 (p-Simplex) The *standard p-simplex* is the closed subset given by

$$\Delta^p := \left\{ (x^1, \ldots, x^p) \in \mathbb{R}^p \;\middle|\; \sum_{i=1}^{p} x^i \leq 1, x^i \geq 0 \,\forall i \right\} \subset \mathbb{R}^p.$$

The interior of Δ^p is a p-dimensional manifold. A smooth differential form on Δ^p is by definition the restriction to Δ^p of a smooth differential form defined on an open neighborhood of Δ^p in $\mathbb{R}^p$. Let $\omega \in \Omega^{p-1}(\Delta^p)$. Explicitly, let

$$\omega = \sum_{j=1}^{p} \omega^j \, dx^1 \wedge \cdots \wedge \widehat{dx^j} \wedge \cdots \wedge dx^p.$$

Then we have

$$d\omega = \sum_{j=1}^{p}(-1)^{j+1}\partial_j\omega^j d^p x$$

and

$$\int_{\Delta^p} d\omega = \sum_{j=1}^{p}(-1)^{j+1}\int_{\Delta^p}\partial_j\omega^j d^p x.$$

Using *Fubini's theorem* and the *fundamental theorem of analysis*, we get

$$\int_{\Delta^p}\partial_j\omega^j dx^j = \omega^j\big|_{x^j=1-\sum_{\substack{i=1\\i\neq j}}x^i} - \omega^j\big|_{x^j=0}.$$

Hence, we have

$$\int_{\Delta^p} d\omega = \sum_{j=1}^{p}(-1)^{j+1}\int_{\Delta^p\cap\{\sum_{i=1}^{p}x^i=1\}}\omega^j dx^1 \wedge \cdots \wedge \widehat{dx^j} \wedge \cdots \wedge dx^p +$$

$$+ \sum_{j=1}^{p}(-1)^j\int_{\Delta^p\cap\{x^j=0\}}\omega^j dx^1 \wedge \cdots \wedge \widehat{dx^j} \wedge \cdots \wedge dx^p. \qquad (2.28)$$

We can rewrite this in another way if we regard the faces on which we integrate as images of $(p-1)$-simplices. Namely, for $i = 0, \ldots, p$, we define smooth maps

$$k_i^{p-1}: \Delta^{p-1} \to \Delta^p,$$

by

$$k_0^{p-1}(a^1, \ldots, a^{p-1}) = \left(1 - \sum_{i=1}^{p-1}a^i, a^1, \ldots, a^{p-1}\right)$$

and

$$k_j^{p-1}(a^1, \ldots, a^{p-1}) = (a^1, \ldots, a^{j-1}, 0, a^j, \ldots, a^{p-1}), \quad \forall j > 0.$$

The j-th integral in the second line of (2.28) is just the integral on Δ^{p-1} of the pullback of ω by k_j^{p-1}. In fact, we have

$$(k_j^{p-1})^*\omega = (k_j^{p-1})^* \sum_{i=1}^{p} \omega^i \, dx^1 \wedge \cdots \wedge \widehat{dx^i} \wedge \cdots \wedge dx^p$$

$$= \omega^j(a^1, \ldots, a^{j-1}, 0, a^j, \ldots, a^{p-1}) d^{p-1}a \qquad (2.29)$$

We integrate over Δ^{p-1} with the standard orientation and rename variables $x^i = a^i$ for $i < j$ and $x^i = a^{i+1}$ for $i > j$. Note that the j-th integral is given by the integral over Δ^{p-1} of the pullback of $(-1)^{j+1}\omega^j dx^1 \wedge \cdots \wedge \widehat{dx^j} \wedge \cdots \wedge dx^p$ by k_0^{p-1}. In particular, we have

$$(k_0^{p-1})^*\omega^j dx^1 \wedge \cdots \wedge \widehat{dx^j} \wedge \cdots \wedge dx^p$$

$$= -\omega^j \left(1 - \sum_i a^i, a^1, \ldots, a^{p-1}\right) \sum_i da^i \wedge da^1 \wedge \cdots \wedge \widehat{da^{j-1}} \wedge \cdots \wedge da^{p-1}$$

$$= (-1)^{j+1}\omega^j \left(1 - \sum_i a^i, a^1, \ldots, a^{p-1}\right) d^{p-1}a. \qquad (2.30)$$

Summing everything up, we get *Stokes' theorem for a simplex*:

$$\int_{\Delta^p} d\omega = \sum_{j=0}^{p} (-1)^j \int_{\Delta^{p-1}} (k_j^{p-1})^*\omega, \qquad (2.31)$$

where the $j = 0$ term corresponds to the whole sum in the first line of (2.28) and each other term corresponds to a term in the second line. Consider a map

$$\sigma: \Delta^p \to M,$$

where M is a manifold. For a p-form ω on M, we define

$$\int_\sigma \omega := \int_{\Delta^p} \sigma^*\omega.$$

If we define $\sigma^j := \sigma \circ k_j^{p-1}: \Delta^{p-1} \to M$, then, by (2.31), we get

$$\int_\sigma d\omega = \sum_{j=0}^{p} (-1)^j \int_{\sigma^j} \omega.$$

Definition 2.5.13 (p-Chains) A p-*chain* with real coefficients in a manifold M is a finite linear combination $\sum_k a_k \sigma_k$, where $a_k \in \mathbb{R}$ for all k, of maps $\sigma_k : \Delta^p \to M$. If ω is a p-form on M, we define

$$\int_{\sum_k a_k \sigma_k} \omega := \sum_k a_k \int_{\sigma_k} \omega. \tag{2.32}$$

Theorem 2.5.14 (Stokes' Theorem for Chains) *We have*

$$\boxed{\int_\sigma d\omega = \int_{\partial\sigma} \omega}$$

where

$$\partial\sigma := \sum_{j=0}^p (-1)^j \sigma^j.$$

Let $\Omega_p(M, \mathbb{R})$ denote the vector space of p-chains in M with *real* coefficients and extend ∂ to it by linearity. Then we get that ∂ is an endomorphism of degree -1 for the graded vector space

$$\Omega_\bullet(M, \mathbb{R}) := \bigoplus_{j=0}^\infty \Omega_j(M, \mathbb{R}),$$

i.e. for all $1 \leq p \leq \dim M$ it is a map of the form

$$\partial : \Omega_p(M, \mathbb{R}) \to \Omega_{p-1}(M, \mathbb{R}).$$

Exercise 2.5.15 Show that $\partial^2 := \partial \circ \partial = 0$.

For $\sigma \in \Omega_p(M, \mathbb{R})$ and $\omega \in \Omega^p(M)$ we can define

$$\langle \sigma, \omega \rangle := \int_\sigma \omega. \tag{2.33}$$

Exercise 2.5.16 Show that $\langle \ , \ \rangle$ as defined in (2.33) is a bilinear map $\Omega_p(M, \mathbb{R}) \times \Omega^p(M) \to \mathbb{R}$.

Note that, by Stokes' theorem for chains, we get

$$\langle \sigma, d\omega \rangle = \langle \partial\sigma, \omega \rangle.$$

In particular, we have a sequence

$$0 \longrightarrow \Omega_n(M, \mathbb{R}) \xrightarrow{\partial} \cdots \xrightarrow{\partial} \Omega_p(M, \mathbb{R}) \xrightarrow{\partial} \cdots \xrightarrow{\partial} \Omega_0(M, \mathbb{R}) \longrightarrow 0$$

If we consider a map $F : M \to N$ between manifolds, it induces a graded linear map

$$F_* : \Omega_\bullet(M, \mathbb{R}) \to \Omega_\bullet(N, \mathbb{R}), \tag{2.34}$$
$$\sigma \mapsto F \circ \sigma.$$

Hence, if $\dim M = \dim N = n$, we get a *chain complex*

$$
\begin{array}{ccccccccccc}
0 & \longrightarrow & \Omega_n(M, \mathbb{R}) & \xrightarrow{\partial} & \cdots & \xrightarrow{\partial} & \Omega_p(M, \mathbb{R}) & \xrightarrow{\partial} & \Omega_{p-1}(M, \mathbb{R}) & \xrightarrow{\partial} & \cdots & \xrightarrow{\partial} & \Omega_0(M, \mathbb{R}) & \longrightarrow 0 \\
 & & \downarrow{\scriptstyle F_*} & & & & \downarrow{\scriptstyle F_*} & & \downarrow{\scriptstyle F_*} & & & & \downarrow{\scriptstyle F_*} & \\
0 & \longrightarrow & \Omega_n(N, \mathbb{R}) & \xrightarrow{\partial} & \cdots & \xrightarrow{\partial} & \Omega_p(N, \mathbb{R}) & \xrightarrow{\partial} & \Omega_{p-1}(N, \mathbb{R}) & \xrightarrow{\partial} & \cdots & \xrightarrow{\partial} & \Omega_0(N, \mathbb{R}) & \longrightarrow 0
\end{array}
$$

Exercise 2.5.17 Show that each square of the diagram commutes, i.e. $\partial \circ F_* = F_* \circ \partial$.

Definition 2.5.18 (Singular Homology Groups) The *p-th singular homology group* on a manifold M with real coefficients is given by

$$H_p(M, \mathbb{R}) := \frac{\ker\left(\partial_{(p)} : \Omega_p(M, \mathbb{R}) \to \Omega_{p-1}(M)\right)}{\operatorname{im}\left(\partial_{(p+1)} : \Omega_{p+1}(M, \mathbb{R}) \to \Omega_p(M, \mathbb{R})\right)},$$

Remark 2.5.19 Elements of $\ker \partial_{(p)} := \{\sigma \in \Omega_p(M, \mathbb{R}) \mid \partial_{(p)}\sigma = 0\}$ are usually called *p-cycles* and elements of $\operatorname{im} \partial_{(p)} := \{\sigma \in \Omega_p(M, \mathbb{R}) \mid \exists \tau \in \Omega_{p+1}(M, \mathbb{R}), \sigma = \partial_{(p+1)}\tau\}$ are usually called *p-boundaries*. The collection of all singular homology groups with real coefficients will be denoted by $H_\bullet(M, \mathbb{R}) := \bigoplus_{p \geq 0} H_p(M, \mathbb{R})$.

Exercise 2.5.20 Using Exercise 2.5.17, show that F_* descends to a graded linear map

$$F_* : H_\bullet(M, \mathbb{R}) \to H_\bullet(N, \mathbb{R}).$$

Thus, we get a chain complex on the level of homology

$$
\begin{array}{ccccccccccc}
0 & \longrightarrow & H_n(M, \mathbb{R}) & \longrightarrow & \cdots & \longrightarrow & H_p(M, \mathbb{R}) & \longrightarrow & H_{p-1}(M, \mathbb{R}) & \longrightarrow & \cdots & \longrightarrow & H_0(M, \mathbb{R}) & \longrightarrow 0 \\
 & & \downarrow{\scriptstyle F_*} & & & & \downarrow{\scriptstyle F_*} & & \downarrow{\scriptstyle F_*} & & & & \downarrow{\scriptstyle F_*} & \\
0 & \longrightarrow & H_n(N, \mathbb{R}) & \longrightarrow & \cdots & \longrightarrow & H_p(N, \mathbb{R}) & \longrightarrow & H_{p-1}(N, \mathbb{R}) & \longrightarrow & \cdots & \longrightarrow & H_0(N, \mathbb{R}) & \longrightarrow 0
\end{array}
$$

Exercise 2.5.21 Show that $\langle F_*\sigma, \omega \rangle = \langle \sigma, F^*\omega \rangle$ for all $\sigma \in \Omega_\bullet(M, \mathbb{R})$ and all $\omega \in \Omega^\bullet(N)$.

2.5.3 de Rham Cohomology and de Rham's Theorem

Let M be an n-dimensional manifold and recall its de Rham complex

$$0 \to C^\infty(M) \xrightarrow{\mathrm{d}} \Omega^1(M) \xrightarrow{\mathrm{d}} \Omega^2(M) \xrightarrow{\mathrm{d}} \cdots \xrightarrow{\mathrm{d}} \Omega^s(M) \xrightarrow{\mathrm{d}} \Omega^{s+1}(M) \xrightarrow{\mathrm{d}} \cdots \xrightarrow{\mathrm{d}} \Omega^n(M) \to 0 \tag{2.35}$$

Definition 2.5.22 (de Rham Cohomology Groups) We define the *s-th de Rham cohomology group* by the quotient

$$H^s(M) := \frac{\ker\left(\mathrm{d}^{(s)} \colon \Omega^s(M) \to \Omega^{s+1}(M)\right)}{\mathrm{im}\left(\mathrm{d}^{(s-1)} \colon \Omega^{s-1}(M) \to \Omega^s(M)\right)}.$$

Remark 2.5.23 We call elements of $\ker \mathrm{d}^{(s)} := \{\omega \in \Omega^s(M) \mid \mathrm{d}^{(s)}\omega = 0\}$ *closed forms* and elements of $\mathrm{im}\, \mathrm{d}^{(s)} := \{\omega \in \Omega^s(M) \mid \exists \eta \in \Omega^{s-1}(M), \omega = \mathrm{d}^{(s-1)}\eta\}$ *exact forms*. The collection of all de Rham cohomology groups will be denoted by $H^\bullet(M) := \bigoplus_{s \geq 0} H^s(M)$.

Example 2.5.24 If $M = \{\mathrm{pt}\}$, we get

$$H^s(M) = \begin{cases} \mathbb{R}, & s = 0 \\ 0, & s \geq 1 \end{cases}$$

For $M = \mathbb{R}$, we can see that $\ker(\mathrm{d}) \cap \Omega^0(\mathbb{R})$ are constant functions and hence we have

$$H^0(\mathbb{R}) = \mathbb{R}.$$

If we take $\Omega^1(\mathbb{R})$, we get that $\ker(\mathrm{d})$ consists of all 1-forms. In particular, if $\omega = g(x)\mathrm{d}x$ is a 1-form, then we get $\mathrm{d}f = g(x)\mathrm{d}x$ with

$$f(x) = \int_0^x g(u)\mathrm{d}u.$$

This shows that every 1-form on $\mathbb{R}$ is exact and thus $H^1(\mathbb{R}) = 0$. If U is a disjoint union of m open intervals on $\mathbb{R}$, we get $H^0(U) = \mathbb{R}^m$ and $H^1(U) = 0$. More generally, we have

$$H^\bullet(\mathbb{R}^n) = \begin{cases} \mathbb{R}, & n = 0 \\ 0, & n \neq 0 \end{cases}$$

This is due to the following lemma.

Lemma 2.5.25 (Poincaré)

$$H^\bullet(\mathbb{R}^n) = H^\bullet(\{pt\}) = \begin{cases} \mathbb{R}, & n = 0 \\ 0, & n \neq 0 \end{cases}$$

Exercise 2.5.26 Using Stokes' theorem for chains (Theorem 2.5.14), show that the bilinear form $\langle\ ,\ \rangle$ defined in (2.33) descends to a bilinear form

$$H_p(M, \mathbb{R}) \times H^p(M) \to \mathbb{R}. \tag{2.36}$$

Theorem 2.5.27 (de Rham[Rha31]) *The bilinear map* (2.36) *is non-degenerate. In particular, we have an isomorphism*

$$(H_p(M, \mathbb{R}))^* \cong H^p(M), \quad \forall p \geq 0.$$

Exercise 2.5.28 (Top Homology Group) Let N be a compact orientable manifold of dimension n. Show that the top homology group $H_n(N, \mathbb{R})$ is non-trivial.

Exercise 2.5.29 Let $\iota \colon S \hookrightarrow M$ be a compact, closed, k-dimensional submanifold of an n-dimensional manifold. Show that there is a unique $\eta_S \in H_k(M, \mathbb{R})$ such that for all $\omega \in \Omega^k(M)$

$$\langle \eta_S, [\omega] \rangle = \int_S \iota^* \omega.$$

For M a compact, connected, orientable manifold of dimension n, $H_n(M, \mathbb{R}) \cong \mathbb{R}$ and a generator is called a *fundamental class* of M.

Exercise 2.5.30 Show that a choice of volume on S naturally induces a fundamental class $[S] \in H_k(S, \mathbb{R})$. What is the relationship between $\iota_*[S] \in H_k(M, \mathbb{R})$ and η_S.

2.6   Hodge Theory for Real Manifolds

2.6.1   Riemannian Manifolds

Definition 2.6.1 (Riemannian Metric) A *Riemannian metric* on a manifold M is a section $g \in \Gamma(T^*M \otimes T^*M)$ with $g(X, Y) \geq 0$ and $g(X, Y) = g(Y, X)$ for all $X, Y \in \Gamma(TM)$. In local coordinates we write

$$g = g_{ij} dx^i \otimes dx^j.$$

Definition 2.6.2 (Riemannian Manifold) A manifold M endowed with a Riemannian metric g is called a *Riemannian manifold*. We usually write it as a tuple (M, g).

Theorem 2.6.3 *Every manifold M has a Riemannian metric.*

Proof Let $\{(U_\alpha, \phi_\alpha)\}_{\alpha \in I}$ be an atlas on M. Define a Riemannian metric g_α on U_α using local coordinates by setting $g_\alpha(X, Y) = \sum_i X^i Y^i$ for $X = \sum_i X^i \partial_i$ and $Y = \sum_j Y^j \partial_j$. Moreover, let $\{\rho_\alpha\}_{\alpha \in I}$ be a partition of unity subordinate to $\{U_\alpha\}_{\alpha \in I}$. Note that each point q has a neighborhood U_q where only finitely many ρ_α are non-vanishing because of the local finiteness for $\{\text{supp}\,\rho_\alpha\}_{\alpha \in I}$. Thus, $\sum_{\alpha \in I} \rho_\alpha g_\alpha$ is a finite sum on U_q. It is not hard to see that at each point q, the sum $\sum_{\alpha \in I} \rho_\alpha g_\alpha$ is an inner product on $T_q M$. We need to show that $\sum_{\alpha \in I} \rho_\alpha g_\alpha$ is smooth. For this, let X and Y be smooth vector fields on M. Then, since $\sum_{\alpha \in I} \rho_\alpha g_\alpha(X, Y)$ is a finite sum of smooth functions on U_q, it is smooth on U_q. Now, since q was arbitrary, $\sum_{\alpha \in I} \rho_\alpha g_\alpha(X, Y)$ is smooth on M.   □

Exercise 2.6.4 (Killing Vector Fields) Let (M, g) be a Riemannian manifold. A vector field X on M is called a *Killing vector field* for g if $L_X g = 0$.

(1) Let $\eta = \sum_{i=1}^{3} dx^i \otimes dx^i$ be the Euclidean metric on $\mathbb{R}^3$. Which of the following are Killing vector fields for η?

  (a) $P_i := \frac{\partial}{\partial x^i}, i = 1, 2, 3$.

  (b) $R_i := \sum_{j,k=1}^{3} \varepsilon_{ijk} x^j \frac{\partial}{\partial x^k}, i = 1, 2, 3$. Here we have denoted by ε_{ijk} the *Levi-Civita tensor* defined by

  $$\varepsilon_{ijk} := \begin{cases} 1, & (i, j, k) = (1, 2, 3), (2, 3, 1), (3, 1, 2), \\ -1, & (i, j, k) = (3, 2, 1), (1, 3, 2), (2, 1, 3), \\ 0, & i = j \vee j = k \vee k = i. \end{cases}$$

  (c) $S := \sum_{i=1}^{3} x^i \frac{\partial}{\partial x^i}$.

(2) Consider the Riemannian metric $g = \frac{\sum_{i=1}^{3} dx^i \otimes dx^i}{\sum_{i=1}^{3} (x^i)^2}$ on $\mathbb{R}^3 \setminus \{0\}$. Which of the above are Killing vector fields for g?

(3) Recall that if we have a diffeomorphism $F \colon M \to N$ between two manifolds M and N and a density σ on N, then $\int_N \sigma = \int_M F^* \sigma$.

  (a) Let M be a compact manifold, X a vector field on M, σ a density on M and $U \subseteq M$ an open subset. Show that

  $$\frac{\partial}{\partial t}\Big|_{t=0} \int_{\Phi_t^X(U)} \sigma = \int_U L_X \sigma.$$

  (b) Show that a Killing vector field X for g is *volume-preserving*, i.e.

  $$\frac{\partial}{\partial t} \text{Vol}_g(\Phi_t^X(U)) = 0$$

  for any open subset $U \subseteq M$.

(c) Find an example of a Riemannian manifold (M, g) and a volume-preserving vector field for g which is not a Killing vector field for g.

2.6.2 Hodge Dual

If we have a Riemannian metric g on an orientable n-dimensional manifold M, we can consider the volume form in $\Omega^n(M)$ given by

$$\mathrm{dVol} = \sqrt{\det(g_{ij})}\mathrm{d}x^1 \wedge \cdots \wedge \mathrm{d}x^n.$$

Definition 2.6.5 (Hodge Star) Let M be some connected, closed, orientable Riemannian manifold of dimension n. Define the *Hodge star* $*\colon \Omega^k(M) \to \Omega^{n-k}(M)$ through the pairing

$$\alpha \wedge *\beta = g(\alpha, \beta)\mathrm{dVol}.$$

Remark 2.6.6 Locally, when choosing coordinates around $x \in M$ with orthonormal basis $e_1, \ldots, e_n$ of $\Omega^1(M)$ in a trivialization around $x \in M$, we get

$$e_{\sigma(1)} \wedge \cdots \wedge e_{\sigma(n)} = \mathrm{sign}(\sigma)e_{\sigma(k+1)} \wedge \cdots \wedge e_{\sigma(n)}.$$

Moreover, note that $*^2 = (-1)^{k(n-k)}$. Thus, $*$ is an isomorphism and hence invertible.

Definition 2.6.7 (Hodge Inner Product) The *Hodge inner product* on $\Omega^k(M)$ is defined by

$$\langle \alpha, \beta \rangle_* := \int_M \alpha \wedge *\beta = \int_M g(\alpha, \beta)\mathrm{dVol}.$$

Note that by Stokes' theorem (Theorem 2.4.28), when assuming that M is closed, we get $\int_M \mathrm{d}\omega = 0$ for all $\omega \in \Omega^{n-1}(M)$. Then, for $\alpha \in \Omega^k(M)$ and $\beta \in \Omega^{k+1}(M)$, we get

$$0 = \int_M \mathrm{d}(\alpha \wedge *\beta)$$

$$= \int_M \left(\mathrm{d}\alpha \wedge *\beta - (-1)^{k+1}\alpha \wedge \mathrm{d} * \beta \right)$$

$$= \int_M \mathrm{d}\alpha \wedge *\beta - \int_M \alpha \wedge *\left(*^{-1} (-1)^{k+1}\mathrm{d} * \beta \right)$$

$$= \langle \mathrm{d}\alpha, \beta \rangle_* - \left\langle \alpha, (-1)^{k+1} *^{-1} \mathrm{d} * \beta \right\rangle_*.$$

Definition 2.6.8 (de Rham Codifferential) We will define the *de Rham codifferential* by

$$d^* := (-1)^{k+1} *^{-1} d* \colon \Omega^k(M) \to \Omega^{k-1}(M).$$

Definition 2.6.9 (Coclosed/Coexact) A differential form $\alpha \in \Omega^k(M)$ is called *coclosed* if $d^*\alpha = 0$. It is called *coexact* if there is some $\beta \in \Omega^{k+1}(M)$ such that $\alpha = d^*\beta$.

Definition 2.6.10 (Hodge Laplacian) The *Hodge Laplacian* is defined as

$$\Delta := dd^* + d^*d \colon \Omega^k(M) \to \Omega^k(M).$$

Example 2.6.11 Let $f \in \Omega^0(\mathbb{R}^3)$. Then we get

$$\begin{aligned}
\Delta f &= (dd^* + d^*d)\,f \\
&= 0 + *d\nabla f\,(dx, dy, dz) \\
&= -*^{-1} d\nabla f\,(dy \wedge dz, dz \wedge dx, dx \wedge dy) \\
&= -*^{-1} \nabla\nabla f\,dx \wedge dy \wedge dz \\
&= -\nabla\nabla f \\
&=: -\nabla^2 f,
\end{aligned}$$

which is the usual Laplacian $\Delta := \frac{\partial^2}{\partial x^2} + \frac{\partial^2}{\partial y^2} + \frac{\partial^2}{\partial z^2}$.

Proposition 2.6.12 *For a closed, compact, oriented Riemannian manifold (M, g) of dimension n, we have*

(1) Δ is self-adjoint,
(2) Δ is positive semi-definite,
(3) $\alpha \in \ker(\Delta)$ if and only if $\alpha \in \ker(d)$ and $\alpha \in \ker(d^*)$.

Proof We can see that for $\alpha, \beta \in \Omega^k(M)$ we get

$$\langle \Delta\alpha, \beta \rangle_* = \langle dd^*\alpha, \beta \rangle_* + \langle d^*d\alpha, \beta \rangle_* = \langle d\alpha, d^*\beta \rangle_* + \langle d\alpha, d\beta \rangle_*.$$

For (1), note that we have

$$\langle \alpha, dd^*\beta \rangle_* + \langle \alpha, d^*d\beta \rangle_* = \langle \alpha, \Delta\beta \rangle_*.$$

For (2) and (3), note that if $\alpha = \beta$, we get

$$\langle \Delta\alpha, \alpha \rangle_* = \|d^*\alpha\|_*^2 + \|d\alpha\|_*^2 \geq 0,$$

where we have equality if and only if $d\alpha = d^*\alpha = 0$. Here we have denoted by $\|\ \|_*$ the norm induced by $\langle\ ,\ \rangle_*$. □

2.6.3 Hodge Decomposition

Theorem 2.6.13 (Hodge Decomposition) *Let M be a closed, orientable, Riemannian manifold of dimension n. Then for each $0 \le k \le n$, we have a decomposition of vector spaces*

$$\Omega^k(M) = \operatorname{im}(d^{(k)}) \oplus \ker(d^*_{(k)}) \oplus \operatorname{Harm}^k_\Delta(M),$$

where $\operatorname{Harm}^k_\Delta(M)$ denotes the space of harmonic k-forms, i.e. elements $\alpha \in \Omega^k(M)$ such that $\Delta\alpha = 0$.

Theorem 2.6.14 (Hodge) *Let M be a closed, orientable, Riemannian manifold of dimension n. Then there is a canonical isomorphism between the space of harmonic k-forms and the k-th de Rham cohomology group:*

$$\operatorname{Harm}^k_\Delta(M) \cong H^k(M), \quad \forall k \ge 0.$$

Corollary 2.6.15 *Let M be a closed, orientable, Riemannian manifold of dimension n. Then the dimension of each de Rham cohomology group is finite:*

$$\dim H^k(M) < \infty, \quad \forall k \ge 0.$$

2.7 Lie Groups and Lie Algebras

The study of groups within differential geometry ant theoretical physics is an important thing in order to deal with symmetries and geometric object which are invariant with respect to a certain group action. Usually, we like to consider finite, or at least finitely generated, groups. However, most of the groups which describe certain geometric symmetries do not have one of these properties, such as e.g. the group $SO_3(\mathbb{R})$ consisting of rotations in 3-dimensional Euclidean space is neither finite nor finitely generated. Studying the concept of *Lie groups* will lead to a way of dealing with such a problem. These are groups which also carry a smooth manifold structure. The key point is to consider its tangent space at the identity element, its *Lie algebra*. The main result of the study of Lie groups is that most of their properties can be completely described by the properties of their Lie algebras.

2.7.1 Lie Groups

Definition 2.7.1 (Lie Group) A *Lie group* is a set G endowed with a group structure and a manifold structure which agree with each other, i.e. the group multiplication map $G \times G \to G$, $(g, h) \mapsto gh$ and taking the inverse map $G \to G$, $g \mapsto g^{-1}$ are smooth maps.

Example 2.7.2 The Euclidean space $\mathbb{R}^n$ endowed with vector addition is a Lie group.

Example 2.7.3 The unit circle $S^1 \subset \mathbb{C} \setminus \{0\}$ is a Lie group with multiplication induced by $\mathbb{C} \setminus \{0\}$.

Example 2.7.4 The manifold $\mathrm{GL}_n(\mathbb{R})$ of $n \times n$ non-singular matrices is a Lie group with respect to the matrix multiplication.

Definition 2.7.5 (Morphism of Lie Groups) A *morphism of Lie groups* is a smooth map which also preserves the group operations. Namely, if G and H are Lie groups, then $f : G \to H$ is a morphism of Lie groups if for all $g, h \in G$ we have $f(gh) = f(g) f(h)$ and $f(\mathrm{id}_G) = \mathrm{id}_H$.

Definition 2.7.6 (Lie Subgroup) Let G be a Lie group. A subgroup $H \subset G$ is called *Lie subgroup* of G if H is a Lie group, a submanifold of G and $\varphi : H \hookrightarrow G$ is a group homomorphism.

Exercise 2.7.7 (Hyperbolic Geometry) Let G be a Lie group and $g \in G$. Denote by $L_g : G \to G$ multiplication with g from the left. A *left-invariant Riemannian metric* on G is a Riemannian metric η such that $L_g^* \eta = \eta$ for all $g \in G$.

A *proper affine function* $f : \mathbb{R} \to \mathbb{R}$ is a function of the form $f(t) = yt + x$ for some $y > 0$ and $x \in \mathbb{R}$. The set of proper affine functions on $\mathbb{H}$ can be given a manifold structure by seeing it as the open subset $\{(x, y) \in \mathbb{R}^2 \mid y > 0\} \subset \mathbb{R}^2$.

(1) Show that $\mathbb{H}$ has a Lie group structure given by composition of functions.
(2) Show that the left-invariant metric η on $\mathbb{H}$, which at the neutral element coincides with the pullback of the Euclidean metric, is given by $\eta_{(x,y)} = \frac{1}{y^2}(dx \otimes dx + dy \otimes dy)$. This metric is called the *Lobachevsky* or *Poincaré* or *Hyperbolic* metric on the upper half-plane.
(3) Compute the length of a vertical line segment between (x, y_1) and (x, y_2) for $y_1 \leq y_2$. What happens in the limit $y_1 \to 0$?
(4) Identifying $(x, y) = z = x + iy$, define an action of $\mathrm{SL}_2(\mathbb{R})$ on $\mathbb{H}$ by

$$\begin{pmatrix} a & b \\ c & d \end{pmatrix} \cdot z = \frac{az + b}{cz + d}.$$

Show that this action is *isometric*, i.e. for every $g \in \mathrm{SL}_2(\mathbb{R})$, the map $\Psi_g : z \mapsto g \cdot z$ is an *isometry* (i.e. $\Psi_g^* \eta = \eta$).

2.7.2 Lie Algebras

Definition 2.7.8 (Lie Algebra) A *Lie algebra* is a vector space $\mathfrak{g}$ together with a *Lie bracket* [,], i.e. a bilinear map [,]: $\mathfrak{g} \times \mathfrak{g} \to \mathfrak{g}$ such that

(1) $[X, Y] = -[Y, X]$, $\forall X, Y \in \mathfrak{g}$ (anti-symmetry)
(2) $[X, [Y, Z]] + [Y, [Z, X]] + [Z, [X, Y]] = 0$, $\forall X, Y, Z \in \mathfrak{g}$ (Jacobi identity)

Example 2.7.9 The vector space of smooth vector fields on some manifold M is a Lie algebra where the Lie bracket is the commutator of vector fields

$$[X, Y] = XY - YX, \quad X, Y \in \mathfrak{X}(M).$$

Example 2.7.10 The vector space $\mathfrak{gl}_n(\mathbb{R})$ of all $n \times n$ real matrices is a Lie algebra with the Lie bracket given by the commutator of matrices

$$[A, B] = AB - BA, \quad A, B \in \mathfrak{gl}_n(\mathbb{R}).$$

Exercise 2.7.11 Let ω be a closed 2-form on some manifold M. Show that

$$\mathfrak{X}_\omega(M) := \{X \in \mathfrak{X}(M) \mid \iota_X \omega = 0\}$$

is a Lie algebra.

Remark 2.7.12 When considering a local basis $(e_i)_{1 \le i \le n}$ of some n-dimensional Lie algebra $\mathfrak{g}$ over $\mathbb{R}$, we can write the Lie bracket as

$$[e_i, e_j] = \sum_{1 \le k \le n} c_{ij}^k e_k.$$

The constants $c_{ij}^k \in \mathbb{R}$ are called the *structure constants* of the Lie algebra ($\mathfrak{g}$, [,]).

Definition 2.7.13 (Morphism of Lie Algebras) Let ($\mathfrak{g}$, [,]$_\mathfrak{g}$) and ($\mathfrak{h}$, [,]$_\mathfrak{h}$) be two Lie algebras. A linear map $f : \mathfrak{g} \to \mathfrak{h}$ is called a *Lie algebra morphism* if

$$f([g_1, g_2]_\mathfrak{g}) = [f(g_1), f(g_2)]_\mathfrak{h}, \quad \forall g_1, g_2 \in \mathfrak{g}.$$

Definition 2.7.14 (Lie Subalgebra) Let $\mathfrak{g}$ be a Lie algebra. A subspace $\mathfrak{h} \subset \mathfrak{g}$ is called a *Lie subalgebra* if $[X, Y] \in \mathfrak{h}$ if $X, Y \in \mathfrak{h}$, where [,] is the Lie bracket induced from $\mathfrak{g}$.

Definition 2.7.15 (Left Multiplication) Let G be a Lie group. For $g \in G$, define the *left multiplication* by g as the map

$$L_g : G \to G,$$

$$a \mapsto g \cdot a.$$

Definition 2.7.16 (Left-invariant Vector Field) A *vector field* X on G is called *left-invariant* if

$$(L_g)_* X = X, \quad \forall g \in G.$$

Remark 2.7.17 There are similar *right* notions.

Let $\mathfrak{g}$ be the vector space of all left-invariant vector fields on G. Together with the Lie bracket $[\ ,\ ]$ of vector fields, $\mathfrak{g}$ forms a Lie algebra. It is called the *Lie algebra of the Lie group* G. We will sometimes also write $\mathrm{Lie}(G)$ for the Lie algebra of G.

Exercise 2.7.18 Show that the map

$$\mathfrak{g} \to T_e G,$$
$$X \mapsto X_e,$$

where e is the identity element in G, is an isomorphism of vector spaces.

Example 2.7.19 (General Linear Group) Recall that $\mathfrak{gl}_n(\mathbb{R})$ is a real vector space of dimension n^2. As already mentioned in Example 2.7.10, $\mathfrak{gl}_n(\mathbb{R})$ is indeed a Lie algebra with Lie bracket $[A, B] = AB - BA$. The general linear group $\mathrm{GL}_n(\mathbb{R})$ is a manifold where we consider open sets of $\mathfrak{gl}_n(\mathbb{R})$ where the determinant does not vanish and, as already mentioned in Example 2.7.4, it is a Lie group with respect to matrix multiplication. Let now x_{ij} be global coordinate functions on $\mathfrak{gl}_n(\mathbb{R})$ assigning to each matrix its ij-entry. For $g_1, g_2 \in \mathrm{GL}_n(\mathbb{R})$, we have $x_{ij}(g_1(g_2)^{-1})$ is a rational function of $(x_{kl}(g_1))$ and $(x_{kl}(g_2))$ of non-vanishing denominator, proving that the map $(g_1, g_2) \mapsto g_1(g_2)^{-1}$ is smooth and thus it is a Lie group. Now let $\mathfrak{g}$ be the Lie algebra of $\mathrm{GL}_n(\mathbb{R})$. Denote by $\alpha \colon T_e\mathfrak{gl}_n(\mathbb{R}) \to \mathfrak{gl}_n(\mathbb{R})$ the identification of the tangent space of $\mathfrak{gl}_n(\mathbb{R})$ at the identity matrix e with $\mathfrak{gl}_n(\mathbb{R})$ itself. Hence, for $v \in T_e\mathfrak{gl}_n(\mathbb{R})$, we have

$$\alpha(v)_{ij} = v(x_{ij}).$$

Now since $T_e\, \mathrm{GL}_n(\mathbb{R}) = T_e\mathfrak{gl}_n(\mathbb{R})$, we have a map $\beta \colon \mathfrak{g} \to \mathfrak{gl}_n(\mathbb{R})$ given by

$$\beta(X) = \alpha(X_e), \quad X \in \mathfrak{g}.$$

Proposition 2.7.20 β *is a Lie algebra isomorphism.*

Proof It is clear that β is a vector space isomorphism, so it remains to show that

$$\beta([X, Y]) = [\beta(X), \beta(Y)], \quad \forall X, Y \in \mathfrak{g}.$$

First note that

$$(x_{ij} \circ L_{g_1})(g_2) = x_{ij}(g_1 g_2) = \sum_k x_{ik}(g_1) x_{kj}(g_2).$$

Since Y is a left-invariant vector field, we get

$$(Y(x_{ij}))(g_1) = dL_{g_1}(Y_e)(x_{ij}) = Y_e(x_{ij} \circ L_{g_1})$$
$$= \sum_k x_{ik}(g_1) Y_e(x_{ki}) = \sum_k x_{ik}(g_1) \alpha(Y_e)_{kj}$$
$$= \sum_k x_{ik}(g_1) \beta(Y)_{kj}.$$

Using this, we can compute the ij-entry of $\beta([X, Y])$. We get

$$\beta([X, Y])_{ij} = [X, Y]_e(x_{ij}) = X_e(Y(x_{ij})) - Y_e(X(x_{ij}))$$
$$= \sum_k \left(X_e(x_{ik}) \beta(Y)_{kj} - Y_e(x_{ik}) \beta(X)_{kj} \right)$$
$$= \sum_k \left(\beta(X)_{ik} \beta(Y)_{kj} - \beta(Y)_{ik} \beta(X)_{kj} \right)$$
$$= [\beta(X), \beta(Y)]_{ij}.$$

This proves the claim.   □

2.7.3 The Exponential Map

Let G be a Lie group and let $\mathfrak{g}$ be its Lie algebra. For $X \in \mathfrak{g}$, we note that the map

$$\lambda \frac{d}{dt} \mapsto \lambda X$$

is a morphism of the Lie algebra $\mathbb{R}$ to $\mathfrak{g}$. By the fact that the real line $\mathbb{R}$ is simply connected, there exists a unique 1-parameter subgroup $\exp_X : \mathbb{R} \to G$ such that

$$d\exp_X \left(\lambda \frac{d}{dt} \right) = \lambda X.$$

This means that the map $t \mapsto \exp_X(t)$ is the unique 1-parameter subgroup of G whose tangent vector at zero is X_e.

Definition 2.7.21 (Exponential Map) The *exponential map* is defined as the map

$$\exp: \mathfrak{g} \to G,$$
$$X \mapsto \exp_X(1). \tag{2.37}$$

Remark 2.7.22 There is a reason why we denote this map by an *exponential*, namely, we will see that for the general linear group it is indeed given by the matrix exponential.

Theorem 2.7.23 *Let $\mathfrak{g}$ be the Lie algebra of a Lie group G and let $X \in \mathfrak{g}$. Then*

(1) $\exp(tX) = \exp_X(t)$ *for all $t \in \mathbb{R}$.*
(2) $\exp((t_1 + t_2)X) = \exp(t_1 X)\exp(t_2 X)$ *for all $t_1, t_2 \in \mathbb{R}$.*
(3) $\exp(-tX) = \exp(tX)^{-1}$ *for all $t \in \mathbb{R}$.*
(4) $\exp: \mathfrak{g} \to G$ *is smooth and $\mathrm{d}\exp: \mathfrak{g}_0 \to T_eG$ is the identity map. Thus $\exp$ defines a diffeomorphism in a neighborhood of zero in $\mathfrak{g}$ onto a neighborhood of e in G.*
(5) $L_g \circ \exp_X$ *is the unique integral curve of X taking the value of g at zero. This means that left-invariant vector fields are always complete.*
(6) *the 1-parameter group of diffeomorphisms X_t for the left-invariant vector field X is given by*

$$X_t = R_{\exp_X(t)}.$$

Theorem 2.7.24 *Let $\varphi: H \to G$ be a morphism of Lie groups. Then the following diagram is commutative:*

$$
\begin{array}{ccc}
H & \xrightarrow{\ \varphi\ } & G \\
{\scriptstyle \exp}\big\uparrow & & \big\uparrow{\scriptstyle \exp} \\
\mathfrak{h} & \xrightarrow[\ \mathrm{d}\varphi\ ]{} & \mathfrak{g}
\end{array}
$$

Proof Let $X \in \mathfrak{h}$. The map $t \mapsto \varphi(\exp(tX))$ is a smooth curve in G with tangent map at zero being $\mathrm{d}\varphi(X_e)$. In fact, this is also a 1-parameter subgroup of G since φ is a morphism of Lie groups. However, $t \mapsto \exp(t(\mathrm{d}\varphi(X)))$ is the unique 1-parameter subgroup of G with tangent map $\mathrm{d}_e\varphi(X)$. Hence,

$$\varphi(\exp(tX)) = \exp(t(\mathrm{d}\varphi(X))),$$

and therefore

$$\varphi(\exp(X)) = \exp(\mathrm{d}\varphi(X)).$$

$\square$

Example 2.7.25 Let $G = T := \mathbb{R}/\mathbb{Z}$ be the 1-dimensional torus and let $\mathfrak{g} \cong \mathbb{R}$ be its Lie algebra. Consider the exponential map

$$\exp\colon \mathbb{R} \to S^1,$$

$$t \mapsto \exp(2\pi i t).$$

It is easy to see that the group homomorphism is given with respect to addition on $\mathfrak{g}$. The torus T in this case can be identified with $S^1 = \{z \in \mathbb{C} \mid |z| = 1\}$ and the exponential map to be $\theta \mapsto e^{i\theta}$. Note that they differ by their period which in the former case is 1 and in the latter case is 2π, thus, usually, one considers the 1-dimensional torus $\mathbb{R}/2\pi\mathbb{Z}$. More generally, if we consider $G = T^n = (S^1)^n$ with Lie algebra $\mathfrak{t}^n \cong \mathbb{R}^n$, the exponential map is given by

$$\exp\colon \mathfrak{t}^n \cong \mathbb{R}^n \to \mathbb{R}^n/(2\pi\mathbb{Z})^n \cong T^n,$$

which is given by the covering $\mathbb{R}^n \to T^n$.

Proposition 2.7.26 *Let (H, φ) be a Lie subgroup of a Lie group G and let $X \in \mathfrak{g} = \mathrm{Lie}(G)$. If $X \in d\varphi(\mathfrak{h})$, then $\exp(tX) \in \varphi(H)$ for all $t \in \mathbb{R}$. Conversely, if $\exp(tX) \in \varphi(H)$ for t in some open interval, then $X \in d\varphi(\mathfrak{h})$.*

Proof Let $X \in d\varphi(\mathfrak{h})$. Then $\exp(tX) \in \varphi(H)$ for all t by Theorem 2.7.24. Now if $\exp(tX) \in \varphi(H)$ for t in some interval $I \subset \mathbb{R}$, then the map $t \mapsto \exp(tX)$ can be written in terms of a composition $\varphi \circ \alpha$ with $\alpha\colon I \hookrightarrow H$ being a smooth map. For $t_0 \in I$ and $\widetilde{X}$ the left-invariant vector field on H determined by $\frac{d\alpha}{dt}(t_0)$, we get $d\varphi(\widetilde{X}) = X$. $\qquad\square$

Theorem 2.7.27 *Consider some subgroup H of a Lie group G and let $\mathfrak{h}$ be some subspace of the Lie algebra $\mathfrak{g}$ of G. Let U be some neighborhood of zero in $\mathfrak{g}$ which is diffeomorphic to the exponential map with some neighborhood V of $e \in G$. Moreover, assume that $\exp(U \cap \mathfrak{h}) = H \cap V$. Then H is a Lie subgroup of G with respect to the relative topology and $\mathfrak{h}$ is a subalgebra of $\mathfrak{g}$ given by the Lie algebra of H.*

Proof We need to show that for the relative topology, the subgroup H carries a differentiable structure such that $H \subset G$ is a submanifold. Then it will follow that $H \subset G$ is a Lie subgroup and hence that the Lie algebra of H is $\mathfrak{h}$. Define the map

$$\varphi := \exp|_{U \cap \mathfrak{h}}\colon U \cap \mathfrak{h} \to V \cap H.$$

Then we get the differentiability property on H by taking the maximal collection of smoothly overlapping coordinate systems containing the collection

$$\{(H \cap hV, \varphi^{-1} \circ L_{h^{-1}}) \mid h \in H\}.$$

$\qquad\square$

Example 2.7.28 (General Linear Group) We want to show that the exponential map

$$\exp \colon \mathfrak{gl}_n(\mathbb{C}) \to \mathrm{GL}_n(\mathbb{C})$$

is given by the matrix exponential. Let $I \in \mathfrak{gl}_n(\mathbb{C})$ be the identity matrix. The matrix exponential is defined by

$$\mathrm{e}^A := \sum_{n \geq 0} \frac{1}{n!} A^n, \quad A \in \mathfrak{gl}_n(\mathbb{C}),$$

where $A^0 := I$. Note that this expression indeed does converge since the right-hand-side converges uniformly for A in some bounded region of $\mathfrak{gl}_n(\mathbb{C})$. In particular, for some bounded region $\Omega \subset \mathfrak{gl}_n(\mathbb{C})$, there is some $\mu > 0$ such that for any matrix $A = (A_{ij}) \in \Omega$, we have $|A_{ij}| \leq \mu$. By induction, we get $|(A^k)^{ij}| \leq n^{(k-1)}\mu^k$. Using the Weierstrass M-test, the series

$$\sum_{k \geq 0} \frac{1}{k!} (A^k)^{ij}, \quad 1 \leq i \leq n,\ 1 \leq j \leq n,$$

does uniformly converge for $A \in \Omega$. Now define $S_k(A)$ to be the k-th partial sum of the matrix exponential, i.e.

$$S_k(A) = \sum_{n=0}^{k} \frac{1}{n!} A^n$$

and let $B \in \mathfrak{gl}_n(\mathbb{C})$. Note that the map $C \mapsto BC$ is a continuous map of $\mathfrak{gl}_n(\mathbb{C})$ to itself and thus

$$B \left(\lim_{k \to \infty} S_k(A) \right) = \lim_{k \to \infty} (B S_k(A)).$$

Actually, if $B \in \mathrm{GL}_n(\mathbb{C})$, then $B\left(\lim_{k \to \infty} S_k(A)\right) B^{-1} = \lim_{k \to \infty} B S_k(A) B^{-1}$. This implies that

$$B \mathrm{e}^A B^{-1} = \mathrm{e}^{BAB^{-1}}.$$

There is a $B \in \mathrm{GL}_n(\mathbb{C})$ such that the entries of BAB^{-1} are all zero below the diagonal. Let $\lambda_1, \ldots, \lambda_n$ be the diagonal entries of BAB^{-1}. Then $\mathrm{e}^{BAB^{-1}}$ is of the same form with diagonal entries $\mathrm{e}^{\lambda_1}, \ldots, \mathrm{e}^{\lambda_n}$. Since $\det \mathrm{e}^{BAB^{-1}} \neq 0$, we get that $\mathrm{e}^A \in \mathrm{GL}_n(\mathbb{C})$ for any $A \in \mathfrak{gl}_n(\mathbb{C})$. Moreover, we have that $\det \mathrm{e}^A = \mathrm{e}^{\mathrm{Tr}\,A}$.

Exercise 2.7.29 Show that if $AB = BA$, then

$$e^{A+B} = e^A e^B.$$

for all $A, B \in \mathfrak{gl}_n(\mathbb{C})$.

2.7.3.1 Matrix Lie Groups

If we look at different subalgebras of $\mathfrak{gl}_n(\mathbb{C})$, we can take their exponential and obtain some Lie subgroups of $\mathrm{GL}_n(\mathbb{C})$. For a matrix A, we will denote by A^T its transpose.

(1) Unitary group:

$$\mathrm{U}_n := \{A \in \mathrm{GL}_n(\mathbb{C}) \mid A^{-1} = \overline{A^T}\}.$$

(2) Special unitary group:

$$\mathrm{SU}_n := \{A \in \mathrm{U}_n \mid \det A = 1\}.$$

(3) Special linear group:

$$\mathrm{SL}_n(\mathbb{C}) := \{A \in \mathrm{GL}_n(\mathbb{C}) \mid \det A = 1\}.$$

(4) Complex orthogonal group:

$$\mathrm{O}_n(\mathbb{C}) := \{A \in \mathrm{GL}_n(\mathbb{C}) \mid A^{-1} = A^T\}.$$

The corresponding Lie algebras are given by

(1) Skew-Hermitian matrices:

$$\mathfrak{u}_n := \{A \in \mathfrak{gl}_n(\mathbb{C}) \mid \overline{A} + A^T = 0\}.$$

(2) Skew-Hermitian matrices with vanishing trace:

$$\mathfrak{su}_n := \{A \in \mathfrak{u}_n \mid \mathrm{Tr}\, A = 0\}.$$

(3) Matrices with vanishing trace:

$$\mathfrak{sl}_n(\mathbb{C}) := \{A \in \mathfrak{gl}_n(\mathbb{C}) \mid \mathrm{Tr}\, A = 0\}.$$

(4) Skew-symmetric matrices:

$$\mathfrak{o}_n(\mathbb{C}) := \{A \in \mathfrak{gl}_n(\mathbb{C}) \mid A + A^T = 0\}.$$

2.7.4 Smooth Actions

Definition 2.7.30 (Representation) A *representation* of a Lie group G on a vector space V is a group homomorphism

$$G \to GL(V).$$

Definition 2.7.31 (Action) Let M be a manifold and denote by

$$\mathrm{Diff}(M) := \{\varphi \colon M \overset{\sim}{\to} M \mid \varphi \text{ diffeomorphism}\}$$

the *diffeomorphism group* of M. An *action* of a Lie group G on M is a group homomorphism

$$\Psi \colon G \to \mathrm{Diff}(M),$$

$$g \mapsto \Psi_g.$$

Remark 2.7.32 We will only consider *left actions* where Ψ is a homomorphism. A *right action* is defined with Ψ being an anti-homomorphism.

Definition 2.7.33 (Evaluation Map) The *evaluation map* associated with an action

$$\Psi \colon G \to \mathrm{Diff}(M)$$

is given by

$$\mathrm{ev}_\Psi \colon M \times G \to M,$$

$$(q, g) \mapsto \Psi_g(q).$$

Remark 2.7.34 The action Ψ is *smooth* if ev_Ψ is smooth.

2.7.5 Adjoint and Coadjoint Representations

Note that any Lie group G acts on itself by *conjugation*:

$$G \to \mathrm{Diff}(M),$$

$$g \mapsto \Psi_g$$

where $\Psi_g(a) = g \cdot a \cdot g^{-1}$. The derivative at the identity of Ψ_g is an invertible linear map

$$\mathrm{Ad}_g : \mathfrak{g} \to \mathfrak{g}.$$

Note that we have identified the Lie algebra $\mathfrak{g}$ with the tangent space $T_e G$.

Definition 2.7.35 (Adjoint Representation) The *adjoint representation* (or *adjoint action*) of G on $\mathfrak{g}$ is given by

$$\mathrm{Ad} : G \to \mathrm{GL}(\mathfrak{g}),$$

$$g \mapsto \mathrm{Ad}_g .$$

Exercise 2.7.36 Check that for matrix Lie groups

$$\frac{d}{dt} \mathrm{Ad}_{\exp(tX)} Y \bigg|_{t=0} = [X, Y], \quad \forall X, Y \in \mathfrak{g}.$$

Hint: use that for a matrix group G (i.e. a subgroup of $\mathrm{GL}_n(\mathbb{R})$ for some n) we have

$$\mathrm{Ad}_g(Y) = gYg^{-1}, \quad \forall g \in G, \forall Y \in \mathfrak{g},$$

and

$$[X, Y] = XY - YX, \quad \forall X, Y \in \mathfrak{g}.$$

Let $\langle \, , \, \rangle$ be the natural pairing between $\mathfrak{g}^*$ and $\mathfrak{g}$ defined as

$$\langle \, , \, \rangle : \mathfrak{g}^* \times \mathfrak{g} \to \mathbb{R},$$

$$(\xi, X) \mapsto \langle \xi, X \rangle := \xi(X).$$

For $\xi \in \mathfrak{g}^*$, we define $\mathrm{Ad}_g^* \xi$ by the property

$$\langle \mathrm{Ad}_g^* \xi, X \rangle = \langle \xi, \mathrm{Ad}_{g^{-1}} X \rangle, \quad \forall X \in \mathfrak{g}.$$

Definition 2.7.37 (Coadjoint Representation) The collection of maps Ad_g^* forms the *coadjoint representation* (or *coadjoint action*) of G on $\mathfrak{g}^*$:

$$\mathrm{Ad}^* : G \to \mathrm{GL}(\mathfrak{g}^*),$$

$$g \mapsto \mathrm{Ad}_g^* .$$

Exercise 2.7.38 Show that for all $g, h \in G$, we have

$$\mathrm{Ad}_g \circ \mathrm{Ad}_h = \mathrm{Ad}_{gh}, \qquad \mathrm{Ad}_g^* \circ \mathrm{Ad}_h^* = \mathrm{Ad}_{gh}^* .$$

Definition 2.7.39 (Adjoint Action) The *adjoint action* of a Lie algebra $\mathfrak{g}$ is defined as

$$\mathrm{ad} : \mathfrak{g} \times \mathfrak{g} \to \mathfrak{g},$$
$$(X, Y) \mapsto \mathrm{ad}_X(Y) := [X, Y].$$

Remark 2.7.40 In other words, we have a map

$$\mathrm{ad} : \mathfrak{g} \to \mathrm{GL}(\mathfrak{g}),$$
$$X \mapsto \mathrm{ad}_X.$$

Exercise 2.7.41 Show that the differential of the adjoint representation Ad of a Lie group G at its identity element is given by the adjoint action ad on its Lie algebra $\mathfrak{g}$.

Definition 2.7.42 (Killing Form) The *Killing form* of a Lie algebra $\mathfrak{g}$ is defined to be the bilinear form induced by its adjoint action:

$$K : \mathfrak{g} \times \mathfrak{g} \to \mathfrak{g},$$
$$(X, Y) \mapsto K(X, Y) := \mathrm{Tr}(\mathrm{ad}_X \circ \mathrm{ad}_Y)$$

2.7.6 Principal Bundles

Let G be a Lie group and let B be a manifold.

Definition 2.7.43 (Principal G-bundle) A *principal G-bundle over B* is a manifold P with a smooth map $\pi : P \to B$ satisfying:

(1) G acts freely on P (from the left),
(2) B is the orbit space for this action and π is the point-orbit projection, and
(3) there is an open covering of B such that to each set U in that covering corresponds a map $\phi_U : \pi^{-1}(U) \to U \times G$ with

$$\phi_U(q) = (\pi(q), s_U(q)), \quad s_U(g \cdot q) = g \cdot s_U(q), \quad \forall q \in \pi^{-1}(U).$$

The G-valued maps s_U are determined by the corresponding ϕ_U. Condition (3) is called the property of being *locally trivial*.

Remark 2.7.44 The manifold B is usually called the *base*, the manifold P is called the *total space*, the Lie group G is called the *structure group*, and the map π is called the *projection*. We can represent a principal G-bundle by the following diagram:

$$
\begin{array}{ccc}
G & \lhook\joinrel\longrightarrow & P \\
 & & \downarrow{\scriptstyle \pi} \\
 & & B
\end{array}
$$

Example 2.7.45 (Product G-bundles) A simple example of a principal G-bundle over some manifold M is given by the product G-bundle $M \times G \to M$. Here, the identity map serves as a trivialization.

Definition 2.7.46 (Complex Projective Space) Similarly as the real projective space (Example 2.1.16), we can construct a complex $(n + 1)$-manifold $\mathbb{C}P^n$, the *complex projective space*, by

$$
\mathbb{C}P^n := (\mathbb{C}^{n+1} \setminus \{0\})/ \sim
$$

where $z \sim \lambda z$ for all $\lambda \in \mathbb{C} \setminus \{0\}$ with $z \in \mathbb{C}^{n+1}$.

Example 2.7.47 (Hopf Fibration) Consider the circle group $S^1 := \{z \in \mathbb{C} \mid |z| = 1\}$ acting on the vector space $\mathbb{C}^{n+1}$ by left-multiplication. This action induces an action of S^1 on the sphere $S^{2n+1} \subset \mathbb{C}^{n+1}$. One can define the complex projective space $\mathbb{C}P^n$ as the orbit space of S^{2n+1} for S^1. Then the projection $S^{2n+1} \to \mathbb{C}P^n$ with fiber S^1 turns out to be a principal S^1 bundle. The *Hopf fibration* (or *Hopf bundle*) is given for the case where $n = 1$, i.e. $S^3 \to \mathbb{C}P^1$ with fiber S^1. Diagrammatically, we have

$$
\begin{array}{ccc}
S^1 & \lhook\joinrel\longrightarrow & S^3 \\
 & & \downarrow \\
 & & \mathbb{C}P^1 \cong S^2
\end{array}
$$

2.7.7 Lie Algebra Cohomology

If we consider a Lie algebra, there is a natural way of extracting a (co)homology theory out of it. In particular, we can extract a cochain complex as follows.

2.7.7.1 The Chevalley–Eilenberg Complex

Let V be a vector space together with a Lie algebra homomorphism $\rho : \mathfrak{g} \to \mathrm{End}(V)$ for some Lie algebra $(\mathfrak{g}, [\ ,\ ])$. In particular, ρ is a linear map such that

$$
\rho([X, Y]) = \rho(X)\rho(Y) - \rho(Y)\rho(X), \quad \forall X, Y \in \mathfrak{g}.
$$

We define the action of some $X \in \mathfrak{g}$ on some $v \in V$ by

$$X \cdot v = \rho(X)(v).$$

Definition 2.7.48 (Chevalley–Eilenberg Complex) The *Chevalley–Eilenberg complex* of $\mathfrak{g}$ with coefficients in V is given by

$$\cdots \to CE^{k-1}(\mathfrak{g}, \rho) \xrightarrow{\delta} CE^k(\mathfrak{g}, \rho) \xrightarrow{\delta} CE^{k+1}(\mathfrak{g}, \rho) \to \cdots$$

where

$$CE^k(\mathfrak{g}, \rho) := \left(\bigwedge^k \mathfrak{g}^* \right) \otimes V, \quad k \geq 0.$$

An element of $\theta \in CE^k(\mathfrak{g}, \rho)$ is a linear map $\bigwedge^k \mathfrak{g} \to V$. The *Chevalley–Eilenberg differential* $\delta: CE^k(\mathfrak{g}, \rho) \to CE^{k+1}(\mathfrak{g}, \rho)$ is defined by

$$(\delta\theta)(X_1, \ldots, X_{k+1}) = \sum_i (-1)^{i+1} \rho(X_i) \theta(X_1, \ldots, \widehat{X}_i, \ldots, X_{k+1})$$

$$+ \sum_{i<j} (-1)^{i+j} \theta([X_i, X_j], X_1, \ldots, \widehat{X}_i, \ldots, \widehat{X}_j, \ldots, X_{k+1}),$$

where the $\widehat{}$ means that the corresponding element is omitted. The fact that δ is a differential follows immediately from the Jacobi identity of $[\ ,\ ]$.

Definition 2.7.49 (Lie Algebra Cohomology) The *Lie algebra cohomology* with respect to the Lie algebra $\mathfrak{g}$ and with coefficients in V is given by the cohomology induced by the Chevalley–Eilenberg complex. The cohomology groups are defined as

$$H^k(\mathfrak{g}, \rho) = H^k(\mathfrak{g}, V) := \frac{\ker\left(\delta: CE^k(\mathfrak{g}, \rho) \to CE^{k+1}(\mathfrak{g}, \rho)\right)}{\mathrm{im}\left(\delta: CE^{k-1}(\mathfrak{g}, \rho) \to CE^k(\mathfrak{g}, \rho)\right)}$$

Remark 2.7.50 Let G be a connected Lie group with Lie algebra $\mathfrak{g}$. If we consider the space of left-invariant de Rham forms on G, denoted by $\Omega_L^\bullet(G)$, it will form a subcomplex of the de Rham complex of G, which is isomorphic to the Chevalley–Eilenberg complex $CE^\bullet(\mathfrak{g}, \mathbb{R})$, where we consider the trivial action of $\mathfrak{g}$ on $\mathbb{R}$. This implies in particular that

$$H_L^\bullet(G) \cong H^\bullet(\mathfrak{g}, \mathbb{R}).$$

This isomorphism associates to each form in $\Omega_L^\bullet(G)$ its value at the identity $e \in G$ when identifying $\mathfrak{g}^*$ with T_e^*G. Additionally, whenever G is compact, the map

$$\alpha \mapsto \int_G L_g^* \alpha dg, \quad \alpha \in \Omega^\bullet(G)$$

induces another isomorphism $H^\bullet(G) \cong H_L^\bullet(G)$ and hence $H^\bullet(G) \cong H^\bullet(\mathfrak{g}, \mathbb{R})$.

Definition 2.7.51 (Semisimple Lie Algebra) A Lie algebra $\mathfrak{g}$ is called *semisimple* if $\mathfrak{g}$ has no non-zero abelian ideals.

Theorem 2.7.52 (Whitehead) *If $\mathfrak{g}$ is a semisimple Lie algebra and V is a finite-dimensional $\mathfrak{g}$-module, then $H^1(\mathfrak{g}, V) = 0$ and $H^2(\mathfrak{g}, V) = 0$.*

Theorem 2.7.53 (Whitehead) *If $\mathfrak{g}$ is a semisimple Lie group and V is a finite-dimensional $\mathfrak{g}$-module such that $V^\mathfrak{g} = 0$, with $V^\mathfrak{g} := \{v \in V \mid X \cdot v = 0, \forall X \in \mathfrak{g}\}$, then*

$$H^k(\mathfrak{g}, V) = 0, \quad \forall k \geq 0.$$

Remark 2.7.54 If $\mathfrak{g}$ is a simple Lie algebra, then $\dim H^3(\mathfrak{g}, \mathbb{K}) = 1$ for some field $\mathbb{K}$.

Remark 2.7.55 Combining Theorems 2.7.53 and 2.7.52, we get that for a semisimple Lie algebra $\mathfrak{g}$ over some field $\mathbb{K}$ and a finite-dimensional $\mathfrak{g}$-module V we get

$$H^\bullet(\mathfrak{g}, V) = H^\bullet(\mathfrak{g}, \mathbb{K}) \otimes V^\mathfrak{g} = \bigoplus_{\substack{k \geq 0 \\ k \neq 1,2}} H^k(\mathfrak{g}, \mathbb{K}) \otimes V^\mathfrak{g}.$$

Remark 2.7.56 If V is given by a smooth Fréchet module of some compact Lie group G and $\mathfrak{g}$ is the Lie algebra of G, one can check that it still holds that $H^\bullet(\mathfrak{g}, V) = H^\bullet(\mathfrak{g}, \mathbb{R}) \otimes V^\mathfrak{g}$ by a result of Ginzburg. If G is a compact Lie group acting on a smooth manifold M, we get that $C^\infty(M)$ is a smooth Fréchet G-module and thus we have

$$H^\bullet(\mathfrak{g}, C^\infty(M)) = H^\bullet(\mathfrak{g}, \mathbb{R}) \otimes (C^\infty(M))^\mathfrak{g}.$$

2.8   Connections and Curvature on Vector Bundles

2.8.1   The Affine Case

If we consider a general manifold not embedded into some Euclidean space, the notion of a *directional derivative* of a function $f \in C^\infty(M)$ is defined through the direction $X_q \in T_q M$ as

$$\nabla_{X_q} f := X_q(f).$$

If we want to consider the directional derivative of another vector field $Y \in \mathfrak{X}(M)$, there is no canonical way of doing this. Hence, we have to introduce the notion of a *connection*.

Definition 2.8.1 (Affine Connection) An *affine connection* on a manifold M is an $\mathbb{R}$-bilinear map

$$\nabla \colon \mathfrak{X}(M) \times \mathfrak{X}(M) \to \mathfrak{X}(M),$$

where we denote $\nabla_X Y := \nabla(X, Y)$, such that it satisfies the following properties:

(1) $\nabla_X Y$ is $C^\infty(M)$-linear in X,
(2) $\nabla_X Y$ satisfies the *Leibniz rule* in Y, i.e. for all $f \in C^\infty(M)$, we have

$$\nabla_X(fY) = X(f)Y + f\nabla_X Y.$$

Definition 2.8.2 (Torsion) Let ∇ be an affine connection on some manifold M. We define the *torsion* of the connection ∇ to be given by

$$T(X, Y) := \nabla_X Y - \nabla_Y X - [X, Y], \quad X, Y \in \mathfrak{X}(M).$$

Remark 2.8.3 We say that an affine connection is *torsion-free*, if its torsion tensor vanishes for all vector fields, i.e.

$$T(X, Y) := \nabla_X Y - \nabla_Y X - [X, Y] = 0, \quad \forall X, Y \in \mathfrak{X}(M).$$

Definition 2.8.4 (Curvature) Let ∇ be an affine connection on some manifold M. We define the *curvature* of the connection ∇ to be given by

$$R(X, Y) := [\nabla_X, \nabla_Y] - \nabla_{[X,Y]}$$
$$= \nabla_X \nabla_Y - \nabla_Y \nabla_X - \nabla_{[X,Y]} \in \mathrm{End}(\mathfrak{X}(M)).$$

Definition 2.8.5 (Compatible Connection) An affine connection ∇ on a Riemannian manifold (M, g) is said to be *compatible with the metric g* if

$$v(g(X, Y)) = g(\nabla_v X, Y) + g(X, \nabla_v Y), \quad \forall X, Y \in \mathfrak{X}(M), \forall v \in T_q M, q \in M.$$

where we considered v to be a derivation applied to the smooth map $q \mapsto g_q(X_q, Y_q)$ for $q \in M$.

Theorem 2.8.6 (Levi-Civita) *For any smooth Riemannian manifold (M, g) with metric g there exists a unique affine connection ∇ which is torsion-free and compatible with the metric g.*

Proof We will first show uniqueness. Let $X, Y, Z \in \mathfrak{X}(M)$. Then, the torsion-free condition gives us the following three equations:

$$\nabla_X Y - \nabla_Y X = [X, Y], \tag{2.38}$$

$$\nabla_Y Z - \nabla_Z Y = [Y, Z], \tag{2.39}$$

$$\nabla_Z X - \nabla_X Z = [Z, X]. \tag{2.40}$$

Moreover, the condition of being compatible with the metric g gives us the following three equations:

$$g(\nabla_X Y, Z) + g(Y, \nabla_X Z) = Xg(Y, Z), \tag{2.41}$$

$$g(\nabla_Y Z, X) + g(Z, \nabla_Y X) = Yg(Z, X), \tag{2.42}$$

$$g(\nabla_Z X, Y) + g(X, \nabla_Z Y) = Zg(X, Y). \tag{2.43}$$

Now take the sum of (2.41) and (2.42), and then subtract (2.43). Moreover, apply the torsion-free conditions (2.38)–(2.40) to obtain

$$2g(\nabla_X Y, Z) = Xg(Y, Z) + Yg(Z, X) - Zg(X, Y)$$
$$+ g(Z, [X, Y]) + g(Y, [Z, X]) + g(X, [Z, Y]). \tag{2.44}$$

It is not hard to see that (2.44) does indeed completely describe $\nabla_X Y$ and thus we have shown uniqueness. Next we prove existence. We need to check that (2.44) does indeed define a connection with the correct properties. In order to do this, let us define the right-hand-side of (2.44) by $L(X, Y, Z)$. Let us first show that L is smooth in the X and Z argument. Indeed, for $f, h \in C^\infty(M)$, we have

$$L(fX, Y, hZ) = fX(hg(Y, Z)) + Y(fhg(Z, X)) - hZ(fg(X, Y))$$
$$+ hg(Z, [fX, Y]) + g(Y, [hZ, fX]) + fg(X, [hZ, Y])$$
$$= fhXg(Y, Z) + fhYg(Z, X) - fhZg(X, Y)$$
$$+ f(Xh)g(Y, Z) + fY(h)g(Z, X)$$
$$+ hY(f)g(Z, X) - hZ(f)g(X, Y)$$
$$+ hfg(Z, [X, Y]) + hfg(Y, [Z, X]) + fhg(X, [Z, Y])$$
$$- hY(f)g(Z, X) - fX(h)g(Y, Z)$$
$$+ hZ(f)g(X, Y) - fY(h)g(X, Z)$$
$$= fhL(X, Y, Z).$$

Hence, the map $q \mapsto L(X, Y, Z)_q$ does only depend on Y and on X_q and Z_q. Moreover, we have

$$L(X, fY, Z) = fL(X, Y, Z) + X(f)g(Y, Z) - Z(f)g(X, Y) + X(f)g(Z, Y) + Z(f)g(X, Y)$$
$$= fL(X, Y, Z) + 2X(f)g(Y, Z).$$

Thus, when we define $\nabla_X Y$ by setting $g(\nabla_X Y, Z) = \frac{1}{2}L(X, Y, Z)$, we get that $(\nabla_X Y)_q$ does only depend on X_q and Y, and

$$g(\nabla_X(fY), Z) = fg(\nabla_X Y, Z) + X(f)g(Y, Z) = g(f\nabla_X Y + X(f)Y, Z).$$

Hence, we have shown that ∇ satisfies the Leibniz rule and therefore is indeed a connection. $\qquad\square$

Remark 2.8.7 The connection ∇ is defined through its way how it is applied to basis elements of the tangent space. In particular, for a connection ∇ we have

$$\nabla_{\partial_i}\partial_j = \Gamma_{ij}^k \partial_k,$$

where the functions $\Gamma_{ij}^k \in C^\infty(M)$ are called the *Christoffel symbols* and they fully determine the connection ∇. When considering the Levi-Civita connection, we can apply Eq. (2.44) to the case where X, Y, Z are given by $\partial_i, \partial_j, \partial_k$ to obtain

$$\Gamma_{ij}^k = \frac{1}{2}g^{k\ell}(\partial_i g_{j\ell} + \partial_j g_{i\ell} - \partial_\ell g_{ij}),$$

where g^{ij} are the components of the inverse matrix of (g_{ij}).

2.8.2 Generalization to Vector Bundles

Definition 2.8.8 (Connection on a Vector Bundle) Let $E \to M$ be a vector bundle over M. A *connection* on E is a map

$$\nabla : \mathfrak{X}(M) \times \Gamma(E) \to \Gamma(E)$$

such that for $X \in \mathfrak{X}$ and $\sigma \in \Gamma(E)$ we get

(1) $\nabla_X \sigma$ is $C^\infty(M)$-linear in X and $\mathbb{R}$-linear in σ,
(2) $\nabla_X \sigma$ satisfies the *Leibniz rule* in σ, i.e. for all $f \in C^\infty(M)$, we have

$$\nabla_X(f\sigma) = X(f)\sigma + f\nabla_X\sigma.$$

Moreover, since $X(f) = \mathrm{d}f(X)$, we can also express it as

$$\nabla_X(f\sigma) = \mathrm{d}f(X)\sigma + f\nabla_X\sigma.$$

Example 2.8.9 Any affine connection on a manifold M is a connection on the tangent bundle $TM \to M$.

Example 2.8.10 Let $M \hookrightarrow \mathbb{R}^n$ be a submanifold of $\mathbb{R}^n$ and let $E := T\mathbb{R}^n|_M$ be the restriction of the tangent bundle of $\mathbb{R}^n$ to M. Define the directional derivative

$$D : \mathfrak{X}(M) \times \Gamma(T\mathbb{R}^n|_M) \to \Gamma(T\mathbb{R}^n|_M),$$

$$(X, Y) \mapsto D_X Y := \sum_i X(Y^i)\partial_i.$$

It is then easy to see that this defines a connection on $T\mathbb{R}^n|_M$.

Example 2.8.11 (Connection on Trivial Bundle) Let $E \to M$ be a trivial bundle of rank r over some manifold M. Consider then the trivialization map $\phi \colon E \to M \times \mathbb{R}^r$. This map induces a connection on E as follows: For a chosen basis $b_1, \ldots, b_r$ of $\mathbb{R}^r$, we can consider a global frame for the trivial bundle $M \times \mathbb{R}^r \to M$ through the maps $s_i \colon q \mapsto (q, b_i)$ for $i = 1, \ldots, r$. Thus, the maps $e_i \colon \phi^{-1} \circ s_i$ define a global frame for $E \to M$ for $i = 1, \ldots, r$. Hence, each section $\sigma \in \Gamma(E)$ can be uniquely written as a linear combination

$$\sigma = \sum_{1 \le i \le r} \sigma^i e_i, \quad \sigma^i \in C^\infty(M).$$

Therefore, we can define a connection ∇ on E by considering the sections e_i to be *flat* for all i, i.e. $\nabla_X e_i = 0$ for any X, and apply the Leibniz rule and $\mathbb{R}$-linearity to define

$$\nabla_X \sigma := \nabla_X \left(\sum_{1 \le i \le r} \sigma^i e_i \right) = \sum_{1 \le i \le r} X(\sigma^i) e_i. \tag{2.45}$$

Definition 2.8.12 (Flat Section) Let $E \to M$ be a vector bundle and let ∇ be a connection on E. A section $\sigma \in \Gamma(E)$ is called *flat* (or *covariantly constant*), if its covariant derivative vanishes, i.e. for all $X \in \mathfrak{X}(M)$ we have $\nabla_X \sigma = 0$.

Exercise 2.8.13 Prove that (2.45) does indeed define a connection on the trivial bundle.

An important Proposition is the following:

Proposition 2.8.14 *Every smooth vector bundle $E \to M$ over some manifold M has a connection.*

Proof Consider an open cover $\{U_\alpha\}_{\alpha\in I}$ of E together with a partition of unity $\{\rho_\alpha\}_{\alpha\in I}$ subordinate to $\{U_\alpha\}_{\alpha\in I}$. On each open set U_α, we get a trivial bundle $E|_{U_\alpha}$ and thus it has a connection ∇^α by Example 2.8.11. Let $X \in \mathfrak{X}(M)$ and $\sigma \in \Gamma(E)$, and denote by $\sigma_\alpha := \sigma|_{U_\alpha}$ the restriction of the section σ to U_α. Then, we can define

$$\nabla_X \sigma := \sum_{\alpha\in I} \rho_\alpha \nabla^\alpha_X \sigma_\alpha.$$

It can be checked that this sum is indeed finite in a neighborhood of a point $q \in M$, since by local finiteness of $\{\mathrm{supp}\,\rho_\alpha\}_{\alpha\in I}$, there exists an open neighborhood U of q which intersects only finitely many sets of $\{\mathrm{supp}\,\rho_\alpha\}_{\alpha\in I}$. This in fact shows that on U all but finitely many of the maps ρ_α are zero and thus $\sum_{\alpha\in I} \rho_\alpha \nabla^\alpha_X \sigma_\alpha$ is finite on U with the condition $\sum_{\alpha\in I} \rho_\alpha = 1$. We leave it as an exercise to check that $\nabla_X \sigma$ is $C^\infty(M)$-linear in X and $\mathbb{R}$-linear and satisfies the Leibniz rule in σ. We can conclude then that ∇ is a connection on E. $\qquad\square$

The notion of curvature still makes sense in the case of vector bundles. Hence, we can define it for $X, Y \in \mathfrak{X}(M)$ and $\sigma \in \Gamma(E)$ as

$$R(X, Y)\sigma = \nabla_X \nabla_Y \sigma - \nabla_Y \nabla_X \sigma - \nabla_{[X,Y]}\sigma \in \Gamma(E).$$

We call R the *curvature tensor* of ∇.

Definition 2.8.15 (Flat Connection) A connection ∇ is called *flat* if its curvature tensor vanishes.

2.8.3 Interpretation as Differential Forms

We want to understand how a connection on a vector bundle can be described locally. Let $E \to M$ be a vector bundle of rank r and let ∇ be a connection on it. We can restrict the connection ∇ on each open subset $U \subset E$ to a connection

$$\nabla^U : \mathfrak{X}(U) \times \Gamma(E|_U) \to \Gamma(E|_U).$$

Consider now a trivial open set U of E and some frame $e_1, \ldots, e_r$ for E over U. Moreover, let $X \in \mathfrak{X}(U)$. Note that we can write each section $\sigma \in \Gamma(E|_U)$ as a linear combination $\sigma = \sum_j \sigma^j e_j$ and thus we can describe $\nabla_X \sigma$ only by using $\nabla_X e_j$ due to linearity and the Leibniz rule. In particular, locally we can write

$$\nabla_X e_j = \sum_{1\le i\le r} \omega^i_j(X) e_i,$$

where ω^i_j are some coefficients depending on the vector field X. Since $\nabla_X e_j$ is $C^\infty(M)$-linear, it follows that also ω^i_j is $C^\infty(M)$-linear and hence ω^i_j defines a 1-form on U.

Definition 2.8.16 (Connection Forms) The 1-forms $\omega^i_j \in \Omega^1(U)$ are called *connection 1-forms* for the connection ∇ relative to the frame $e_1, \ldots, e_r$.

We can also express the curvature tensor locally in the frame $e_1, \ldots, e_r$. We can write

$$R(X, Y)e_j = \sum_{1 \le j \le r} \Omega^i_j(X, Y)e_i, \quad X, Y \in \mathfrak{X}(U). \tag{2.46}$$

Note that by definition of the curvature tensor, R is alternating and $C^\infty(M)$-bilinear, which implies that also Ω^i_j is alternating and $C^\infty(M)$-bilinear. This implies that Ω^i_j is a 2-form on U.

Definition 2.8.17 (Curvature Forms) The 2-forms $\Omega^i_j \in \Omega^2(U)$ are called *curvature 2-forms* for the connection ∇ relative to the frame $e_1, \ldots, e_r$.

Theorem 2.8.18 *Let ∇ be a connection on a vector bundle $E \to M$ of rank r and let U be a trivializing open set of E together with a frame $e_1, \ldots, e_r$ for E. Then, relative to the frame, the curvature forms Ω^i_j are given by the* second structural equation:

$$\Omega^i_j = d\omega^i_j + \sum_{1 \le k \le r} \omega^i_k \wedge \omega^k_j.$$

Proof Consider two vector fields $X, Y \in \mathfrak{X}(U)$. then we have

$$\nabla_X \nabla_Y e_j = \nabla_X \left(\sum_{1 \le k \le r} \omega^k_j(Y)e_k \right)$$

$$= \sum_{1 \le k \le r} X\omega^k_j(Y)e_k + \sum_{1 \le k \le r} \omega^k_j(Y)\nabla_X e_k$$

$$= \sum_{1 \le i \le r} X\omega^i_j(Y)e_i + \sum_{1 \le i,k \le r} \omega^k_j(Y)\omega^i_k(X)e_i.$$

Thus, we have

$$\nabla_Y \nabla_X e_j = \sum_{1 \le i \le r} Y\omega^i_j(X)e_i + \sum_{1 \le i,k \le r} \omega^k_j(X)\omega^i_k(Y)e_i.$$

Moreover, we have

$$\nabla_{[X,Y]}e_j = \sum_{1 \le i \le r} \omega^i_j([X,Y])e_i.$$

If we put everything together, we get

$$R(X,Y)e_j = \nabla_X \nabla_Y e_j - \nabla_Y \nabla_X e_j - \nabla_{[X,Y]}e_j$$
$$= \left(X\omega^i_j(Y) - Y\omega^i_j(X) - \omega^i_j([X,Y])\right)e_j + \left(\omega^i_k(X)\omega^k_j(Y) - \omega^i_k(Y)\omega^k_j(X)\right)e_j.$$
$$(2.47)$$

If we now use the equations

$$(\alpha \wedge \beta)(X,Y) = \alpha(X)\beta(Y) - \alpha(Y)\beta(X), \qquad (2.48)$$

$$d\alpha(X,Y) = X\alpha(Y) - Y\alpha(X) - \alpha([X,Y]), \qquad (2.49)$$

we get that (2.47) is given by

$$d\omega^i_j(X,Y)e_i + \omega^i_k \wedge \omega^k_j(X,Y)e_i = \left(d\omega^i_j + \omega^i_k \wedge \omega^k_j\right)(X,Y)e_i, \qquad (2.50)$$

where we have used the Einstein summation convention. Comparing with (2.46), we get

$$\Omega^i_j = d\omega^i_j + \sum_{1 \le k \le r} \omega^i_k \wedge \omega^k_j.$$

$\square$

Exercise 2.8.19 Prove Eqs. (2.48) and (2.49).

2.9 Distributions and Frobenius' Theorem

2.9.1 Plane Distributions

Definition 2.9.1 (Plane Distribution) A *k-plane distribution* D, or simply a *k-distribution* or just a *distribution*, on a smooth *n*-dimensional manifold M is a collection $\{D_q\}_{q \in M}$ of linear *k*-dimensional subspaces $D_q \in T_q M$ for all $q \in M$.

Remark 2.9.2 The number k is called the *rank* of the distribution. Obviously, we want $k \le n$.

Definition 2.9.3 (Distribution) A *k-distribution* D on M is called *smooth* if every $q \in M$ has an open neighborhood U and smooth vector fields $X_1, \ldots, X_k$ defined on U such that

$$D_x = \mathrm{span}\{(X_1)_x, \ldots, (X_k)_x\}, \quad \forall x \in U.$$

The vector fields $X_1, \ldots, X_k$ are also called *(local) generators* for D on U.

Remark 2.9.4 In the case when $U = M$, we speak of *global generators*. The existence of local generators is required since there are certain interesting distributions which do not have global generators.

We say that a vector field $X \in \mathfrak{X}(M)$ is a *tangent* to a distribution D if $X_q \in D_q$ for all $q \in M$. Note that any linear combination of tangent vector fields are again tangent.

Definition 2.9.5 (Involutive Distribution) A smooth distribution D on M is called *involutive* if $\Gamma(D)$ is a Lie subalgebra of $\mathfrak{X}(M)$, i.e. when $[X, Y] \in \Gamma(D)$ for all $X, Y \in \Gamma(D)$.

Remark 2.9.6 A distribution generated by vector fields $X_1, \ldots, X_k$ is involutive if and only if $[X_i, X_j]$ is a linear combination over $C^\infty(M)$ of the generators. The only-if side follows from the definition. The if-implication can be obtained by observing that a vector field tangent to the distribution is necessarily a linear combination of the generators. Moreover, we have

$$\left[\sum_i f_i X_i, \sum_j g_j Y_j \right] = \sum_{i,j} \left(\Big(f_i X_i(g_j) - g_i X_i(f_j) \Big) X_j + f_i g_j [X_i, X_j] \right).$$

Remark 2.9.7 Note that the previous discussion directly implies that any distribution of rank 1 is involutive. Locally, it is generated by a single vector field X and by the skew-symmetry of the Lie bracket, we have $[X, X] = 0$.

Definition 2.9.8 (Push-forward of a Distribution) Let D be a distribution on M and let $F \colon M \to N$ be a diffeomorphism. We define the *push-forward* $F_* D$ of D by

$$(F_* D)_y := \mathrm{d}_{F^{-1}(y)} D_{F^{-1}(y)}, \quad \forall y \in N.$$

2.9.2 Frobenius' Theorem

Definition 2.9.9 (Integral Manifold) An immersion $\psi \colon N \to M$ with N connected is called an *integral manifold* for a distribution D on M if

$$\mathrm{d}_n \psi(T_n N) = D_{\psi(n)}, \quad \forall n \in N.$$

Remark 2.9.10 An integral manifold that is not a proper restriction of an integral manifold is called *maximal*.

Remark 2.9.11 If ψ is an embedding, restricting $\psi : N \to \psi(N)$ to its image allows us to rewrite the above condition as

$$\psi_*(TN) = D\big|_{\psi(N)},$$

where the fact that D can be restricted to $\psi(N)$, i.e. $D_q \in T_q\psi(N)$ for all $q \in \psi(N)$, is part of the condition.

Definition 2.9.12 (Integrable Distribution) A smooth distribution D on M is called *integrable* if for all $q \in M$ there is an integral manifold D passing through q.

Lemma 2.9.13 *If a distribution D is integrable, then D is involutive.*

Proof For all $q \in M$, we can find an integral manifold $\psi : N \to M$ with $q \in \psi(N)$. If X and Y are tangent to D, in a neighborhood of q in $\psi(N)$ we can write them as push-forwards of vector fields $\widetilde{X}$ and $\widetilde{Y}$ on N. Since the push-forward preserves the Lie bracket and TN is involutive, we see that in this neighborhood $Z := [X, Y]$ is the push-forward of $[\widetilde{X}, \widetilde{Y}]$ and hence tangent to D. Note also that this is indeed the Lie bracket of X and Y, as they do not have components transverse to $\psi(N)$ by definition. We can compute Z by this procedure at each point in M, which shows that D is involutive. $\square$

Proposition 2.9.14 *Let X be a vector field on a Hausdorff manifold M. Let $q \in M$ be a point such that $X_q \neq 0$. Then there is a chart (U, ϕ_U) with $U \ni q$ such that $(\phi_U)_* X\big|_U$ is the constant vector field $(1, 0, \ldots, 0)$. As a consequence, if γ is an integral curve of X passing through U, then $\phi_U \circ \gamma$ is of the form $\{x \in \phi_U(U) \mid x^1(t) = x_0^1 + t;\ x^j(t) = x_0^j,\ j > 1\}$ where the x_0^is are constants.*

Theorem 2.9.15 (Frobenius[Fro77]) *Let D be an involutive k-distribution on a smooth, Hausdorff n-dimensional manifold M. Then each point $q \in M$ has a chart neighborhood (U, ϕ) such that $\phi_* D = \mathrm{span}\left\{\frac{\partial}{\partial x^1}, \ldots, \frac{\partial}{\partial x^k}\right\}$, where $x^1, \ldots, x^k$ are coordinates on $\phi(U)$.*

Corollary 2.9.16 *On a smooth, Hausdorff manifold a smooth distribution is involutive if and only if it is integrable.*

Proof of Theorem 2.9.15 We will use induction on the rank k of the distribution. For $k = 1$, This is the content of Proposition 2.9.14. We assume that we have proved the theorem for rank $k - 1$. Let $(X_1, \ldots, X_k)$ be generators of the distribution in a neighborhood of q. In particular, they are all not vanishing at q. By Proposition 2.9.14, we can find a chart neighborhood (V, χ) of q with $\chi(q) = 0$ and $\chi_* X_1 = \frac{\partial}{\partial y^1}$, where $y^1, \ldots, y^n$ are coordinates on $\chi(V)$. We can define new generators of $\chi_* D$ by

$$Y_1 := \chi_* X_1 = \frac{\partial}{\partial y^1}$$

and for $i > 1$

$$Y_i := \chi_* X_i - (\chi_* X_i(y^1)) \chi_* X_1.$$

For $i > 1$ we have $Y_i(y^1) = 0$ and thus, for $i, j > 1$, we have $[Y_i, Y_j](y^1) = 0$. This means that the expansion of $[Y_i, Y_j]$ in the Y_ℓs does not contain Y_1. Hence, the distribution D' defined on $S := \{y \in \chi(V) \mid y^1 = 0\}$ as the span of $Y_2, \ldots, Y_k$ is involutive. By the induction assumption, we can find a neighborhood U of 0 in S and a diffeomorphism τ such that $\tau_* Y_i = \frac{\partial}{\partial w^i}$, for $i = 2, \ldots, w$, where $w^2, \ldots, w^n$ are coordinates on $\tau(U)$. Let $\widetilde{U} := U \times (-\varepsilon, \varepsilon)$ for some $\varepsilon > 0$ such that $\widetilde{U} \subset \chi(V)$. We then have the projection $\pi : \widetilde{U} \to U$. Finally, consider the diffeomorphism

$$\widetilde{\tau} : \widetilde{U} \to \tau(U) \times (-\varepsilon, \varepsilon),$$

$$(u, y^1) \mapsto (\tau(u), y^1),$$

and write $x^1 = y^1$, $x^i = \tau^i(y^2, \ldots, y^n) = w^i$ for $i > 1$. We write $Z_i := \widetilde{\tau}_* Y_i$ for $i = 1, \ldots, k$ being the generators of the distribution $\widetilde{D} := \widetilde{\tau}_* \chi_* D$. Now since $\frac{\partial x^i}{\partial y^1}$ is equal to one if $i = 1$ and zero otherwise, we get that $Z_1 = \frac{\partial}{\partial x^1}$. For $i = 2, \ldots, k$ and $j > 1$ we have

$$\frac{\partial}{\partial x^1}(Z_i(x^j)) = Z_1(Z_i(x^j)) = [Z_1, Z_i](x^j) = \sum_{\ell=2}^{k} c_i^\ell Z_\ell(x^j),$$

where the c_i^ℓ are functions that are guaranteed to exist by the involutivity of the distribution. For fixed j and fixed $x^2, \ldots, x^n$ we regard these identities as ODEs in the variable x^1. Note that, for $i = 2, \ldots, k$ and $j > k$, we have $Z_i(x^j) = 0$ at $x^1 = 0$ (since at $x^1 = 0$ we have $Z_i = Y_i$). This means that $Z_i(x^j) = 0$ for $i = 2, \ldots, k$ and $j > k$, is the unique solution with this initial condition. These identities mean that $\widetilde{D}$ is the distribution spanned by the vector fields $\frac{\partial}{\partial x^1}, \ldots, \frac{\partial}{\partial x^k}$.   □

2.10   Connections and Curvature on Principal Bundles

2.10.1   Vertical and Horizontal Subbundles

Consider a Lie group G and its Lie algebra $\mathfrak{g}$. Moreover, let $\pi : P \to M$ be a principal G-bundle.

Definition 2.10.1 (Fundamental Vector Field) The *fundamental vector field* on a principal G-bundle P associated to an element $X \in \mathfrak{g}$ is defined as

$$X_p^\# = \frac{d}{dt}\Big|_{t=0} p \exp(tX) \in T_p P.$$

Proposition 2.10.2 *For every $X \in \mathfrak{g}$, the associated fundamental vector field $X^{\#}$ is smooth on P.*

Definition 2.10.3 (Vertical Subbundle) The *vertical bundle* of a principal bundle P is defined as the subbundle $V \subset TP$ given by

$$V := \ker(\mathrm{d}\pi),$$

where $\mathrm{d}\pi : TP \to TM$ is the tangent map of the bundle map π.

Definition 2.10.4 (Horizontal Distribution) A distribution H on P is called *horizontal*, if it is the complement of the vertical subbundle $V \subset TP$, i.e.

$$T_p P = V_p \oplus H_p, \quad \forall p \in P.$$

If H is a horizontal distribution on P, we can define the map $j_p : G \to P$, $j_p(g) = pg$ for $p \in P$. The vertical space V_p is then canonically identified with the Lie algebra $\mathfrak{g}$ through the isomorphism induced by j_p as $(j_p)_* : \mathfrak{g} \to V_p$. Note also that the fundamental vector field of an element $X \in \mathfrak{g}$ can be obtained by using the tangent map $(j_p)_*$ as

$$(j_p)_* X = \frac{\mathrm{d}}{\mathrm{d}t}\Big|_{t=0} j_p(\exp(tX)) = \frac{\mathrm{d}}{\mathrm{d}t}\Big|_{t=0} p\exp(tX) = X_p^{\#}.$$

2.10.2 Ehresmann Connection and Curvature

In order to understand the notion of a connection on a principal bundle, let us define at first

$$\omega_p := ((j_p)_*)^{-1} \circ \pi_{\mathrm{vert}} : T_p P \xrightarrow{\pi_{\mathrm{vert}}} V_p \xrightarrow{((j_p)_*)^{-1}} \mathfrak{g}, \tag{2.51}$$

where $\pi_{\mathrm{vert}} : T_p P = V_p \oplus H_p \to V_p$ denotes the vertical projection. It is easy to see that ω defines a $\mathfrak{g}$-valued 1-form on P.

Theorem 2.10.5 *If H is a smooth right-invariant horizontal distribution on a principal G-bundle $\pi : P \to M$, the $\mathfrak{g}$-valued 1-form ω on P as defined in (2.51) satisfies the following properties:*

(1) for any $A \in \mathfrak{X}(\mathfrak{g})$ and $p \in P$, we have $\omega_p(X_p^{\#}) = X$,
(2) for any $g \in G$ we have $R_g^* \omega = \mathrm{Ad}_g^{-1} \omega$, where R_g denotes right-multiplication with g,
(3) ω is smooth.

Definition 2.10.6 (Ehresmann Connection) An *Ehresmann connection* (or just a *connection*) on a principal G-bundle $P \to M$ is a $\mathfrak{g}$-valued 1-form ω on P satisfying the properties of Theorem 2.10.5.

Definition 2.10.7 (Curvature on Principal Bundle) Let ω be an Ehresmann connection on a principal G-bundle $P \to M$. The *curvature* of the connection ω is then defined as the $\mathfrak{g}$-valued 2-form

$$\Omega = d\omega + \frac{1}{2}[\omega, \omega].$$

Exercise 2.10.8 Let $P \to M$ be a principal bundle over some manifold M endowed with an Ehresmann connection ω. Let X, Y be horizontal vector fields on P. Show that

(1) $\Omega(X, Y) = -\omega([X, Y])$.
(2) $[X, Y]$ is horizontal if and only if $\Omega(X, Y) = 0$.

2.11 Basics of Category Theory

There is a generalization of the concept of sets and maps between them which turns out to be quite helpful in order to think of certain mathematical structure in a more general setting. In particular, when considering e.g. the two sets $\{1, 2\}$ and $\{2, 3\}$ then they are not equal as sets but they are isomorphic, meaning that in a certain sense the are still the same. The concept of isomorphism can be generalized using the methods of category theory. It also turns out that it provides a suitable language in order to axiomatically define what a quantum field theory is as we will see later on (see Sect. 6.1). It will also provide a way of interpreting certain concepts of deformation quantization from different point of view which helps to understand the underlying structure in a much deeper way (see Sect. 5.5). We will restrict ourselves here only to the basic definitions in order to understand the applications we need later on. For more about category theory we refer e.g. to the excellent reference [Mac71].

2.11.1 Definition of a Category

Definition 2.11.1 (Category) A *category* $\mathscr{C}$ consists of the following data:

(1) A collection of *objects*, usually denoted by $\mathrm{obj}(\mathscr{C})$.
(2) A collection of *morphisms* $\mathrm{Hom}_{\mathscr{C}}(x, y)$ between any two objects $x, y \in \mathrm{obj}(\mathscr{C})$.
(3) A unique *identity morphism* $\mathrm{id}_x \in \mathrm{Hom}_{\mathscr{C}}(x, x)$ for each object $x \in \mathrm{obj}(\mathscr{C})$.

(4) A composition map

$$\circ\colon \mathrm{Hom}_{\mathscr{C}}(y, z) \times \mathrm{Hom}_{\mathscr{C}}(x, y) \to \mathrm{Hom}_{\mathscr{C}}(x, z)$$

For any objects $x, y, z \in \mathrm{obj}(\mathscr{C})$.

Moreover, they satisfy the following axioms:

(i) For any pair of objects $x, y \in \mathrm{obj}(\mathscr{C})$ and any morphism $f \in \mathrm{Hom}_{\mathscr{C}}(x, y)$ we get

$$f \circ \mathrm{id}_x = f,$$

and

$$\mathrm{id}_y \circ f = f.$$

(ii) For any objects $w, x, y, z \in \mathrm{obj}(\mathscr{C})$ and any morphisms $f \in \mathrm{Hom}_{\mathscr{C}}(x, y)$, $g \in \mathrm{Hom}_{\mathscr{C}}(y, z), h \in \mathrm{Hom}_{\mathscr{C}}(z, w)$, we have

$$(h \circ g) \circ f = h \circ (g \circ f).$$

Remark 2.11.2 Usually we will just write $x \in \mathscr{C}$ instead of $x \in \mathrm{obj}(\mathscr{C})$ in order to avoid any cumbersome notation. Moreover, we will often denote the space of morphisms by Hom instead of $\mathrm{Hom}_{\mathscr{C}}$ whenever the underlying category is understood.

Example 2.11.3 (Category of Vector Spaces) The *category of vector spaces* over a field k is denoted by **Vect**$_k$. Objects are vector spaces and morphisms are linear homomorphisms.

Example 2.11.4 (Category of Topological Spaces) The *category of topological spaces* is denoted by **Top**. Objects are given by topological spaces and morphisms are given by continuous maps.

Definition 2.11.5 (Small Category) A category $\mathscr{C}$ is called *small* if the collection of objects $\mathrm{obj}(\mathscr{C})$ is a set and for any two objects $x, y \in \mathscr{C}$ the collection of morphisms $\mathrm{Hom}_{\mathscr{C}}(x, y)$ is a set.

Definition 2.11.6 (Opposite Category) For any category $\mathscr{C}$ we can form its *opposite category* $\mathscr{C}^{op}$ by having the same objects as $\mathscr{C}$ but the morphisms are reversed, i.e. $\mathrm{Hom}_{\mathscr{C}^{op}}(x, y) = \mathrm{Hom}_{\mathscr{C}}(y, x)$.

2.11.2 Functors

Similarly as we are considering morphisms between objects of small categories, which forms a set, we can ask ourselves what a corresponding notion is for a category rather than a set. This is the notion of a *functor* between categories.

Definition 2.11.7 (Functor) A *functor* $F : \mathscr{C} \to \mathscr{D}$ between two categories $\mathscr{C}$ and $\mathscr{D}$ consists of the following data:

(1) A map $\mathrm{obj}(\mathscr{C}) \to \mathrm{obj}(\mathscr{D})$ which is also denoted by F.
(2) For any objects $x, y \in \mathscr{C}$ a map of sets

$$\mathrm{Hom}_{\mathscr{C}}(x, y) \to \mathrm{Hom}_{\mathscr{D}}(F(x), F(y)).$$

Moreover, they satisfy the following axioms:

(i) For any object $x \in \mathscr{C}$ we have $F(\mathrm{id}_x) = \mathrm{id}_{F(x)}$.
(ii) For any objects $x, y, z \in \mathscr{C}$ and morphisms $f \in \mathrm{Hom}_{\mathscr{C}}(x, y)$, $g \in \mathrm{Hom}_{\mathscr{C}}(y, z)$ we have

$$F(g \circ f) = F(g) \circ F(f).$$

Definition 2.11.8 (Faithful) A functor $F : \mathscr{C} \to \mathscr{D}$ is called *faithful* if the map

$$\mathrm{Hom}_{\mathscr{C}}(x, y) \to \mathrm{Hom}_{\mathscr{D}}(F(x), F(y))$$

is injective for any objects $x, y \in \mathscr{C}$.

Definition 2.11.9 (Full) A functor $F : \mathscr{C} \to \mathscr{D}$ is called *full* if the map

$$\mathrm{Hom}_{\mathscr{C}}(x, y) \to \mathrm{Hom}_{\mathscr{D}}(F(x), F(y))$$

is surjective for any objects $x, y \in \mathscr{C}$.

Definition 2.11.10 (Contravariant) A functor F between two categories $\mathscr{C}$ and $\mathscr{D}$ is called *contravariant* if it is a functor $F : \mathscr{C}^{op} \to \mathscr{D}$.

Remark 2.11.11 Categories also form a category themselves, the category of categories, denoted by **Cat** with morphisms given by functors.

Definition 2.11.12 (Isomorphism) A morphism $f \in \mathrm{Hom}(x, y)$ in a category is called an *isomorphism* if there is a morphism $f^{-1} \in \mathrm{Hom}(y, x)$ such that

$$f^{-1} \circ f = \mathrm{id}_x, \qquad f \circ f^{-1} = \mathrm{id}_y.$$

Such a morphism f is then also called *invertible*.

2.11.3 Monoidal Categories

A category which is equipped with a certain type of *product* between its objects is called *monoidal*. Examples of such products can be: the direct product $\times$, the direct sum $\oplus$, or the tensor product $\otimes$. By abuse of notation (and since it is done by many authors), we will use the same symbol $\otimes$ as for the tensor product to denote such a general monoidal product.

Definition 2.11.13 (Strict Monoidal Category) A *strict monoidal category* is a triple $(\mathscr{C}, \otimes, \mathbf{1})$, where $\mathscr{C}$ is a category, $\otimes : \mathscr{C} \times \mathscr{C} \to \mathscr{C}$ an associative bifunctor, i.e. a functor in each argument such that

$$\otimes(\otimes \times I_\mathscr{C}) = \otimes(I_\mathscr{C} \times \otimes) : \mathscr{C} \times \mathscr{C} \times \mathscr{C} \to \mathscr{C},$$

with $I_\mathscr{C}$ the identity functor on $\mathscr{C}$, i.e. the functor that acts on each object $a \in \mathscr{C}$ and endomorphism $a \to a$ as the identity and a distinguished object $\mathbf{1} \in \mathscr{C}$ which is a left and right unit for $\otimes$, i.e.

$$\otimes(\mathbf{1} \times I_\mathscr{C}) = \mathrm{id}_\mathscr{C} = \otimes(I_\mathscr{C} \times \mathbf{1}).$$

Definition 2.11.14 (Monoidal Category) A *monoidal category* is a quintupel $(\mathscr{C}, \otimes, \mathbf{1}, \alpha, \lambda, \rho)$ where $\mathscr{C}$ is a category, $\otimes : \mathscr{C} \times \mathscr{C} \to \mathscr{C}$ a bifunctor, $\mathbf{1} \in \mathscr{C}$ a distinguished object and α, λ, ρ three isomorphisms such that

$$\alpha = \alpha_{a,b,c} : a \otimes (b \otimes c) \cong (a \otimes b) \otimes c, \qquad \forall a, b, c \in \mathscr{C}$$

and the *pentagon diagram*

$$
\begin{array}{ccccc}
a \otimes (b \otimes (c \otimes d)) & \xrightarrow{\ \alpha\ } & (a \otimes b) \otimes (c \otimes d) & \xrightarrow{\ \alpha\ } & ((a \otimes b) \otimes c) \otimes d \\
{\scriptstyle I_\mathscr{C} \otimes \alpha} \downarrow & & & & \uparrow {\scriptstyle \alpha \otimes I_\mathscr{C}} \\
a \otimes ((b \otimes c) \otimes d) & & \xrightarrow{\hspace{5cm}\alpha\hspace{5cm}} & & (a \otimes (b \otimes c)) \otimes d
\end{array}
$$

commutes for all $a, b, c, d \in \mathscr{C}$. Moreover, we have

$$\lambda_a : \mathbf{1} \otimes a \cong a, \qquad \rho_a : a \otimes \mathbf{1} \cong a, \qquad \forall a \in \mathscr{C}$$

and the triangular diagram

$$
\begin{array}{ccc}
a \otimes (\mathbf{1} \otimes c) & \xrightarrow{\ \alpha\ } & (a \otimes \mathbf{1}) \otimes c \\
{\scriptstyle \mathbf{1} \otimes \lambda} \searrow & & \swarrow {\scriptstyle \rho \otimes \mathbf{1}} \\
& a \otimes c &
\end{array}
$$

commutes for all $a, c \in \mathscr{C}$. Also we have

$$\lambda_1 = \rho_1 : \mathbf{1} \otimes \mathbf{1} \to \mathbf{1}.$$

Exercise 2.11.15 Show that the commutative diagrams in Definition 2.11.14 imply that the following diagrams commute:

Exercise 2.11.16 Show that the category of vector spaces $\mathbf{Vect}_k$ over some field k forms a monoidal category with monoidal product given by the tensor product $\otimes_k$ of vector spaces over k. Show that, if R denotes a commutative ring, this is also true for the category of R-modules $\mathbf{Mod}_R$ together with the tensor product $\otimes_R$ of R-modules.

Exercise 2.11.17 Show that the category of topological spaces **Top** forms a monoidal category with monoidal product given by disjoint union $\sqcup$ of topological spaces.

Exercise 2.11.18 Show that the category of sets, denoted by **Set** whose objects are sets and whose morphisms are total functions between sets with composition given by composition of functions, is a monoidal category with monoidal product given by the direct product $\times$ of sets.

Definition 2.11.19 (Symmetric Monoidal Category) A *symmetric monoidal category* is a category $(\mathscr{C}, \otimes, \mathbf{1}, \alpha, \lambda, \rho)$ such that for all $a, b \in \mathscr{C}$ there is an isomorphism $s_{a,b} : a \otimes b \to b \otimes a$ such that the following diagrams commute:

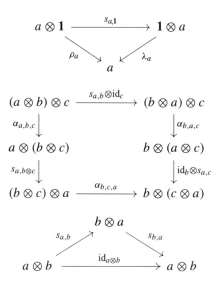

Exercise 2.11.20 Show that $(\mathbf{Vect}_k, \otimes_k, k)$ and $(\mathbf{Top}, \sqcup, \varnothing)$ are both symmetric monoidal categories.

Definition 2.11.21 (Lax Monoidal Functor) A functor $F \colon (\mathscr{C}, \otimes_{\mathscr{C}}, \mathbf{1}_{\mathscr{C}}) \to (\mathscr{D}, \otimes_{\mathscr{D}}, \mathbf{1}_{\mathscr{D}})$ between two monoidal categories is called *lax monoidal* if there is a morphism $\epsilon \colon \mathbf{1}_{\mathscr{D}} \to F(\mathbf{1}_{\mathscr{C}})$ and a natural transformation $\mu_{a,b} \colon F(a) \otimes_{\mathscr{C}} F(b) \to F(a \otimes_{\mathscr{C}} b)$ for all $a, b \in \mathscr{C}$, such that the following diagrams commute:

$$\begin{array}{ccc}
(F(a) \otimes_{\mathscr{D}} F(b)) \otimes_{\mathscr{D}} F(c) & \xrightarrow{\alpha^{\mathscr{D}}_{F(a),F(b),F(c)}} & F(a) \otimes_{\mathscr{D}} (F(b) \otimes_{\mathscr{D}} F(c)) \\
\downarrow{\mu_{a,b} \otimes \mathrm{id}_{F(c)}} & & \downarrow{\mathrm{id}_{F(a)} \otimes \mu_{b,c}} \\
F(a \otimes_{\mathscr{C}} b) \otimes_{\mathscr{D}} F(c) & & F(a) \otimes_{\mathscr{D}} F(a \otimes_{\mathscr{C}} c) \\
\downarrow{\mu_{a \otimes_{\mathscr{C}} b, c}} & & \downarrow{\mu_{a, b \otimes_{\mathscr{C}} c}} \\
F((a \otimes_{\mathscr{C}} b) \otimes_{\mathscr{C}} c) & \xrightarrow{F(\alpha^{\mathscr{C}}_{a,b,c})} & F(a \otimes_{\mathscr{C}} (b \otimes_{\mathscr{C}} c))
\end{array}$$

$$\begin{array}{ccc}
\mathbf{1}_{\mathscr{D}} \otimes_{\mathscr{D}} F(a) & \xrightarrow{\epsilon \otimes \mathrm{id}_{F(a)}} & F(\mathbf{1}_{\mathscr{C}}) \otimes_{\mathscr{D}} F(a) \\
\downarrow{\lambda^{\mathscr{D}}_{F(a)}} & & \downarrow{\mu_{\mathbf{1}_{\mathscr{C}}, a}} \\
F(a) & \xleftarrow{F(\lambda^{\mathscr{C}}_a)} & F(\mathbf{1}_{\mathscr{C}} \otimes_{\mathscr{C}} a)
\end{array}$$

$$\begin{array}{ccc}
F(a) \otimes_{\mathscr{D}} \mathbf{1}_{\mathscr{D}} & \xrightarrow{\mathrm{id}_{F(a)} \otimes \epsilon} & F(a) \otimes_{\mathscr{D}} F(\mathbf{1}_{\mathscr{C}}) \\
\downarrow{\rho^{\mathscr{D}}_{F(a)}} & & \downarrow{\mu_{a, \mathbf{1}_{\mathscr{C}}}} \\
F(a) & \xleftarrow{F(\rho^{\mathscr{C}}_a)} & F(a \otimes_{\mathscr{C}} \mathbf{1}_{\mathscr{C}})
\end{array}$$

Definition 2.11.22 ((Strong) Monoidal Functor) If ϵ and $\mu_{a,b}$ are isomorphisms in Definition 2.11.21, then F is called a *strong monoidal functor*. Many times, the word *strong* is dropped and one only speaks of a *monoidal functor*.

Definition 2.11.23 (Strict Monoidal Functor) If ϵ and $\mu_{a,b}$ in Definition 2.11.21 are even identity morphisms, then F is called a *strict monoidal functor*.

Definition 2.11.24 (Symmetric Monoidal Functor) A monoidal functor $F \colon (\mathscr{C}, \otimes_{\mathscr{C}}, \mathbf{1}_{\mathscr{C}}) \to (\mathscr{D}, \otimes_{\mathscr{D}}, \mathbf{1}_{\mathscr{D}})$ between symmetric monoidal categories is called *symmetric* if for all $a, b \in \mathscr{C}$ the following diagram commutes:

$$\begin{array}{ccc}
F(a) \otimes_{\mathscr{D}} F(b) & \xrightarrow{s^{\mathscr{D}}_{F(a),F(b)}} & F(b) \otimes_{\mathscr{D}} F(a) \\
\downarrow{\mu_{a,b}} & & \downarrow{\mu_{b,a}} \\
F(a \otimes_{\mathscr{C}} b) & \xrightarrow{F(s^{\mathscr{C}}_{a,b})} & F(b \otimes_{\mathscr{C}} a)
\end{array}$$

Chapter 3
Symplectic Geometry

Symplectic geometry is mainly motivated by the study of classical mechanics and dynamical systems. It can only be described for even-dimensional geometries where it serves as a way of measuring 2-dimensional objects. The word *symplectic* comes form the greek word for *complex*, which indicates that this theory is naturally associated to the field of complex numbers. In fact, it describes a generalization of complex geometry since the structure is not as strict as one can associated to each symplectic structure a certain compatible (almost) complex structure that is unique up to homotopy which is not integrable. This means that symplectic structures provide a more flexible structure than complex ones. In classical mechanics, we usually describe the dynamics of a mass particle through the coordinates which are given by position q_i and momentum p_i. Hence, the prototypical space where the particle moves is given by $\mathbb{R}^6$, since we have 3 space coordinates and to each one the corresponding momentum coordinate. In general, one can consider the space $M := \mathbb{R}^{2n}$ for $n \geq 1$. The symplectic structure is then defined as an area 2-form $\omega = \sum_{i=1}^{n} dq_i \wedge dp_i$. The dynamical information is usually encoded in a function $H \in C^\infty(M)$ called the *Hamiltonian*. It is convenient to extract a vector field $X_H \in \mathfrak{X}(M)$ out of H in order to express the dynamics in terms of the flow lines of this vector field, i.e. to consider a differential equation with respect to the change of H. In particular, one would like to consider a map $\omega \colon TM \to T^*M$, or equivalently an element of $T^*M \otimes T^*M$, such that $dH = \iota_{X_H}\omega = \omega(X_H, \ )$. Additionally, one would like the choice of X_H for each H to be unique in this way. It is easy to see that this will require ω to be non-degenerate. Moreover, one would like to have H such that it does not change along flow lines which means that $dH(X_H) = 0$. This would imply that $\omega(X_H, X_H) = 0$ and thus one requires ω to be alternating. This is the reason why we want ω to be a 2-form. Note that this also implies that the underlying space has to be even-dimensional since every skew-symmetric linear map for odd dimensions is singular. Finally, one also would like that ω does not change under

N. Moshayedi, *Kontsevich's Deformation Quantization and Quantum Field Theory*, Lecture Notes in Mathematics 2311, https://doi.org/10.1007/978-3-031-05122-7_3

flow lines. Mathematically, this is expressed as the vanishing of the Lie derivative $L_{X_H}\omega = 0$. Using Cartan's magic formula (2.19), we get

$$L_{X_H}\omega = \mathrm{d}\iota_{X_H}\omega + \iota_{X_H}\mathrm{d}\omega = \mathrm{d}(\mathrm{d}H) + \iota_{X_H}\mathrm{d}\omega = \mathrm{d}\omega(X_H).$$

Hence, if we require $\mathrm{d}\omega(X_H) = 0$ for the vector field induced by different Hamiltonians H, we require ω to be closed, i.e. $\mathrm{d}\omega = 0$. The aim of this chapter is to introduce the most important concepts of symplectic geometry in a precise way by first considering the linear setting and then move to the global case. We will then move to more sophisticated methods which will be relevant to several aspects of quantization as we will see later on.

Some parts of this chapter follow closely the excellent reference [Can08] and use additional material from [Cat18, Arn78, Don96, DK90, MS95, EG98, Wei71, Wei77, Wei81, BGV92, AB84, CF04, Xu91, MX00, Ber01, AN01].

3.1 Symplectic Manifolds

3.1.1 Symplectic Form

Let V be an m-dimensional vector space over $\mathbb{R}$ and let

$$\Omega : V \times V \to \mathbb{R}$$

be a bilinear map.

Definition 3.1.1 (Skew-Symmetric) The map Ω is called *skew-symmetric* if $\Omega(u, v) = -\Omega(v, u)$ for all $u, v \in V$.

Theorem 3.1.2 (Standard Form) *Let Ω be a skew-symmetric bilinear map on V. Then there is a basis $u_1, \ldots, u_k, e_1, \ldots, e_n, f_1, \ldots, f_n$ of V such that*

$$\Omega(u_i, v) = 0, \quad 1 \le i \le k, \forall v \in V, \tag{3.1}$$

$$\Omega(e_i, e_j) = \Omega(f_i, f_j) = 0, \quad 1 \le i, j \le n \tag{3.2}$$

$$\Omega(e_i, f_j) = \delta_{ij}, \quad 1 \le i, j \le n. \tag{3.3}$$

Remark 3.1.3 The basis in Theorem 3.1.2 is not unique even if it is historically called *canonical* basis.

Proof of Theorem 3.1.2 Let $U := \{u \in V \mid \Omega(u, v) = 0, \forall v \in V\}$. Choose a basis $u_1, \ldots, u_k$ of U and choose a complementary space W to U in V, i.e. such that

$$V = W \oplus U.$$

Let $e_1 \in W$ be a non-zero element. Then there is $f_1 \in W$ such that $\Omega(e_1, f_1) \neq 0$. Now assume that $\Omega(e_1, f_1) = 1$. Let W_1 be the span of e_1 and f_1 and let

$$W_1^\Omega := \{w \in W \mid \Omega(w, v) = 0, \ \forall v \in W_1\}.$$

We can then show that $W_1 \cap W_1^\Omega = \{0\}$. Indeed, suppose that $v = ae_1 + bf_1 \in W_1 \cap W_1^\Omega$. Then $0 = \Omega(v, e_1) = -b$ and $0 = \Omega(v, f_1) = a$ which together implies that $v = 0$. Moreover, we have $W = W_1 \oplus W_1^\Omega$. Indeed, suppose that $v \in W$ has $\Omega(v, e_1) = c$ and $\Omega(v, f_1) = d$. Then $v = (-cf_1 + de_1) + (v + cf_1 - de_1)$, where $-cf_1 + de_1 \in W_1$ and $v + cf_1 - de_1 \in W_1^\Omega$. Now let $e_2 \in W_1^\Omega$ be a non-zero element. Then there is $f_2 \in W_1^\Omega$ such that $\Omega(e_2, f_2) \neq 0$. Now assume that $\Omega(e_2, f_2) = 1$. Moreover, let W_2 be given by the span of e_2 and f_2. This construction can be continued until some point since $\dim V < \infty$ and thus we obtain

$$V = U \oplus W_1 \oplus \cdots \oplus W_n.$$

where all the summands are orthogonal with respect to Ω and where W_i has basis e_i, f_i with $\Omega(e_i, f_i) = 1$. $\qquad\square$

Remark 3.1.4 Note that the dimension of the subspace $U \subset V$ does not depend on the choice of basis and hence is an invariant on (V, Ω). Since $\dim U + 2n = \dim V$, we get that n is an invariant of (V, Ω). We call the number $2n$ the *rank* of Ω.

3.1.2 Symplectic Vector Spaces

Let V be an m-dimensional real vector space and let $\Omega \colon V \times V \to \mathbb{R}$ be a bilinear form. Define the map $\tilde{\Omega} \colon V \to V^*$ to be the linear map defined by

$$\tilde{\Omega}(v)(u) := \Omega(v, u).$$

Note that $\ker \tilde{\Omega} = U$.

Definition 3.1.5 (Linear Symplectic Form) A skew-symmetric bilinear form Ω is called *symplectic* (or *non-degenerate*) if $\tilde{\Omega}$ is bijective, i.e. $U = \{0\}$.

Remark 3.1.6 We sometimes also call a symplectic form Ω a *linear symplectic structure*.

Definition 3.1.7 (Symplectic Vector Space) We call a vector space V endowed with a linear symplectic structure Ω *symplectic*.

Exercise 3.1.8 Let Ω be a symplectic structure. Check that the map $\tilde{\Omega}$ is a bijection and that $\dim U = 0$ so $\dim V$ is even. Moreover, check that a symplectic vector space (V, Ω) has a basis $e_1, \ldots, e_n, f_1, \ldots, f_n$ satisfying

$$\Omega(e_i, f_j) = \delta_{ij}, \qquad \Omega(e_i, e_j) = \Omega(f_i, f_j) = 0.$$

We will call such a basis *symplectic*.

Definition 3.1.9 (Symplectic Subspace) A subspace $W \subset V$ is called *symplectic* if $\Omega\big|_W$ is non-degenerate.

Exercise 3.1.10 Show that the subspace given by the span of e_1 and f_1 is symplectic.

Definition 3.1.11 (Isotropic Subspace) A subspace $W \subset V$ is called *isotropic* if $\Omega\big|_W = 0$.

Exercise 3.1.12 Show that the subspace given by the span of e_1 and e_2 is isotropic.

Definition 3.1.13 (Symplectic Orthogonal) Let $W \subset V$ be a subspace of a symplectic vector space (V, Ω). The *symplectic orthogonal* of W is defined as

$$W^\Omega := \{v \in V \mid \Omega(v, u) = 0, \ \forall u \in W\}.$$

Exercise 3.1.14 Show that $(W^\Omega)^\Omega = W$.

Exercise 3.1.15 Show that a subspace $W \subset V$ is isotropic if $W \subseteq W^\Omega$. Moreover, show that if W is isotropic, then $\dim W \leq \frac{1}{2} \dim V$.

Definition 3.1.16 (Coisotropic Subspace) A subspace $W \subset V$ of a symplectic vector space (V, Ω) is called *coisotropic* if

$$W^\Omega \subseteq W.$$

Exercise 3.1.17 Show that every codimension 1 subspace $W \subset V$ is coisotropic.

Definition 3.1.18 (Lagrangian Subspace) An isotropic subspace $W \subset V$ of a symplectic vector space (V, Ω) is called *Lagrangian* if it is maximal, i.e. $\dim W = \frac{1}{2} \dim V$.

Exercise 3.1.19 Show that a subspace $W \subset V$ of a symplectic vector space (V, Ω) is Lagrangian if and only if W is isotropic and coisotropic if and only if $W = W^\Omega$.

Definition 3.1.20 (Standard Symplectic Vector Space) The *standard symplectic vector space* is defined as the vector space $\mathbb{R}^{2n}$ endowed with the linear symplectic structure Ω_0 defined such that the basis

$$
\begin{aligned}
e_1 &= (1, 0, \ldots, 0), \ldots, e_n = (0, \ldots, 0, 1, 0, \ldots, 0), \\
f_1 &= (0, \ldots, 0, 1, 0, \ldots, 0), \ldots, f_n = (0, \ldots, 0, 1),
\end{aligned}
\tag{3.4}
$$

where 1 is at the i-th entry for e_i and on the $(n+i)$-th entry for f_i with $1 \leq i \leq n$, is a symplectic basis.

Remark 3.1.21 We can extend Ω_0 to other vectors by using its values on a basis and bilinearity.

Definition 3.1.22 (Linear Symplectomorphism) A *linear symplectomorphism* φ between symplectic vector spaces (V, Ω) and (V', Ω') is a linear isomorphism $\varphi: V \xrightarrow{\sim} V$ such that

$$\varphi^* \Omega' = \Omega.$$

Remark 3.1.23 By definition we have

$$(\varphi^* \Omega')(u, v) := \Omega'(\varphi(u), \varphi(v)).$$

Definition 3.1.24 (Symplectomorphic Spaces) We call two symplectic vector spaces (V, Ω) and (V', Ω') *symplectomorphic* if there exists a symplectomorphism between them.

Exercise 3.1.25 Show that the relation of being symplectomorphic defines an equivalence relation in the set of all even-dimensional vector spaces. Moreover, show that every $2n$-dimensional symplectic vector space (V, Ω) is symplectomorphic to the standard symplectic vector space $(\mathbb{R}^{2n}, \Omega_0)$.

3.1.3 Symplectic Manifolds

Let ω be a 2-form on a manifold M. Note that for each point $q \in M$, the map

$$\omega_q: T_q M \times T_q M \to \mathbb{R}$$

is skew-symmetric bilinear on the tangent space $T_q M$.

Definition 3.1.26 (Symplectic Form) A 2-form ω is called *symplectic* if ω is closed and ω_q is symplectic for all $q \in M$.

Definition 3.1.27 (Symplectic Manifold) A *symplectic manifold* is a pair (M, ω) where M is a manifold and ω is a symplectic form.

Example 3.1.28 Let $M = \mathbb{R}^{2n}$ with coordinates $q_1, \ldots, q_n, p_1, \ldots, p_n$. Then one can check that the 2-form

$$\omega_0 := \sum_{i=1}^{n} dq_i \wedge dp_i$$

is symplectic and that the set

$$
\left\{ \left. \frac{\partial}{\partial q_1} \right|_q, \ldots, \left. \frac{\partial}{\partial q_n} \right|_q, \left. \frac{\partial}{\partial p_1} \right|_q, \ldots, \left. \frac{\partial}{\partial p_n} \right|_q \right\}
$$

defines a symplectic basis of $T_q M$.

Example 3.1.29 Let $M = \mathbb{C}^n$ with coordinates $z_1, \ldots, z_n$. Then one can check that the 2-form

$$
\omega_0 := \frac{i}{2} \sum_{k=1}^n dz_k \wedge d\bar{z}_k
$$

is symplectic. This is similar to Example 3.1.28 by the identification $\mathbb{C}^n \cong \mathbb{R}^{2n}$ and $z_k = x_k + iy_k$. See also Sect. 3.10 for more constructions on the complex side.

Remark 3.1.30 Let (M, ω) be a symplectic manifold. The closedness condition of the symplectic form ω can be interpreted via Stokes' theorem (Theorem 2.4.28). Namely, for a 3-dimensional submanifold $N \subset M$ with boundary ∂N, we have

$$
\int_{\partial N} \omega = \int_N d\omega = 0.
$$

Thus, if we consider two surfaces S_1 and S_2 inside M, i.e. $\dim S_i = 2$ for $i = 1, 2$, with the property that $\partial N = S_1 \cup S_2$, we get

$$
\int_{S_1} \omega = \int_{S_2} \omega.
$$

3.1.4 Symplectomorphisms

Definition 3.1.31 (Symplectomorphism) Let (M_1, ω_1) and (M_2, ω_2) be $2n$-dimensional symplectic manifolds, and let $\varphi \colon M_1 \to M_2$ be a diffeomorphism. Then φ is called a *symplectomorphism* if

$$
\varphi^* \omega_2 = \omega_1.
$$

Remark 3.1.32 Note that by definition we have

$$
(\varphi^* \omega_2)_q(u, v) = (\omega_2)_{\varphi(q)}(d_q \varphi(u), d_q \varphi(v)), \quad \forall u, v \in T_q M.
$$

The classification of symplectic manifolds up to symplectomorphisms is an interesting problem. The next theorem takes care of this locally. In fact, as any n-dimensional manifold looks locally like $\mathbb{R}^n$, one can show that any $2n$-dimensional symplectic manifold (M, ω) is locally symplectomorphic to $(\mathbb{R}^{2n}, \omega_0)$. In particular, the dimension is the only local invariant of symplectic manifolds up to symplectomorphisms.

Theorem 3.1.33 (Darboux[Dar82]) *Let (M, ω) be a 2n-dimensional symplectic manifold and let $q \in M$. Then there is a coordinate chart $(U, q_1, \ldots, q_n, p_1, \ldots, p_n)$ centered at q such that on $U \subset M$ we have*

$$\omega = \sum_{i=1}^{n} dq_i \wedge dp_i.$$

Definition 3.1.34 (Darboux Chart) A local coordinate chart $(U, q_1, \ldots, q_n, p_1, \ldots, p_n)$ is called a *Darboux chart*.

Remark 3.1.35 The proof of Theorem 3.1.33 will be given later as an exercise (Exercise 3.5.11) for a simple application of another important theorem (Moser's relative theorem).

3.2 The Cotangent Bundle as a Symplectic Manifold

Let N be an n-dimensional manifold and let $M := T^*N$ be its cotangent bundle. Consider coordinate charts $(U, q_1, \ldots, q_n)$ on N with $q_i : U \to \mathbb{R}$. Then at any $q \in U$ we have a basis of T_q^*N defined by the linear maps

$$d_q q_1, \ldots, d_q q_n.$$

In particular, if $p \in T_q^*N$, then $p = \sum_{i=1}^{n} p_i d_q q_i$ for some $p_1, \ldots, p_n \in \mathbb{R}$. Note that this induces a map

$$T^*U \to \mathbb{R}^{2n},$$

$$(q, p) \mapsto (q_1, \ldots, q_n, p_1, \ldots, p_n). \tag{3.5}$$

The chart $(T^*U, q_1, \ldots, q_n, p_1, \ldots, p_n)$ is a coordinate chart of T^*N. The transition functions on the overlaps are smooth; given two charts $(U, q_1, \ldots, q_n)$ and $(U', q_1' \ldots, q_n')$ and $q \in U \cap U'$ with $p \in T_q^*N$, then

$$p = \sum_{i=1}^{n} p_i d_q q_i = \sum_{1 \le i, j \le n} p_i \left(\frac{\partial q_i}{\partial q_j} \right) d_q q_j' = \sum_{j=1}^{n} p_j' d_q q_j',$$

where $p'_j = \sum_i p_i \left(\frac{\partial q_i}{\partial q_j} \right)$ is smooth. Hence, $M = T^*N$ is a $2n$-dimensional manifold.

3.2.1 Tautological and Canonical Forms

Consider a coordinate chart $(U, q_1, \ldots, q_n)$ for N with associated cotangent coordinates $(T^*U, q_1, \ldots, q_n, p_1, \ldots, p_n)$. Define a 2-form ω on T^*U by

$$\omega := \sum_{i=1}^{n} dq_i \wedge dp_i.$$

we want to show that this expression is independent of the choice of coordinates. Indeed, consider the 1-form

$$\alpha := \sum_{i=1}^{n} p_i dq_i,$$

and note that $\omega = -d\alpha$. Let $(U, q_1, \ldots, q_n, p_1, \ldots, p_n)$ and $(U', q'_1, \ldots, q'_n, p'_1, \ldots, p'_n)$ be two coordinate charts on T^*N. As we have seen, on the intersection $U \cap U'$ they are related by $p'_j = \sum_i p_i \left(\frac{\partial q_i}{\partial q_j} \right)$. Since $dq'_j = \sum_i \left(\frac{\partial q'_j}{\partial q_i} \right) dq_i$, we get

$$\alpha = \sum_i p_i dq_i = \sum_j p'_j dq'_j = \alpha'.$$

Hence, since α is intrinsically defined, so is ω. This finishes the claim.

Definition 3.2.1 (Tautological Form) The 1-form α is called *tautological form*.

Remark 3.2.2 The tautological form is sometimes also called *Liouville 1-form* and $\omega = -d\alpha$ is often called *canonical symplectic form*.

Let us denote by

$$\pi : M := T^*N \to N,$$

$$(q, p) \mapsto N \tag{3.6}$$

the natural projection for $p \in T^*_q N$. We define the tautological 1-form α point-wise as

$$\alpha_{(q,p)} = (d_{(q,p)}\pi)^* p \in T^*_q M.$$

where $(d_{(q,p)}\pi)^*$ denotes the transpose of $d_{(q,p)}\pi$, i.e. $(d_{(q,p)}\pi)^* p = p \circ d_{(q,p)}\pi$. We have the three maps

$$\pi : M := T^*N \to N, \tag{3.7}$$

$$d_{(q,p)}\pi : T_{(q,p)}M \to T_q N, \tag{3.8}$$

$$(d_{(q,p)}\pi)^* : T_q^*N \to T_{(q,p)}^*M. \tag{3.9}$$

In fact, we have

$$\alpha_{(q,p)}(v) = p\left((d_{(q,p)}\pi)v\right), \quad \forall v \in T_{(q,p)}M.$$

The canonical symplectic form is then defined by

$$\omega = -d\alpha.$$

Exercise 3.2.3 Show that the tautological form α is uniquely characterized by the property that, for every 1-form $\mu : N \to T^*N$ we have

$$\mu^*\alpha = \mu.$$

3.2.2   Symplectic Volume

Let V be a vector space of dimension $\dim V < \infty$. Any skew-symmetric bilinear map $\Omega \in \bigwedge^2 V^*$ is of the form

$$\Omega = e_1^* \wedge f_1^* + \cdots + e_n^* \wedge f_n^*,$$

where $u_1^*, \ldots, u_k^*, e_1^*, \ldots, e_n^*, f_1^*, \ldots, f_n^*$ is a basis of V^* dual to the standard basis. Here we have set $\dim V = k + 2n$. If Ω is also non-degenerate, i.e. a symplectic form on a vector space V with $\dim V = 2n$, then the n-th exterior power of

$$\Omega^n := \underbrace{\Omega \wedge \cdots \wedge \Omega}_{n}$$

does not vanish.

Exercise 3.2.4 Show that this also holds for the n-th exterior power ω^n of a symplectic form ω on a $2n$-dimensional symplectic manifold (M, ω). Deduce that it defines a volume form on M.

Exercise 3.2.4 shows that any symplectic manifold (M, ω) can be canonically oriented by the symplectic structure.

Definition 3.2.5 (Symplectic Volume) Let (M, ω) be a symplectic manifold. Then the form

$$\frac{\omega^n}{n!}$$

is called the *symplectic volume* of (M, ω).

Remark 3.2.6 The symplectic volume is sometimes also called *Liouville volume*.

Exercise 3.2.7 Show that if, conversely, a given 2-form $\Omega \in \bigwedge^2 V^*$ satisfies $\Omega^n \neq 0$, then Ω is symplectic.

Exercise 3.2.8 Let (M, ω) be a $2n$-dimensional symplectic manifold. Show that, if M is compact, the de Rham cohomology class $[\omega^n] \in H^{2n}(M)$ is non-zero. *Hint: Use Stokes' theorem (Theorem 2.4.28)*. Conclude then that $[\omega]$ is not exact and show that for $n > 1$, there are no symplectic structures on the sphere S^{2n}.

3.3 Lagrangian Submanifolds

3.3.1 Lagrangian Submanifolds of Cotangent Bundles

Definition 3.3.1 (Lagrangian Submanifold) Let (M, ω) be a $2n$-dimensional symplectic manifold. A submanifold $L \subset M$ is called *Lagrangian* if for any point $q \in L$ we get that $T_q L$ is a Lagrangian subspace of $T_q M$.

Remark 3.3.2 Recall that this is equivalent to say that $L \subset M$ is Lagrangian if and only if $\omega_q|_{T_q L} = 0$ and $\dim T_q L = \frac{1}{2} \dim T_q M$ for all $q \in L$. Equivalently, if $i : L \hookrightarrow M$ denotes the inclusion of L into M, then L is Lagrangian if and only if $i^*\omega = 0$ and $\dim L = \frac{1}{2} \dim M$.

Let N be an n-dimensional manifold and let $M := T^*N$ be its cotangent bundle. Consider coordinates $q_1, \ldots, q_n$ on $U \subseteq N$ with cotangent coordinates $q_1, \ldots, q_n, p_1, \ldots, p_n$ on T^*U, then the tautological 1-form on T^*N is given by

$$\alpha = \sum_i p_i \, dq_i$$

and the canonical 2-form on T^*N is

$$\omega = -d\alpha = \sum_i dq_i \wedge dp_i.$$

3.3.2 Conormal Bundle

Let S be any k-dimensional submanifold of N.

Definition 3.3.3 (Conormal Space) The *conormal space* at $q \in S$ is defined by

$$N_q^* S := \{p \in T_q^* N \mid p(v) = 0, \; \forall v \in T_q S\}.$$

Definition 3.3.4 (Conormal Bundle) The *conormal bundle* of S is

$$N^* S = \{(q, p) \in T^* N \mid q \in S, \; p \in N_q^* S\}.$$

Exercise 3.3.5 Show that the conormal bundle $N^* S$ is an n-dimensional submanifold of $T^* N$.

Proposition 3.3.6 *Let* $i \colon N^* S \hookrightarrow T^* N$ *be the inclusion, and let* α *be the tautological 1-form on* $T^* N$. *Then*

$$i^* \alpha = 0.$$

Definition 3.3.7 (Adapted Coordinate Chart) A coordinate chart $(U, q_1, \ldots, q_n)$ on N is said to be *adapted* to a k-dimensional submanifold $S \subset N$ if $S \cap U$ is described by $q_{k+1} = \cdots = q_n = 0$.

Proof of Proposition 3.3.6 Let $(U, q_1, \ldots, q_n)$ be a coordinate system on N centered at $q \in S$ and adapted to S, so that $U \cap S$ is described by $q_{k+1} = \cdots = q_n = 0$. Let $(T^* U, q_1, \ldots, q_n, p_1, \ldots, p_n)$ be the associated cotangent coordinate system. The submanifold $N^* S \cap T^* U$ is then described by

$$q_{k+1} = \cdots = q_n = 0,$$
$$p_1 = \cdots = p_k = 0.$$

Since $\alpha = \sum_i p_i dq_i$ on $T^* U$, we conclude that for $(q, p) \in N^* S$ we get

$$(i^* \alpha)_{(q,p)} = \alpha_{(q,p)} \big|_{T_{(q,p)}(N^* S)} = \sum_{i > k} p_i dq_i \Big|_{\mathrm{span}\left(\frac{\partial}{\partial q_i}\right)_{i \leq k}} = 0.$$

$\square$

Definition 3.3.8 (Zero Section) The *zero section* of $T^* N$ is defined by

$$N_0 := \{(q, p) \in T^* N \mid p = 0 \in T_q^* N\}.$$

Exercise 3.3.9 Let $i_0 \colon N_0 \hookrightarrow T^*N$ be the inclusion of the zero section and let $\omega = -d\alpha$ be the canonical symplectic form on T^*N. Show that $i_0^*\omega$ and N_0 are Lagrangian submanifolds of T^*N.

Corollary 3.3.10 *For any submanifold $S \subset N$, the conormal bundle N^*S is a Lagrangian submanifold of T^*N.*

Exercise 3.3.11 Prove Corollary 3.3.10.

Remark 3.3.12 If $S = \{q\} \subset N$, then the conormal bundle is given by a cotangent fiber $N^*S = T_q^*N$. If $S = N$, then the conormal bundle is the zero section of T^*N, i.e. $N^*S = N_0$.

3.3.3 Graphs and Symplectomorphisms

Let (M_1, ω_1) and (M_2, ω_2) be two $2n$-dimensional symplectic manifolds. Given a diffeomorphism $\varphi \colon M_1 \xrightarrow{\sim} M_2$, when is it a symplectomorphism? Equivalently, when do we have

$$\varphi^*\omega_2 = \omega_1 ?$$

Consider two projection maps

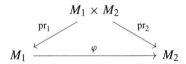

Then

$$\omega := (\mathrm{pr}_1)^*\omega_1 + (\mathrm{pr}_2)^*\omega_2$$

is a 2-form on $M_1 \times M_2$ which is closed:

$$d\omega = (\mathrm{pr}_1)^* \underbrace{d\omega_1}_{=0} + (\mathrm{pr}_2)^* \underbrace{d\omega_2}_{=0} = 0,$$

and symplectic:

$$\omega^{2n} = \binom{2n}{n} \left((\mathrm{pr}_1)^*\omega_1\right)^n \wedge \left((\mathrm{pr}_2)^*\omega_2\right)^n \neq 0.$$

Note that for all $\lambda_1, \lambda_2 \in \mathbb{R}$ we get that

$$\lambda_1 (\mathrm{pr}_1)^* \omega_1 + \lambda_2 (\mathrm{pr})^* \omega_2$$

is a symplectic form on $M_1 \times M_2$. The *twisted product form* on $M_1 \times M_2$ is obtained by taking $\lambda_1 = 1$ and $\lambda_2 = -1$. Namely,

$$\tilde{\omega} = (\mathrm{pr}_1)^* \omega_1 - (\mathrm{pr}_2)^* \omega_2.$$

For a diffeomorphism $\varphi \colon M_1 \xrightarrow{\sim} M_2$, we define its *graph* by

$$\Gamma_\varphi := \mathrm{graph}\, \varphi = \{(q, \varphi(q)) \mid q \in M_1\} \subset M_1 \times M_2. \tag{3.10}$$

Remark 3.3.13 Note that $\dim \Gamma_\varphi = 2n$ and that Γ_φ is the embedded image of M_1 in $M_1 \times M_2$. The embedding is given by the map

$$\gamma \colon M_1 \to M_1 \times M_2,$$

$$q \mapsto (q, \varphi(q)).$$

Proposition 3.3.14 *A diffeomorphism φ is a symplectomorphism if and only if Γ_φ is a Lagrangian submanifold of $(M_1 \times M_2, \tilde{\omega})$.*

Proof The graph Γ_φ is Lagrangian if and only if $\gamma^* \tilde{\omega} = 0$. But we have

$$\gamma^* \tilde{\omega} = \gamma^* (\mathrm{pr}_1)^* \omega_1 - \gamma^* (\mathrm{pr}_2)^* \omega_2 = (\mathrm{pr}_1 \circ \gamma)^* \omega_1 - (\mathrm{pr}_2 \circ \gamma)^* \omega_2,$$

where $\mathrm{pr}_1 \circ \gamma$ is the identity map on M_1 whereas $\mathrm{pr}_2 \circ \gamma = \varphi$. Hence

$$\gamma^* \tilde{\omega} = 0 \iff \varphi^* \omega_2 = \omega_1.$$

$\square$

3.4 Local Theory

3.4.1 Isotopies and Vector Fields

Let M be a manifold and $\rho \colon M \times \mathbb{R} \to M$ a map with $\rho_t(q) := \rho(q, t)$.

Definition 3.4.1 (Isotopy) The map ρ is an *isotopy* if each $\rho_t \colon M \to M$ is a diffeomorphism and $\rho_0 = \mathrm{id}_M$.

Given an isotopy ρ we can construct a *time-dependent* vector field. This means that we get a family of vector fields X_t for $t \in \mathbb{R}$, which at $q \in M$ satisfy

$$X_t(q) = \frac{\mathrm{d}}{\mathrm{d}s}\rho_s(p)\bigg|_{s=t}, \qquad \forall p = \rho_t^{-1}(q).$$

Basically, this means

$$\frac{\mathrm{d}\rho_t}{\mathrm{d}t} = X_t \circ \rho_t. \tag{3.11}$$

Conversely, let X_t be a time-dependent vector field and assume either that M is compact or X_t is compactly supported for all t. Then there exists an isotopy ρ satisfying (3.11).

Moreover, if M is compact, we have a one-to-one correspondence

$$\{\text{isotopies of } M\} \longleftrightarrow \{\text{time-dependent vector fields on } M\},$$

$$(\rho_t)_{t\in\mathbb{R}} \longleftrightarrow (X_t)_{t\in\mathbb{R}}.$$

We want to recall the notion of an exponential map as in Sect. 2.7.3.

Definition 3.4.2 (Exponential Map) If a vector field $X_t = X$ is time-independent, we call the associated isotopy the *exponential map* of X and we denote it by $\exp(tX)$.

Remark 3.4.3 Note that the family $\{\exp(tX): M \to M \mid t \in \mathbb{R}\}$ is the unique smooth family of diffeomorphisms satisfying the Cauchy problem

$$\exp(tX)\big|_{t=0} = \mathrm{id}_M,$$

$$\frac{\mathrm{d}}{\mathrm{d}t}(\exp(tX))(q) = X(\exp(tX)(q)).$$

Remark 3.4.4 The exponential map is the same as the *flow* of a vector field (see Sect. 2.2, Definition 2.2.31). The flow of a time-dependent vector field is given by the corresponding isotopy.

Exercise 3.4.5 Show that for a time-dependent vector field X_t and $\omega \in \Omega^s(M)$ we have

$$\frac{\mathrm{d}}{\mathrm{d}t}\rho_t^*\omega = \rho_t^* L_{X_t}\omega, \tag{3.12}$$

where ρ is the (local) isotopy generated by X_t.

Proposition 3.4.6 *For a smooth family* $(\omega_t)_{t\in\mathbb{R}}$ *of s-forms, we have*

$$\boxed{\frac{d}{dt}\rho_t^*\omega_t = \rho_t^*\left(L_{X_t}\omega_t + \frac{d}{dt}\omega_t\right).}$$ (3.13)

Remark 3.4.7 Equation (3.13) will turn out to be very useful for the proof of Moser's theorem (see Sect. 3.5, Theorem 3.5.6)

Proof of Proposition 3.4.6 If $f(x, y)$ is a real function of two variables, we can use the chain rule to get

$$\frac{d}{dt}f(t, t) = \frac{d}{dx}f(x, t)\Big|_{x=t} + \frac{d}{dy}f(t, y)\Big|_{y=t}.$$

Hence, we get

$$\frac{d}{dt}\rho_t^*\omega_t = \underbrace{\frac{d}{dx}\rho_x^*\omega_t\Big|_{x=t}}_{\rho_x^*L_{X_t}\omega_t\big|_{x=t}\ \text{by (3.12)}} + \underbrace{\frac{d}{dy}\rho_t^*\omega_y\Big|_{y=t}}_{\rho_t^*\frac{d}{dy}\omega_y\big|_{y=t}} = \rho_t^*\left(L_{X_t}\omega_t + \frac{d}{dt}\omega_t\right).$$

$\square$

3.4.2 Tubular Neighborhood Theorem

Let M be an n-dimensional manifold and let $S \subset M$ be a k-dimensional submanifold and consider the inclusion map

$$i: S \hookrightarrow M.$$

By the differential of the inclusion $d_q i : T_q S \hookrightarrow T_q M$, we have an inclusion of the tangent space of X at a point $q \in S$ into the tangent space of M at the point q.

Definition 3.4.8 (Normal Space) The *normal space* to S at the point $q \in S$ is given by the $(n - k)$-dimensional vector space defined by the quotient

$$N_q S := T_q M / T_q S.$$

Definition 3.4.9 (Normal Bundle) The *normal bundle* is then given by

$$NS := \{(q, p) \mid q \in S, \ p \in N_q S\}.$$

Remark 3.4.10 Using the natural projection, NS is a vector bundle over S of rank $n - k$ and hence as a manifold it is n-dimensional. The zero section of NS

$$i_0 \colon S \hookrightarrow NS,$$

$$q \mapsto (q, 0),$$

embeds S as a closed submanifold of NS.

Definition 3.4.11 (Convex Neighborhood) A neighborhood U_0 of the zero section S in NS is called *convex* if the intersection $U_0 \cap N_q S$ with each fiber is convex.

Theorem 3.4.12 (Tubular Neighborhood Theorem) *There exists a convex neighborhood U_0 of S in NS, a neighborhood U of S in M, and a diffeomorphism $\varphi \colon U_0 \to U$ such that the following diagram commutes:*

Remark 3.4.13 Restricting to the subset $U_0 \subseteq NS$, we obtain a submersion $U_0 \overset{\pi_0}{\to} S$ with all fibers $\pi_0^{-1}(q)$ convex. We can extend this fibration to U by setting $\pi :=$ $\pi_0 \circ \varphi^{-1}$ such that if $NS \supseteq U_0 \overset{\pi_0}{\to} S$ is a fibration, then $M \supseteq U \overset{\pi}{\to} S$ is a fibration. This is called the *tubular neighborhood fibration*.

3.4.3 Homotopy Formula

Let U be a tubular neighborhood of a submanifold $S \subset M$. The restriction of de Rham cohomology groups

$$i^* \colon H^s(U) \to H^s(X)$$

by the inclusion map is surjective. By the tubular neighborhood fibration, i^* is also injective since the de Rham cohomology is homotopy invariant. In fact, we have the following corollary:

Corollary 3.4.14 *For any degree s we have*

$$H^s(U) \cong H^s(S).$$

Remark 3.4.15 Corollary 3.4.14 says that if ω is a closed s-form on U and $i^*\omega$ is exact on S, then ω is exact.

Proposition 3.4.16 *If a closed s-form ω on U has restriction $i^*\omega = 0$, then ω is exact, i.e. $\omega = d\mu$ for some $\mu \in \Omega^{s-1}(U)$. Moreover, we can choose μ such that $\mu_q = 0$ for all $q \in S$.*

Proof By using the map $\varphi: U_0 \xrightarrow{\sim} U$, we can work over U_0. For $t \in [0, 1]$, define a map

$$\rho_t: U_0 \to U_0,$$
$$(q, p) \mapsto (q, tp).$$

This is well-defined since U_0 is convex. The map ρ_1 is the identity and $\rho_0 = i_0 \circ \pi_0$. Moreover, each ρ_t fixes S, i.e. $\rho_t \circ i_0 = i_0$. Hence, we say that the family $(\rho_t)_{t \in [0,1]}$ is a *homotopy* from $i_0 \circ \pi_0$ to the identity fixing S. The map π_0 is called *retraction* because $\pi_0 \circ i_0$ is the identity. The submanifold S is then called a *deformation retract* of U.

A (de Rham) *homotopy operator* between $\rho_0 = i_0 \circ \pi_0$ and $\rho_1 = \mathrm{id}$ is a linear map

$$Q: \Omega^s(U_0) \to \Omega^{s-1}(U_0)$$

satisfying the *homotopy formula*

$$\boxed{\mathrm{id} - (i_0 \circ \pi_0)^* = dQ + Qd.} \tag{3.14}$$

When $d\omega = 0$ and $i_0^*\omega = 0$, the operator Q gives $\omega = dQ\omega$, so that we can take $\mu = Q\omega$. A concrete operator Q is given by the formula

$$Q\omega = \int_0^1 \rho_t^*(\iota_{X_t}\omega)dt, \tag{3.15}$$

where X_t, at the point $p = \rho_t(q)$, is the vector tangent to the curve $\rho_s(q)$ at $s = t$. We claim that the operator (3.15) satisfies the homotopy formula. Indeed, we compute

$$Qd\omega + dQ\omega = \int_0^1 \rho_t^*(\iota_{X_t}d\omega)dt + d\int_0^1 \rho_t^*(\iota_{X_t}\omega)dt = \int_0^1 \rho_t^*(\underbrace{\iota_{X_t}d\omega + d\iota_{X_t}\omega}_{=L_{X_t}\omega})dt.$$

Hence, the result follows from (3.12) and the fundamental theorem of analysis:

$$Qd\omega + dQ\omega = \int_0^1 \frac{d}{dt}\rho_t^*\omega dt = \rho_1^*\omega - \rho_0^*\omega.$$

This completes the proof since, for our case, we have that $\rho_t(q) = q$ is a constant curve for all $q \in S$ and for all t, so X_t vanishes at all q and all t. Hence, $\mu_q = 0$. $\quad\square$

3.5   Moser's Theorem

3.5.1   Equivalences for Symplectic Structures

Let M be a $2n$-dimensional manifold with two symplectic forms ω_0 and ω_1, so that (M, ω_0) and (M, ω_1) are two symplectic manifolds.

Definition 3.5.1 (Symplectomorphic) (M, ω_0) and (M, ω_1) are *symplectomorphic* if there is a diffeomorphism $\varphi\colon M \to M$ with $\varphi^*\omega_1 = \omega_0$.

Definition 3.5.2 (Strongly Isotopic) (M, ω_0) and (M, ω_1) are *strongly isotopic* if there is an isotopy $\rho_t\colon M \to M$ such that $\rho_1^*\omega_1 = \omega_0$.

Definition 3.5.3 (Deformation-Equivalent) (M, ω_0) and (M, ω_1) are *deformation-equivalent* if there is a smooth family ω_t of symplectic forms joining ω_0 to ω_1.

Definition 3.5.4 (Isotopic) (M, ω_0) and (M, ω_1) are *isotopic* if they are deformation-equivalent with the de Rham cohomology classes $[\omega_t]$ independent of t.

Remark 3.5.5 We have *strongly isotopic* $\Longrightarrow$ *symplectomorphic*, and *isotopic* $\Longrightarrow$ *deformation-equivalent*. Moreover, we have *strongly isotopic* $\Longrightarrow$ *isotopic*, because if $\rho_t\colon M \to M$ is an isotopy such that $\rho_1^*\omega_1 = \omega_0$, then the set $\omega_t := \rho_t^*\omega_1$ is a smooth family of symplectic forms joining ω_1 to ω_0 and $[\omega_t] = [\omega_1]$ for all t by the homotopy invariance of the de Rham cohomology.

3.5.2   Moser's Trick

Consider the following problem: Let M be a $2n$-dimensional manifold and let $S \subset M$ be a k-dimensional submanifold. Moreover, consider neighborhoods U_0 and U_1 of S and symplectic forms ω_0 and ω_1 on U_0 and U_1 respectively. Does there exist a symplectomorphism preserving S? More precisely, does there exist a diffeomorphism $\varphi\colon U_0 \to U_1$ with $\varphi^*\omega_1 = \omega_0$ and $\varphi(S) = S$?

  We want to consider the extreme case when $S = M$ and we consider M to be compact with symplectic forms ω_0 and ω_1. So the question will change to: are (M, ω_0) and (M, ω_1) symplectomorphic, i.e. does there exist a diffeomorphism $\varphi\colon M \to M$ such that $\varphi^*\omega_1 = \omega_0$?

Moser's question was whether we can find such a φ which is homotopic to the identity on M. A necessary condition is

$$[\omega_0] = [\omega_1] \in H^2(M)$$

because if $\varphi \sim \mathrm{id}_M$ then, by the homotopy formula (3.14), there exists a homotopy operator Q such that

$$(\mathrm{id}_M)^* \omega_1 - \varphi^* \omega_1 = dQ\omega_1 + Q \underbrace{d\omega_1}_{=0}.$$

Thus, we get

$$\omega_1 = \varphi^* \omega_1 + dQ\omega_1$$

and hence

$$[\omega_1] = [\varphi^* \omega_1] = [\omega_0].$$

So we ask ourselves whether, if $[\omega_0] = [\omega_1]$, there exist a diffeomorphism φ homotopic to id_M such that $\varphi^* \omega_1 = \omega_0$. In [Mos65], *Moser* proved that, with certain assumptions, this is true. Later, *McDuff* showed that, in general, this is not true by constructing a counterexample [MS95, Example 7.23].

Theorem 3.5.6 (Moser (Version I)[Mos65]) *Suppose that M is compact, $[\omega_0] = [\omega_1]$ and that the 2-form $\omega_t = (1 - t)\omega_0 + t\omega_1$ is symplectic for all $t \in [0, 1]$. Then there exists an isotopy $\rho : M \times \mathbb{R} \to M$ such that $\rho_t^* \omega_t = \omega_0$.*

Remark 3.5.7 In particular, $\varphi := \rho_t : M \to M$ satisfies $\varphi^* \omega_1 = \omega_0$.

The argument for the proof is known as *Moser's trick*.

Proof of Theorem 3.5.6 Suppose that there exists an isotopy $\rho : M \times \mathbb{R} \to M$ such that $\rho_t^* \omega_t = \omega_0$ for $t \in [0, 1]$. Let

$$X_t := \frac{d\rho_t}{dt} \circ \rho_t^{-1}, \quad \forall t \in \mathbb{R}.$$

Then we have

$$0 = \frac{d}{dt}(\rho_t^* \omega_t) = \rho_t^* \left(L_{X_t} \omega_t + \frac{d}{dt} \omega_t \right),$$

which is equivalent to

$$L_{X_t} \omega_t + \frac{d}{dt} \omega_t = 0. \tag{3.16}$$

Suppose conversely that we can find a smooth time-dependent vector field X_t for $t \in \mathbb{R}$, such that (3.16) holds for $t \in [0, 1]$. Since M is compact, we can integrate X_t to an isotopy $\rho : M \times \mathbb{R} \to M$ with

$$\frac{\mathrm{d}}{\mathrm{d}t}(\rho_t^* \omega_t) = 0,$$

which implies that

$$\rho_t^* \omega_t = \rho_0^* \omega_0 = \omega_0.$$

This means that everything reduces to solving (3.16) for X_t. First, note that, from $\omega_t = (1 - t)\omega_0 + t\omega_1$, we get

$$\frac{\mathrm{d}}{\mathrm{d}t}\omega_t = \omega_1 - \omega_0.$$

Second, since $[\omega_0] = [\omega_1]$, there is a 1-form μ such that

$$\omega_1 - \omega_0 = \mathrm{d}\mu.$$

Third, by Cartan's magic formula (2.19), we have

$$L_{X_t}\omega_t = \mathrm{d}\iota_{X_t}\omega_t + \iota_{X_t}\underbrace{\mathrm{d}\omega_t}_{=0}.$$

putting everything together, we need to find X_t such that

$$\mathrm{d}\iota_{X_t}\omega_t + \mathrm{d}\mu = 0.$$

Clearly, it is sufficient to solve

$$\iota_{X_t}\omega_t + \mu = 0.$$

By non-degeneracy of ω_t, we can solve this point-wise to obtain a unique (smooth) X_t.   □

Theorem 3.5.8 (Moser (Version II)[Mos65]) *Let M be a compact manifold with symplectic forms ω_0 and ω_1. Suppose that $(\omega_t)_{t \in [0,1]}$ is a smooth family of closed 2-forms joining ω_0 and ω_1 and satisfying:*

(1) (cohomology assumption) $[\omega_t]$ is independent of t, i.e.

$$\frac{\mathrm{d}}{\mathrm{d}t}[\omega_t] = \left[\frac{\mathrm{d}}{\mathrm{d}t}\omega_t\right] = 0,$$

(2) (non-degeneracy condition) ω_t is non-degenerate for all $t \in [0, 1]$.

Then there exists an isotopy $\rho\colon M \times \mathbb{R} \to M$ *such that*

$$\rho_t^*\omega_t = \omega_0, \quad \forall t \in [0, 1].$$

Proof (Moser's Trick) We have the following implications: Condition (1) implies that there exists a family of 1-forms μ_t such that

$$\frac{\mathrm{d}}{\mathrm{d}t}\omega_t = \mathrm{d}\mu_t, \quad \forall t \in [0, 1].$$

Indeed, we can find a smooth family of 1-forms μ_t such that $\frac{\mathrm{d}}{\mathrm{d}t}\omega_t = \mathrm{d}\mu_t$. The argument uses a combination of the *Poincaré lemma* for compactly-supported forms and the *Mayer–Vietoris sequence* in order to use induction on the number of charts in a good cover of M (see [BT82] for a detailed discussion of the Poincaré lemma and the Mayer–Vietoris construction). Condition (2) implies that there exists a unique family of vector fields X_t such that

$$\boxed{\iota_{X_t}\omega_t + \mu_t = 0.} \qquad (3.17)$$

Equation (3.16) is called *Moser's equation*. We can extend X_t to all $t \in \mathbb{R}$. Let ρ be the isotopy generated by X_t (ρ exists by compactness of M). Then, using Cartan's magic formula and Moser's equation, we indeed have

$$\frac{\mathrm{d}}{\mathrm{d}t}(\rho_t^*\omega_t) = \rho_t^*\left(L_{X_t}\omega_t + \frac{\mathrm{d}}{\mathrm{d}t}\omega_t\right) = \rho_t^*(\mathrm{d}\iota_{X_t}\omega_t + \mathrm{d}\mu_t) = 0.$$

$\square$

Remark 3.5.9 Note that we have used compactness of M to be able to integrate X_t for all $t \in \mathbb{R}$. If M is not compact, we need to check the existence of a solution ρ_t for the differential equation

$$\frac{\mathrm{d}}{\mathrm{d}t}\rho_t = X_t \circ \rho_t, \quad \forall t \in [0, 1].$$

Theorem 3.5.10 (Moser (Relative Version)[Mos65]) *Let M be a manifold, S a compact submanifold of M, $i\colon S \hookrightarrow M$ the inclusion map, ω_0 and ω_1 two symplectic forms on M. Then, if $\omega_0|_q = \omega_1|_q$ for all $q \in S$, we get that there exist neighborhoods U_0, U_1 of S in M and a diffeomorphism $\varphi\colon U_0 \to U_1$ such that $\varphi^*\omega_1 = \omega_0$ and the following diagram commutes:*

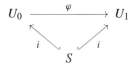

Proof Choose a tubular neighborhood U_0 of S. The 2-form $\omega_1 - \omega_0$ is closed on U_0 and $(\omega_1 - \omega_0)_q = 0$ for all $q \in S$. By the homotopy formula on the tubular neighborhood, there exists a 1-form μ on U_0 such that $\omega_1 - \omega_0 = d\mu$ and $\mu_q = 0$ at all $q \in S$. Consider a family $\omega_t = (1-t)\omega_0 + t\omega_1 = \omega_0 + t d\mu$ of closed 2-forms on U_0. Shrinking U_0 if necessary, we can assume that ω_t is symplectic for $t \in [0, 1]$. Then we can solve Moser's equation $\iota_{X_t} \omega_t = -\mu$ and note that $X_t|_S = 0$. Shrinking U_0 again if necessary, there is an isotopy $\rho \colon U_0 \times [0, 1] \to M$ with $\rho_t^* \omega_t = \omega_0$ for all $t \in [0, 1]$. Since $X_t|_S = 0$, we have $\rho_t|_S = \mathrm{id}_S$. Then we can set $\varphi := \rho_1$ and $U_1 := \rho_1(U_0)$.   □

Exercise 3.5.11 Prove Darboux's theorem (Theorem 3.1.33) by using the relative version of Moser's theorem for $S = \{q\}$.

3.6 Weinstein's Tubular Neighborhood Theorem

3.6.1 Weinstein's Lagrangian Neighborhood Theorem

Theorem 3.6.1 (Weinstein's Lagrangian Neighborhood Theorem[Wei71]) *Let M be a $2n$-dimensional manifold, S a compact n-dimensional submanifold $i \colon S \hookrightarrow M$ the inclusion map, and ω_0 and ω_1 symplectic forms on M such that $i^*\omega_0 = i^*\omega_1 = 0$, i.e. S is a Lagrangian submanifold of both (M, ω_0) and (M, ω_1). Then there exist neighborhoods U_0 and U_1 of S in M and a diffeomorphism $\varphi \colon U_0 \to U_1$ such that $\varphi^*\omega_1 = \omega_0$ and the following diagram commutes:*

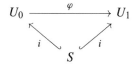

Theorem 3.6.2 (Whitney's Extension Theorem[Whi34]) *Let M be an n-dimensional manifold and S a k-dimensional submanifold with $k < n$. Suppose that at each $q \in S$ we are given a linear isomorphism $L_q \colon T_q M \xrightarrow{\sim} T_q M$ such that $L_q|_{T_q S} = \mathrm{id}_{T_q S}$ and L_q depends smoothly on q. Then there exists an embedding $h \colon W \to M$ of some neighborhood W of S in M such that $h|_S = \mathrm{id}_S$ and $d_q h = L_q$ for all $q \in S$.*

Proof of Theorem 3.6.1 Let g be a *Riemannian metric* on M, i.e. at each $q \in M$ we get that g_q is a positive-definite inner product. Fix $q \in S$ and let $V := T_q M$, $U := T_q S$ and $W := U^\perp$ be the orthogonal complement of U in V relative to g_q. Since $i^*\omega_0 = i^*\omega_1 = 0$, the space U is a Lagrangian subspace of both $(V, \omega_0|_q)$ and $(V, \omega_1|_q)$. By the symplectic linear algebra, we canonically get from $U^\perp$ a linear isomorphism $L_q \colon T_q M \to T_q M$ such that $L_q|_{T_q S} = \mathrm{id}_{T_q S}$ and $L_q^* \omega_1|_p = \omega_0|_q$. Note that L_q varies smoothly with respect to q since the construction is canonical.

By the Whitney extension theorem, there exists a neighborhood W of S and an embedding $h \colon W \hookrightarrow S$ with $h|_S = \mathrm{id}_S$ and $\mathrm{d}_q h = L_q$ for $q \in S$. Hence, at any $q \in S$, we have

$$(h^*\omega_1)_q = (\mathrm{d}_q h)^* \omega_1|_q = L_q^* \omega_1|_q = \omega_0|_q.$$

applying the relative version of Moser's theorem to ω_0 and $h^*\omega_1$, we find a neighborhood U_0 of S and an embedding $f \colon U_0 \hookrightarrow W$ such that $f|_S = \mathrm{id}_S$ and $f^*(h^*\omega_1) = \omega_0$ on U_0. Then we can set $\varphi := h \circ f$. $\qquad \square$

Theorem 3.6.3 (Coisotropic Embedding Theorem) *Let M be a $2n$-dimensional manifold, S a k-dimensional submanifold with $k < n$, $i \colon S \hookrightarrow M$ the inclusion map and ω_0 and ω_1 two symplectic forms on M such that $i^*\omega_0 = i^*\omega_1$ with S being coisotropic for both (M, ω_0) and (M, ω_1). Then there exist neighborhoods U_0 and U_1 of S in M and a diffeomorphism $\varphi \colon U_0 \to U$ such that $\varphi^*\omega_1 = \omega_0$ and the following diagram commutes:*

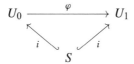

Exercise 3.6.4 Prove Theorem 3.6.3. *Hint: See [Wei77] for some inspiration and [GS77, Got82] for a proof.*

3.6.2 Weinstein's Tubular Neighborhood Theorem

Theorem 3.6.5 (Weinstein's Tubular Neighborhood Theorem[Wei71]) *Let (M, ω) be a symplectic manifold, S a compact Lagrangian submanifold of M, ω_0 the canonical symplectic form on T^*S, $i_0 \colon S \hookrightarrow T^*S$ the Lagrangian embedding as the zero section, and $i \colon S \hookrightarrow M$ the Lagrangian embedding given by the inclusion. Then there exist neighborhoods U_0 of S in T^*S, U of S in M and a diffeomorphism $\varphi \colon U_0 \to U$ such that $\varphi^*\omega = \omega_0$ and the following diagram commutes:*

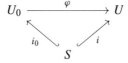

Remark 3.6.6 A similar statement for isotropic submanifolds was also proved by Weinstein in [Wei77, Wei81].

Proof of Theorem 3.6.5 The proof uses the tubular neighborhood theorem and Weinstein's Lagrangian neighborhood theorem. We first use the tubular neighborhood theorem. Since $NS \cong T^*S$, we can find a neighborhood W_0 of S in T^*S, a neighborhood W of S in M and a diffeomorphism $\psi: W_0 \to W$ such that the following diagram commutes:

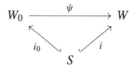

Let ω_0 be the canonical symplectic form on T^*S and $\omega_1 := \psi^*\omega$. Note that ω_0 and ω_1 are symplectic forms on W_0. The submanifold S is Lagrangian for both ω_0 and ω_1. Now we use Weinstein's Lagrangian neighborhood theorem. There exist neighborhoods U_0 and U_1 of S in W_0 and a diffeomorphism $\theta: U_0 \to U_1$ such that $\theta^*\omega_1 = \omega_0$ and the following diagram commutes:

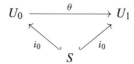

Now we can take $\varphi = \psi \circ \theta$ and $U_1 = \varphi(U_0)$. It is easy to check that $\varphi^*\omega = \theta^* \underbrace{\psi^*\omega}_{\omega_1} = \omega_0$. $\square$

3.6.3   Some Applications

Definition 3.6.7 (C^1-Topology) Let X and Y be two manifolds. A sequence of C^1 maps $f_i: X \to Y$ is said to *converge in the C^1-topology* to a map $f: X \to Y$ if and only if the sequence (f_i) itself and the sequence of the differentials $df_i: TX \to TY$ converge uniformly on compact sets.

Remark 3.6.8 Let (M, ω) be a symplectic manifold. Note that the graph of the identity map, denoted by $\Delta := \Gamma_{\mathrm{id}_M}$ (recall Eq. (3.10) for the definition), is a Lagrangian submanifold of $(M \times M, (\mathrm{pr}_1)^*\omega - (\mathrm{pr}_2)^*\omega)$ (see also Proposition 3.3.14). Moreover, we say that f is C^1-*close* to another map g, if f is in some small neighborhood of g in the C^1-topology.

By Weinstein's tubular neighborhood theorem, there is a neighborhood U of

$$\Delta \subset (M \times M, (\mathrm{pr}_1)^*\omega - (\mathrm{pr}_2)^*\omega)$$

which is symplectomorphic to a neighborhood U_0 of M in (T^*M, ω_0). Let $\varphi: U \to U_0$ be the symplectomorphism satisfying $\varphi(q, q) = (q, 0)$ for all $q \in M$.

Denote by

$$\text{Sympl}(M, \omega) := \{f \colon M \xrightarrow{\sim} M \mid f^*\omega = \omega\}$$

and suppose that $f \in \text{Sympl}(M, \omega)$ is sufficiently C^1-*close* to the identity, i.e. f is in some sufficiently small neighborhood of the identity id_M in the C^1-topology. Then we can assume that the graph of f lies inside of U. Let $j \colon M \hookrightarrow U$ be the embedding as Γ_f and $i \colon M \hookrightarrow U$ the embedding as $\Gamma_{\text{id}_M} = \Delta$. The map j is sufficiently C^1-close to i. By Weinstein's theorem, $U \simeq U_0 \subseteq T^*M$, so the above j and i induce two embeddings: $j_0 \colon M \hookrightarrow U_0$ where $j_0 = \varphi \circ j$ and $i_0 \colon M \hookrightarrow U_0$ embedding as 0-section. Hence, we have

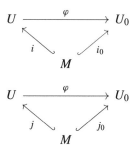

where $i(q) = (q, q)$, $i_0(q) = (q, 0)$, $j(q) = (q, f(q))$ and $j_0(q) = \varphi(q, f(q))$ for $q \in M$. The map j_0 is sufficiently C^1-close to i_0. Thus, the image set $j_0(M)$ intersects each T_q^*M at one point μ_q depending smoothly on q. The image of j_0 is the image of a smooth intersection $\mu \colon M \to T^*M$, that is, a 1-form $\mu = j_0 \circ (\pi \circ j_0)^{-1}$. Therefore,

$$\Gamma_f \simeq \{(q, \mu_q) \mid q \in M, \ \mu_q \in T_q^*M\}. \tag{3.18}$$

Vice-versa, if μ is a 1-form sufficiently C^1-close to the zero 1-form, then we also get (3.18). In fact, we get

$$\Gamma_f \text{ is Lagrangian} \iff \mu \text{ is closed.}$$

Hence, we can conclude that a small C^1-neighborhood of the identity in $\text{Sympl}(M, \omega)$ is homeomorphic to a C^1-neighborhood of zero in the vector space of closed 1-forms on M. Thus, we have

$$T_{\text{id}_M}(\text{Sympl}(M, \omega)) \simeq \{\mu \in \Omega^1(M) \mid d\mu = 0\}.$$

In particular, $T_{\text{id}_M}(\text{Sympl}(M, \omega))$ contains the space of exact 1-forms

$$\{\mu = dh \mid h \in C^\infty(M)\} \simeq C^\infty(M)/\{\text{locally constant functions}\}.$$

Theorem 3.6.9 *Let (M, ω) be a compact symplectic manifold with $H^1(M) = 0$. Then any symplectomorphism of M which is sufficiently C^1-close to the identity has at least two fixed points.*

Proof Suppose that any $f \in \mathrm{Sympl}(M, \omega)$ is sufficiently C^1-close to the identity id_M. Then the graph Γ_f is homotopic to a closed 1-form on M. Now the fact that $\mathrm{d}\mu = 0$ and $H^1(M) = 0$ imply that $\mu = \mathrm{d}h$ for some $h \in C^\infty(M)$. Since M is compact, h has at least two critical points. Note that fixed points of f are equal to critical points of h. Moreover, fixed points of f are equal to the intersection of Γ_f with the diagonal $\Delta \subset M \times M$ which is again equal to $S := \{q \in M \mid \mu_q = \mathrm{d}_q h = 0\}$. Finally, note that the critical points of h are exactly the points in S.   □

We say that a submanifold Y of M is C^1-*close* to another submanifold X of M, when there is a diffeomorphism $X \to Y$ which is, as a map into M, C^1-close to the inclusion $X \hookrightarrow M$.

Theorem 3.6.10 (Lagrangian Intersection Theorem) *Let (M, ω) be a symplectic manifold. Suppose that X is a compact Lagrangian submanifold of M with $H^1(X) = 0$. Then every Lagrangian submanifold of M which is C^1-close to X intersects X in at least two points.*

Exercise 3.6.11 Prove Theorem 3.6.10.

Definition 3.6.12 (Morse Function) A *Morse function* on a manifold M is a function $h \colon M \to \mathbb{R}$ whose critical points are all non-degenerate, i.e. the Hessian at a critical point q is not singular, $\det\left(\frac{\partial^2 h}{\partial x_i \partial x_j}\big|_q\right) \neq 0$.

Let (M, ω) be a symplectic manifold and suppose that $h_t \colon M \to \mathbb{R}$ is a smooth family of functions which is *1-periodic*, i.e. $h_t = h_{t+1}$. Let $\rho \colon M \times \mathbb{R} \to M$ be the isotopy generated by the time-dependent vector field X_t defined by $\omega(X_t, \ ) = \mathrm{d}h_t$. Then we say that f is *exactly homotopic to the identity* if $f = \rho_1$ for such h_t. Equivalently, using the notions which will be introduced in Sect. 3.7, we have the following definition:

Definition 3.6.13 (Exactly Homotopic to the Identity) A symplectomorphism $f \in \mathrm{Sympl}(M, \omega)$ is *exactly homotopic to the identity* when f is the time-1 map of an isotopy generated by some smooth time-dependent 1-periodic Hamiltonian function.

Definition 3.6.14 (Non-degenerate Fixed Point) A fixed point q of a function $f \colon M \to M$ is called *non-degenerate* if $\mathrm{d}_q f \colon T_q M \to T_q M$ is not singular.

Conjecture 3.6.15 (Arnold) *Let (M, ω) be a compact symplectic manifold, and $f \colon M \to M$ a symplectomorphism which is exactly homotopic to the identity. Then*

$$|\{\textit{fixed points of } f\}| \geq \min |\{\textit{critical points of a smooth function on } M\}|.$$

Using the notion of a Morse function *as in Definition 3.6.12, we obtain*

$$|\{non\text{-}degenerate\ fixed\ points\ of\ f\}| \geq \min |\{critical\ points\ of\ a\ Morse\ function\ on\ M\}|$$

$$\geq \sum_{i=0}^{2n} \dim H^i(M, \mathbb{R}).$$

Remark 3.6.16 The Arnold conjecture has been proven by Conley–Zehnder, Floer, Hofer–Salamon, Ono, Fukaya–Ono, Liu–Tian by using *Floer homology*. This is an infinite-dimensional version of *Morse theory*. Sharper bound versions of the Arnold conjecture are still open.

Exercise 3.6.17 Compute the estimates for the number of fixed points on the compact symplectic manifolds S^2, $S^2 \times S^2$ and $T^2 = S^1 \times S^1$.

3.7 Classical Mechanics

3.7.1 Lagrangian Mechanics and Variational Principle

In classical mechanics we are usually interested in the extremal points of certain functions on some infinite-dimensional domain. We will call such a function a *functional*. In order, to compute the extremal points of a functional, we need to describe the notion of differentiation of such an object. In particular, we will consider functionals which are mappings from the space of paths of the form $\gamma = \{(t, q) \mid q(t) = q, \ t_0 \leq t \leq t_1\}$ to $\mathbb{R}$. An example of such a functional is

$$S(\gamma) = \int_{t_0}^{t_1} \sqrt{1 + \dot{q}^2} \mathrm{d}t,$$

where $\dot{x} := \frac{\mathrm{d}q}{\mathrm{d}t}$. Let now $h > 0$ be small. We say that a functional S is *differentiable* if $S(\gamma + h) - S(\gamma) = F(h) + R(h, \gamma)$, where F depends linearly on h and $R(h, \gamma) = O(h^2)$, i.e. for $|h| < \varepsilon$ and $\left|\frac{\mathrm{d}h}{\mathrm{d}t}\right| < \varepsilon$, we have $|R| < C\varepsilon$ for some $C \in \mathbb{R}$. Let us now consider the functional S to be given in form

$$S(\gamma) = \int_{t_0}^{t_1} L(q, \dot{q}, t)\mathrm{d}t, \tag{3.19}$$

where γ denotes again a curve in the (t, q)-plane as before and $L(q, \dot{q}, t)$ denotes a differentiable function of three variables, usually called the *Lagrangian* function of the system. The functional S is usually called the *action functional* in the physics literature.

Theorem 3.7.1 *The action functional* (3.19) *is differentiable and its differential is given by*

$$F(h) = \int_{t_0}^{t_1} \left(\frac{\partial L}{\partial q} - \frac{d}{dt} \frac{\partial L}{\partial \dot{q}} \right) h \, dt + \left(\frac{\partial L}{\partial \dot{q}} h \right) \Big|_{t_0}^{t_1}$$

Proof Indeed, note that we have

$$S(\gamma + h) - S(\gamma) = \int_{t_0}^{t_1} \left(L(q - h, \dot{q} - \dot{h}, t) - L(q, \dot{q}, t) \right) dt \tag{3.20}$$

$$= \int_{t_0}^{t_1} \left(\frac{\partial L}{\partial q} h + \frac{\partial L}{\partial \dot{q}} \dot{h} \right) dt + O(h^2) = F(h) + R, \tag{3.21}$$

where

$$F(h) = \int_{t_0}^{t_1} \left(\frac{\partial L}{\partial q} h + \frac{\partial L}{\partial \dot{q}} \dot{h} \right) dt$$

and $R = O(h^2)$. Using integration by parts, we get

$$\int_{t_0}^{t_1} \frac{\partial L}{\partial \dot{q}} \dot{h} \, dt = - \int_{t_0}^{t_1} h \frac{d}{dt} \left(\frac{\partial L}{\partial \dot{q}} \right) dt + \left(h \frac{\partial L}{\partial \dot{q}} \right) \Big|_{t_0}^{t_1}.$$

□

Similarly as for usual functions where we require the differential to vanish in order to compute extremal points, we require for a functional S that its differential $F(h)$ vanishes for all h at an extremal path γ. Using this property, one can show that γ is an extremal path of S if and only if

$$\boxed{\frac{d}{dt} \frac{\partial L}{\partial \dot{q}} - \frac{\partial L}{\partial q} = 0.} \tag{3.22}$$

Equation (3.22) is called *Euler–Lagrange (EL) equation*. We also denote the variation of the functional S by δS. Hence, the EL equations are equivalent to $\delta S = 0$. In particular, one can regard δ as the de Rham differential on the infinite-dimensional path space in the (t, q)-plane.

Example 3.7.2 (Planar Pendulum) Consider a pendulum in the plane which consists of a particle with mass m on a (massless) rod of length ℓ moving vertically in the plane (see Fig. 3.1). The particle will move on the circle $S_\ell^1 \subset \mathbb{R}^2$ of radius ℓ. The Lagrangian is then given by

$$L(q, \dot{q}, t) = \frac{1}{2} m (\dot{q}_1^2 + \dot{q}_2^2) - m g q_2,$$

Fig. 3.1 Pendulum

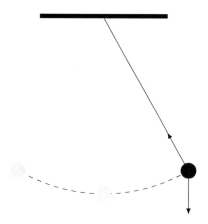

where q_1, q_2 are the coordinates in $\mathbb{R}^2$ and $\dot{q}_1, \dot{q}_2$ are the corresponding coordinates in $T\mathbb{R}^2$. Here g is a constant (gravitational acceleration). Denote the coordinate of S_ℓ^1 by φ.

We can then compute the equations of motion by considering the map

$$S_\ell^1 \to \mathbb{R}^2,$$

$$\varphi \mapsto (\ell \sin(\varphi), -\ell \cos(\varphi)).$$

The corresponding tangent map is given by

$$T S_\ell^1 \to T\mathbb{R}^2,$$

$$(\varphi, \dot{\varphi}) \mapsto (\ell \sin(\varphi), -\ell \cos(\varphi), \ell \cos(\varphi)\dot{\varphi}, \ell \sin(\varphi)\dot{\varphi}).$$

Hence, the constraint Lagrangian is given by

$$L(\varphi, \dot{\varphi}, t) = \frac{1}{2}m(\ell^2 \cos^2(\varphi)\dot{\varphi}^2 + \ell^2 \sin^2(\varphi)\dot{\varphi}^2) + mg\ell \cos(\varphi)$$

$$= \frac{1}{2}m\ell^2\dot{\varphi}^2 + mg\ell \cos(\varphi).$$

The Euler–Lagrange equation is given by

$$m\ell^2 \frac{d\dot{\varphi}}{dt} + mg\ell \sin(\varphi) = 0.$$

Which gives us the following second order differential equation:

$$\ddot{\varphi} = -\frac{g}{\ell} \sin(\varphi).$$

This is exactly the equation of motion for such a pendulum.

3.7.2   Hamiltonian and Symplectic Vector Fields

Let (M, ω) be a symplectic manifold and let $H: M \to \mathbb{R}$ be a smooth function. By non-degeneracy of ω, there is a unique vector field X_H on M such that

$$\iota_{X_H}\omega = \mathrm{d}H.$$

Suppose that M is compact or that X_H is complete. Let $\rho_t: M \to M$ for $t \in \mathbb{R}$ be the 1-parameter family of diffeomorphisms generated by X_H, i.e. satisfying the Cauchy problem

$$\rho_0 = \mathrm{id}_M,$$

$$\frac{\mathrm{d}}{\mathrm{d}t}\rho_t = X_H \circ \rho_t.$$

In fact, each diffeomorphism ρ_t preserves ω, i.e. $\rho_t^*\omega = \omega$ for all t. Indeed, note that

$$\frac{\mathrm{d}}{\mathrm{d}t}\rho_t^*\omega = \rho_t^* L_{X_H}\omega = \rho_t^*(\mathrm{d}\underbrace{\iota_{X_H}\omega}_{=\mathrm{d}H} + \iota_{X_H}\underbrace{\mathrm{d}\omega}_{=0}) = 0.$$

Definition 3.7.3 (Hamiltonian Vector Field and Hamiltonian Function)   A vector field X_H as above is called the *Hamiltonian vector field* with *Hamiltonian function H*.

Exercise 3.7.4 Let X be a vector field on some manifold M. Then there is a unique vector field $X_\sharp$ on the cotangent bundle T^*M whose flow is the lift of the flow of X. Let α be the tautological 1-form on T^*M and let $\omega = -\mathrm{d}\alpha$ be the canonical symplectic form on T^*M. Show that $X_\sharp$ is a Hamiltonian vector field with Hamiltonian function $H := \iota_{X_\sharp}\alpha$.

Remark 3.7.5 If X_H is Hamiltonian, we get

$$L_{X_H} H = \iota_{X_H}\mathrm{d}H = \iota_{X_H}\iota_{X_H}\omega = 0.$$

Hence, Hamiltonian vector fields preserve their Hamiltonian functions and each integral curve $(\rho_t(x))_{t\in\mathbb{R}}$ of X_H must be contained in a level set of H, i.e.

$$H(x) = (\rho_t^* H)(x) = H(\rho_t(x)), \quad \forall t.$$

Definition 3.7.6 (Symplectic Vector Field)   A vector field X on a symplectic manifold (M, ω) preserving ω, i.e. $L_X\omega = 0$, is called *symplectic*.

Remark 3.7.7 Note that X is *symplectic* if and only if $\iota_X\omega$ is closed and X is *Hamiltonian* if and only if $\iota_X\omega$ is exact.

Example 3.7.8 Let $M = T^2 := \mathbb{R}^2/\mathbb{Z}^2$ be the 2-dimensional torus and let (θ_1, θ_2) denote the coordinates of T^2 in $\mathbb{R}^2$. Moreover, endow M with the symplectic form $\omega = d\theta_1 \wedge d\theta_2$. Consider the basis vector field $\frac{\partial}{\partial\theta_2}$ and note that

$$\iota_{\frac{\partial}{\partial\theta_2}} \omega = -d\theta_1,$$

which is closed but not exact since θ_1 is not a function on M. Thus $\frac{\partial}{\partial\theta_2}$ is a symplectic but not a Hamiltonian vector field.

Exercise 3.7.9 Let (M, ω) be a symplectic manifold and consider a symplectic vector field $X \in \mathfrak{X}(M)$ together with a path γ in M. Let S_t be the surface swept out after times t by the flow along X starting at γ. We define the flux

$$\varphi_\gamma(X) := \lim_{t\to 0} \frac{1}{t} \int_{S_t} \omega.$$

Prove that X is Hamiltonian if and only if $\varphi_\gamma(X) = 0$ for all $\gamma \in H_1(M, \mathbb{R})$.

Remark 3.7.10 Locally, on every contractible set, every symplectic vector field is Hamiltonian. If $H^1(M) = 0$, then globally every symplectic vector field is Hamiltonian. In general, $H^1(M)$ measures the obstruction for symplectic vector fields to be Hamiltonian.

3.7.3 Hamiltonian Mechanics

Consider the Euclidean space $\mathbb{R}^{2n}$ with coordinates $(q_1, \ldots, q_n, p_1, \ldots, p_n)$ and $\omega_0 = \sum_{1 \le j \le n} dq_j \wedge dp_j$. If the *Hamilton equations*

$$\begin{aligned}
\frac{dq_i}{dt}(t) &= \frac{\partial H}{\partial p_i}, \\
\frac{dp_i}{dt}(t) &= -\frac{\partial H}{\partial q_i},
\end{aligned} \tag{3.23}$$

are satisfied, then the curve $\rho_t = (q(t), p(t))$ is an integral curve for X_H. Indeed, let

$$X_H = \sum_{i=1}^n \left(\frac{\partial H}{\partial p_i} \frac{\partial}{\partial q_i} - \frac{\partial H}{\partial q_i} \frac{\partial}{\partial p_i} \right).$$

Then we get

$$\iota_{X_H}\omega = \sum_{j=1}^{n}\iota_{X_H}(\mathrm{d}q_j \wedge \mathrm{d}p_j) = \sum_{j=1}^{n}\left[(\iota_{X_H}\mathrm{d}q_j) \wedge \mathrm{d}p_j - \mathrm{d}q_j \wedge (\iota_{X_H}\mathrm{d}p_j)\right]$$

$$= \sum_{j=1}^{n}\left(\frac{\partial H}{\partial p_j}\mathrm{d}p_j + \frac{\partial H}{\partial q_j}\mathrm{d}q_j\right) = \mathrm{d}H.$$

Consider the case when $n = 3$. By Newton's second law, a particle with mass $m \in \mathbb{R}_{>0}$ moving in *configuration space* $\mathbb{R}^3$ with coordinates $q := (q_1, q_2, q_3)$ in a potential $V(q)$, moves along a curve $q(t)$ satisfying

$$m\frac{\mathrm{d}^2 q}{\mathrm{d}t^2} = -\nabla V(q).$$

One then introduces momentum coordinates $p_i := m\frac{\mathrm{d}q_i}{\mathrm{d}t}$ for $i = 1, 2, 3$ and a total energy function

$$H(q, p) = \text{kinetic energy} + \text{potential energy}$$

$$= \frac{1}{2m}\|p\|^2 + V(q).$$

The *phase space* is then given by $T^*\mathbb{R}^3 \cong \mathbb{R}^6$ with coordinates $(q_1, q_2, q_3, p_1, p_2, p_3)$. Newton's second law in $\mathbb{R}^3$ is equivalent to the Hamilton equations in $\mathbb{R}^6$:

$$\frac{\mathrm{d}q_i}{\mathrm{d}t} = \frac{1}{m}p_i = \frac{\partial H}{\partial p_i},$$

$$\frac{\mathrm{d}p_i}{\mathrm{d}t} = m\frac{\mathrm{d}^2 q_i}{\mathrm{d}t^2} = -\frac{\partial V}{\partial q_i} = -\frac{\partial H}{\partial q_i}.$$

Remark 3.7.11 The total energy H is *conserved* by the physical motion, i.e. $\frac{\mathrm{d}}{\mathrm{d}t}H = 0$.

3.7.4   Relation of Lie Brackets and Further Structure

Recall that vector fields are differential operators on functions. In particular, if X is a vector field and $f \in C^\infty(M)$, with $\mathrm{d}f$ being the corresponding 1-form, then

$$X(f) = \mathrm{d}f(X) = L_X f.$$

In fact, for two vector fields X, Y, we have

$$L_{[X,Y]}f = L_X(L_Y f) - L_Y(L_X f).$$

Proposition 3.7.12 *If X and Y are symplectic vector fields on a symplectic manifold (M, ω), then $[X, Y]$ is a Hamiltonian vector field with Hamiltonian function $\omega(Y, X)$.*

Proof We have

$$
\begin{aligned}
\iota_{[X,Y]}\omega &= L_X \iota_Y \omega - \iota_Y L_X \omega \\
&= d\iota_X \iota_Y \omega + \iota_X \underbrace{d\iota_Y \omega}_{=0} - \iota_Y \underbrace{d\iota_X \omega}_{=0} - \iota_Y \iota_X \underbrace{d\omega}_{=0} \\
&= d(\omega(Y, X)).
\end{aligned}
$$

$\square$

Denote by $\mathfrak{X}_H(M)$ the space of *Hamiltonian vector fields* and by $\mathfrak{X}_S(M)$ the space of *symplectic vector fields* on a manifold M.

Corollary 3.7.13 *The inclusions*

$$(\mathfrak{X}_H(M), [\ ,\ ]) \subseteq (\mathfrak{X}_S(M), [\ ,\ ]) \subseteq (\mathfrak{X}(M), [\ ,\ ])$$

are inclusions of Lie algebras.

Definition 3.7.14 (Poisson Bracket) The *Poisson bracket* of two functions $f, g \in C^\infty(M)$ on a symplectic manifold (M, ω) is defined by

$$\{f, g\} := \omega(X_f, X_g). \tag{3.24}$$

Remark 3.7.15 Note that we have $X_{\{f,g\}} = -[X_f, X_g]$ since $X_{\omega(X_f,X_g)} = [X_g, X_f]$.

Exercise 3.7.16 Show that the Poisson bracket $\{\ ,\ \}$ satisfies the *Jacobi identity*, i.e.

$$\{f, \{g, h\}\} + \{g, \{h, f\}\} + \{h, \{f, g\}\} = 0, \quad \forall f, g, h \in C^\infty(M).$$

Definition 3.7.17 (Poisson Algebra) A *Poisson algebra* is a commutative associative algebra $\mathcal{A}$ with a Lie bracket $\{\ ,\ \} : \mathcal{A} \times \mathcal{A} \to \mathcal{A}$ satisfying the *Leibniz rule*

$$\{f, gh\} = \{f, g\}h + g\{f, h\}, \quad \forall f, g, h \in \mathcal{A}.$$

Exercise 3.7.18 Show that the Poisson bracket defined as in (3.24) satisfies the Leibniz rule and deduce that if (M, ω) is a symplectic manifold, then $(C^\infty(M), \{\ ,\ \})$ is a Poisson algebra.

Remark 3.7.19 Note that we have a Lie algebra *anti-homomorphism*

$$C^\infty(M) \to \mathfrak{X}(M),$$

$$H \mapsto X_H$$

such that the Poisson bracket $\{\ ,\ \}$ will correspond to $-[\ ,\ ]$.

3.7.5 Integrable Systems

Definition 3.7.20 (Hamiltonian System) A *Hamiltonian system* is a triple (M, ω, H), where (M, ω) is a symplectic manifold and $H \in C^\infty(M)$ is a *Hamiltonian function*.

Theorem 3.7.21 *We have $\{f, H\} = 0$ if and only if f is constant along integral curves of X_H.*

Proof Let ρ_t be the flow of X_H. Then we have

$$\frac{\mathrm{d}}{\mathrm{d}t}(f \circ \rho_t) = \rho_t^* L_{X_H} f = \rho_t^* \iota_{X_H} \mathrm{d}f = \rho_t^* \iota_{X_H} \iota_{X_f} \omega = \rho_t^* \omega(X_f, X_H) = \rho_t^* \{f, H\}.$$
(3.25)

$\square$

Remark 3.7.22 Given a Hamiltonian system (M, ω, H), a function f satisfying $\{f, H\} = 0$ is called an *integral of motion* or a *constant of motion*. In general, Hamiltonian system do not admit integrals of motion which are *independent* of the Hamiltonian function.

Definition 3.7.23 (Independent Functions) We say that functions $f_1, \ldots, f_n$ on a manifold M are *independent* if their differentials $\mathrm{d}_q f_1, \ldots, \mathrm{d}_q f_n$ are linearly independent at all points $q \in M$ in some open dense subset of M.

Definition 3.7.24 ((Completely) Integrable System) A Hamiltonian system (M, ω, H) is *(completely) integrable* if it has $n = \frac{1}{2} \dim M$ independent integrals of motion $f_1 = H, f_2 \ldots, f_n$, which are pairwise in *involution* with respect to the Poisson bracket, i.e.

$$\{f_i, f_j\} = 0, \quad \forall i, j.$$

Let (M, ω, H) be an integrable system of dimension $2n$ with integrals of motion $f_1 = H, f_2, \ldots, f_n$. Let $c \in \mathbb{R}^n$ be a regular value of $f := (f_1, \ldots, f_n)$. Note that the corresponding *level set*, $f^{-1}(c)$, is a Lagrangian submanifold, because it is n-dimensional and its tangent bundle is isotropic.

Lemma 3.7.25 *If the Hamiltonian vector fields $X_{f_1}, \ldots, X_{f_n}$ are complete on the level set $f^{-1}(c)$, then the connected components of $f^{-1}(c)$ are* homogeneous spaces *for $\mathbb{R}^n$, i.e. are of the form $\mathbb{R}^{n-k} \times T^k$ for some k with $0 \leq k \leq n$, where T^k denotes the k-torus.*

Exercise 3.7.26 Prove Lemma 3.7.25. *Hint: follow the flows to obtain coordinates.*

Note that any compact component of $f^{-1}(c)$ must be a torus. These components, when they exist, are called *Liouville tori.*

Theorem 3.7.27 (Arnold–Liouville[Arn78, Lio55]) *Let (M, ω, H) be an integrable system of dimension $2n$ with integrals of motion $f_1 = H, f_2, \ldots, f_n$. Let $c \in \mathbb{R}^n$ be a regular value of $f := (f_1, \ldots, f_n)$. The corresponding level set $f^{-1}(c)$ is a Lagrangian submanifold of M.*

(1) If the flows of $X_{f_1}, \ldots, X_{f_n}$ starting at a point $q \in f^{-1}(c)$ are complete, then the connected component of $f^{-1}(c)$ containing q is a homogeneous space for $\mathbb{R}^n$. With respect to this affine structure, that component has coordinates $\phi_1, \ldots, \phi_n$, known as angle coordinates, *in which the flows of the vector fields $X_{f_1}, \ldots, X_{f_n}$ are linear.*

(2) There are coordinates $\psi_1, \ldots, \psi_n$, known as action coordinates, *complementary to the angle coordinates such that the ψ_is are integrals of motion and*

$$\phi_1, \ldots, \phi_n, \psi_1, \ldots, \psi_n$$

form a Darboux chart.

Definition 3.7.28 (Superintegrable System) Let (M, ω) be a $2n$-dimensional connected symplectic manifold and let $(C^\infty(M), \{\ ,\ \})$ be the Poisson algebra of smooth functions on M. A subset $f = (f_1, \ldots, f_k)$ for $n \leq k \leq 2n$ with $f_i \in C^\infty(M)$ is called *superintegrable system* if the following hold:

(1) All maps f_i are independent, i.e. the k-form $\bigwedge^k df_i$ is nowhere vanishing on M. This implies that $f : M \to \mathbb{R}^k$ is a submersion, which implies that $f : M \to f(M)$ is a fibered manifold over some $N \subset \mathbb{R}^k$ endowed with the coordinates (x_i) such that $x_i \circ f = f_i$.

(2) There are smooth functions s_{ij} on N such that

$$\{f_i, f_j\} = s_{ij} \circ f, \quad 1 \leq i, j \leq k.$$

(3) The matrix $S = (s_{ij})$ is of constant corank $m = 2n - k$ at each point of N.

The generalization of the Arnold–Liouville theorem for superintegrable systems is covered by the *Mishchenko–Fomenko theorem.*

Theorem 3.7.29 (Mishchenko–Fomenko[MF78]) *Consider the Hamiltonian vector fields X_{f_i} of the functions f_i and assume that they are complete for all $1 \leq i \leq n$. Moreover, let the fibers of the fibered manifold $f : M \to f(M)$ be connected and mutually diffeomorphic. Then the following hold:*

(1) The fibers of f are diffeomorphic to the toroidal cylinder $\mathbb{R}^{m-r} \times T^r$ with $m = 2n - k$ and $0 \leq r \leq m$.

(2) For a fiber M of f, there is an open neighborhood U_M which is a trivial principal bundle

$$U_M = N_M \times \mathbb{R}^{m-r} \times T^r \xrightarrow{f} N_M$$

with structure group $\mathbb{R}^{m-r} \times T^r$ with $m = 2n - k$ and $0 \leq r \leq m$.

(3) The neighborhood U_M is provided with generalized action-angle coordinates.

Example 3.7.30 (Kepler System) Let us consider a 3-dimensional example, the 2-body problem of planetary motion considered by Kepler, which was historically one of the first examples of an integrable system. Denote for a vector $\vec{x} \in \mathbb{R}^3$ by $x := \|\vec{x}\|$ its norm where $\| \ \|$ is the norm induced by the standard inner product on $\mathbb{R}^3$. The Hamiltonian is given by

$$H = \frac{1}{2}(p_1^2 + p_2^2 + p_3^2) - \frac{k}{r}$$

with $r = \sqrt{q_1^2 + q_2^2 + q_3^2}$ and $k \in \mathbb{R}$ some constant. It is easy to see that this system is invariant under rotations and that the angular momentum $\vec{J} := \vec{q} \times \vec{p}$ is conserved. Consider the following two conserved quantities

$$H = \frac{1}{2}\left(p_r^2 + \frac{p_\theta^2}{r^2} + \frac{p_\phi^2}{r^2 \sin^2(\theta)}\right) + V(r), \quad J^2 = p_\theta^2 + \frac{p_\phi^2}{\sin^2(\theta)}, \quad J_3 = p_\phi.$$

Let us consider, w.l.o.g. that $\vec{J} = (0, 0, J_3)$, i.e. we restrict the motion to the plane with $\theta = \frac{\pi}{2}$. Using the Hamilton equations, we get

$$\dot{\phi} = \{H, \phi\} = \frac{p_\phi}{r^2 \sin^2(\theta)}$$

and thus $p_\theta = r^2 \dot{\phi}$. This shows that the angular momentum is indeed conserved as it was discovered by Kepler. The solution is constructed as follows: The momenta p_r and p_ϕ are obtained in terms of the energy E and angular momentum J

$$p_r = \sqrt{2(E - V) - \frac{J^2}{r^2}}, \quad p_\phi = J_3 = J.$$

Then we can construct the generating function

$$S = \int \sqrt{2(H - V) - \frac{J^2}{r^2}}\, dr + J \int d\phi.$$

The associated angle variables are then given by

$$\psi_E = \frac{\partial S}{\partial E}, \quad \psi_J = \frac{\partial S}{\partial J},$$

with corresponding equations of motion

$$\dot{\psi}_E = 1, \quad \dot{\psi}_J = 0,$$

such that $\psi_E = t$. Now, if we go back to spherical coordinates, we plug into ψ_E and obtain

$$t = \int \frac{dr}{\sqrt{2(E - V) - \frac{J^2}{r^2}}}.$$

Using the equation of motion for ψ_J, we get

$$\phi = \int \frac{J\, dr}{r^2 \sqrt{2(E - V) - \frac{J^2}{r^2}}} = \arccos\left(\frac{\frac{J}{r} - \frac{k}{J}}{\sqrt{2E + \frac{k^2}{J^2}}}\right).$$

In order to be able to distinguish between the different types of orbits, we define

$$p = \frac{J^2}{k}, \quad e = \sqrt{1 + \frac{2EJ^2}{k^2}}.$$

This gives us

$$r = \frac{p}{1 + e\cos(\theta)},$$

which is called the *focal equation of a conic section*. When $e < 1$, we get that the conic section defined through the equation is an ellipse. Another conserved quantity of the system is given by the *Runge–Lenz vector*

$$\vec{A} = \vec{v} \times \vec{J} - k\frac{\vec{r}}{r}.$$

Exercise 3.7.31 Show that the integrable system in Example 3.7.30 is superintegrable.

3.8 Moment Maps

3.8.1 Symplectic and Hamiltonian Actions

Let (M, ω) be a symplectic manifold, and G a Lie group. Let $\Psi: G \to \mathrm{Diff}(M)$ be a (smooth) action.

Definition 3.8.1 (Symplectic Action) The action Ψ is a *symplectic action* if

$$\Psi: G \to \mathrm{Sympl}(M, \omega) \subset \mathrm{Diff}(M),$$

i.e., *G acts by symplectomorphisms.*

Remark 3.8.2 It is easy to show that there is a one-to-one correspondence between *complete vector fields* on a manifold M and *smooth actions* of $\mathbb{R}$ on M. One can show this by associating to a complete vector field X its exponential map $\exp(tX)$ and, vice-versa, to a smooth action Ψ its derivative $\frac{\mathrm{d}\Psi_t(q)}{\mathrm{d}t}\big|_{t=0} =: X_q$. Thus, we also get a one-to-one correspondence between *complete symplectic vector fields* on M and *symplectic actions* of $\mathbb{R}$ on M.

Definition 3.8.3 (Hamiltonian Action I) A symplectic action Ψ of S^1 or $\mathbb{R}$ on (M, ω) is *Hamiltonian* if the vector field generated by Ψ is Hamiltonian. Equivalently, an action Ψ of S^1 or $\mathbb{R}$ on (M, ω) is *Hamiltonian* if there is a function $H: M \to \mathbb{R}$ with $\iota_X \omega = \mathrm{d}H$, where X is the vector field generated by Ψ.

Example 3.8.4 Let $G = T^n := \mathbb{R}^n / \mathbb{Z}^n$ be the n-dimensional torus. The Lie algebra of G is then given by $\mathfrak{g} \cong \mathbb{R}^n$ endowed with the zero Lie bracket. Assume we have a Hamiltonian action of G on a symplectic manifold (M, ω). Then, if $X_1, \ldots, X_n$ is a basis of $\mathfrak{g}$, we get a collection of corresponding Hamiltonians $H_1, \ldots, H_n$ on M. Since $[X_i, X_j] = 0$ for all $1 \leq i, j \leq n$, we get that the corresponding Hamiltonian vector fields X_{H_i} satisfy

$$X_{\{H_i, H_j\}} = [X_{H_i}, X_{H_j}] = 0.$$

Thus, we have $\{H_i, H_j\}$ is constant for $1 \leq i, j \leq n$. It is not hard to see that actually we get

$$\{H_i, H_j\} = 0, \quad 1 \leq i, j \leq n.$$

Remark 3.8.5 When G is not a product of copies of S^1 or $\mathbb{R}$, the solution is to use an upgraded Hamiltonian function, which is called a *moment map*.

3.8.2 Hamiltonian Actions II and Moment Maps

Let (M, ω) be a symplectic manifold, G a Lie group, and $\Psi: G \to \mathrm{Sympl}(M, \omega)$ a (smooth) symplectic action, i.e. a group homomorphism such that the evaluation map $\mathrm{ev}_\Psi(g, q) := \Psi_g(q)$ is smooth. Moreover, let $\mathfrak{g}$ be the Lie algebra of G and $\mathfrak{g}^*$ its dual space.

Definition 3.8.6 (Hamiltonian Action II) The action Ψ is called *Hamiltonian* if there exists a map

$$\mu: M \to \mathfrak{g}^*$$

such that for each $X \in \mathfrak{g}$, the map

$$\mu^X: M \to \mathbb{R},$$
$$q \mapsto \mu^X(q) := \langle \mu(q), X \rangle,$$

is the component of μ along X and $X^\#$ is the fundamental vector field on M generated by the 1-parameter subgroup $(\exp(tX))_{t \in \mathbb{R}} \subseteq G$. Then

$$\mathrm{d}\mu^X = \iota_{X^\#}\omega,$$

i.e., μ^X is a Hamiltonian function for the vector field $X^\#$. Moreover, μ is *equivariant* with respect to the given action Ψ of G on M and the coadjoint action Ad^* of G on $\mathfrak{g}^*$:

$$\mu \circ \Psi_g = \mathrm{Ad}_g^* \circ \mu, \quad \forall g \in G.$$

Definition 3.8.7 (Moment Map) A map μ as in Definition 3.8.6 is called a *moment map*.

Definition 3.8.8 (Hamiltonian G-Space) The quadruple (M, ω, G, μ) is called a *Hamiltonian G-space*.

Definition 3.8.9 (Comoment Map) Let G be a connected Lie group. A *comoment map* is a map

$$\mu^*: \mathfrak{g} \to C^\infty(M),$$

such that

(1) $\mu^*(X) := \mu^X$ is a Hamiltonian function for the vector field $X^\#$,
(2) μ^* is a Lie algebra homomorphism:

$$\mu^*[X, Y] = \{\mu^*(X), \mu^*(Y)\},$$

where $\{\ ,\ \}$ denotes the Poisson bracket on $C^\infty(M)$.

Remark 3.8.10 The condition for an action to be Hamiltonian can be equivalently rephrased by using the notion of a comoment map instead of a moment map.

Example 3.8.11 (Torus Action) Let M be a compact manifold and consider the group given by the torus T with the corresponding action. Let $\mathfrak{t}$ be its Lie algebra and choose a basis $X_1, \ldots, X_k$ of $\mathfrak{t}$. The existence of a moment map is then equivalent to a primitive $\widetilde{\mu}^{X_i}$ for $\iota_{X_i}\omega$ for any $1 \le i \le k$. Hence, to get $\widetilde{\mu}$, we can define

$$\widetilde{\mu}^X = \sum_{1 \le i \le k} \lambda_i \widetilde{\mu}^{X_i}, \quad X = \sum_{1 \le i \le k} \lambda_i X_i.$$

E.g., assume that $H^1(M) = 0$ which is the case if M is simply connected. Then we get that any T-action is Hamiltonian. The moment map μ is then indeed well-defined up to some addition of a constant element in $\mathfrak{t}^*$.

Example 3.8.12 (Circle Action) Consider the circle S^1 acting on $\mathbb{C}^n$ by $u \cdot (z_1, \ldots, z_n) = (uz_1, \ldots, uz_n)$ with $z_i = z_i + iy_i$. Then we can consider the fundamental vector field associated to $\frac{\partial}{\partial\theta} \in T_1 S^1$ which is given by

$$X_H = \sum_{1 \le j \le n} \left(-y_j \frac{\partial}{\partial x_j} + x_j \frac{\partial}{\partial y_j} \right).$$

The moment map μ for this action is given by the Hamiltonian

$$H = \frac{1}{2} \sum_{1 \le j \le n} |z_j|^2.$$

3.9 Symplectic Reduction

3.9.1 Quotient Manifold by Group Action

Let G be a group and let M be a manifold. Denote by $\Psi : G \to \mathrm{Diff}(M)$ any action of G on the diffeomorphism group of M.

Definition 3.9.1 (Orbit) The *orbit* of G through $q \in M$ is

$$\mathcal{O}_q := \{\Psi_g(q) \mid g \in G\}.$$

Definition 3.9.2 (Stabilizer) The *stabilizer* (or *isotropy*) of $q \in M$ is the subgroup

$$G_q := \{g \in G \mid \Psi_g(q) = q\}.$$

Exercise 3.9.3 Show that if p is in the orbit of q, then G_p and G_q are conjugate subgroups.

Definition 3.9.4 (Transitive/Free/Locally Free) The action Ψ is called

- *transitive* if there is just one orbit,
- *free* if all stabilizers are trivial, i.e. only consist of the identity element $\{e\}$,
- *locally free* if all stabilizers are discrete.

Let $\sim$ be the orbit equivalence relation, i.e. for $q, p \in M$, we define $q \sim p$ if and only if q and p are on the same orbit.

Definition 3.9.5 (Orbit Space) The space $M/\!\!\sim \, =: M/G$ is called the *orbit space*.

Remark 3.9.6 We can endow M/G with the *quotient topology* with respect to the projection

$$\pi : M \to M/G,$$

$$q \mapsto \mathcal{O}_q.$$

Theorem 3.9.7 *If a compact Lie group G acts freely on a manifold M, then M/G is a manifold and the map $\pi : M \to M/G$ is a principal G-bundle.*

Proof First, we will show that, for any $q \in M$, the G-orbit through q is a compact embedded submanifold of M diffeomorphic to G. Note that the evaluation map

$$\mathrm{ev} : G \times M \to M,$$

$$(g, q) \mapsto \mathrm{ev}(g, q) := g \cdot q$$

is smooth since the action is smooth. We claim that, for $q \in M$, the map ev_q provides the desired embedding. Note that the image of ev_q is the G-orbit through q. Since the action of G is free, we get that ev_q is injective. Clearly, the map ev_q is proper, since a compact (and hence closed) subset N of M has inverse image $(\mathrm{ev}_q)^{-1}(N)$ being a closed subset of a compact Lie group G, hence compact. We still have to show that ev_q is an immersion. Note that for $X \in \mathfrak{g} \cong T_e G$ we have

$$d_e \, \mathrm{ev}_q(X) = 0 \Longleftrightarrow X_q^\# = 0 \Longleftrightarrow X = 0,$$

since the action is free. Hence, we can conclude that $d_e \, ev_q$ is injective. Thus, at any point $g \in G$, for $X \in T_g G$, we have

$$d_g \, ev_q(X) = 0 \iff d_e(ev_q \circ R_g) \circ d_g R_{g^{-1}}(X) = 0,$$

where $R_g \colon G \to G$ denotes *right multiplication* by g. On the other hand, $ev_q \circ R_g = ev_{g \cdot q}$ has an injective differential at the identity e, and $d_g R_{g^{-1}}$ is an isomorphism. This implies that $d_g \, ev_q$ is always injective.

In fact, one can show that even if the action is not free, the G-orbit through q is a compact embedded submanifold of M. In that case, the orbit is diffeomorphic to the quotient of G by the isotropy of q, i.e.

$$\mathcal{O}_q \cong G/G_q.$$

Let S be a transverse section to $\mathcal{O}_q$ at q. We call S a *slice*. Choose coordinates $x_1, \ldots, x_n$ centered at q such that

$$\mathcal{O}_q \cong G \colon x_1 = \cdots = x_k = 0,$$
$$S \colon x_{k+1} = \cdots = x_n = 0.$$

Let $S_\epsilon := S \cap B_\epsilon(0, \mathbb{R}^n)$, where $B_\epsilon(0, \mathbb{R}^n)$ denotes the ball of radius ϵ centered at 0 in $\mathbb{R}^n$. Let $\eta \colon G \times S \to M$, $\eta(g, s) = g \cdot s$. Then we can apply the following equivariant version of the tubular neighborhood theorem:

Theorem 3.9.8 (Slice Theorem) *Let G be a compact Lie group G acting on a manifold M such that G acts freely at $q \in M$. For sufficiently small ϵ, $\eta \colon G \times S_\epsilon \to M$ maps $G \times S_\epsilon$ diffeomorphically onto a G-invariant neighborhood U of the G-orbit through q.*

We will use the following corollaries of the Slice theorem:

Corollary 3.9.9 *If the action of G is free at q, then the action is free on U.*

Corollary 3.9.10 *The set of points where G acts freely is open.*

Corollary 3.9.11 *The set $G \times S_\epsilon \cong U$ is G-invariant. Hence, the quotient*

$$U/G \cong S_\epsilon$$

is smooth.

Now we can conclude the proof that M/G is a manifold and $\pi \colon M \to M/G$ is a smooth fiber map. For $q \in M$, let $p = \pi(q) \in M/G$. Choose a G-invariant neighborhood U of q as in the slice theorem: $U \cong G \times S_\epsilon$. Then $\pi(U) = U/G =: \mathcal{V}$ is an open neighborhood of p in M/G. By the slice theorem, we get that $S_\epsilon \xrightarrow{\sim} \mathcal{V}$ is a homeomorphism. We will use such neighborhoods $\mathcal{V}$ as charts on M/G. We want to show that the transition functions associated with these charts are smooth.

For this, consider two G-invariant open sets U_1, U_2 in M and corresponding slices S_1, S_2 of the G-action. Then $S_{12} := S_1 \cap U_2$ and $S_{21} := S_2 \cap U_1$ are both slices for the G-action on $U_1 \cap U_2$. To compute the transition map $S_{12} \to S_{21}$, consider the diagram

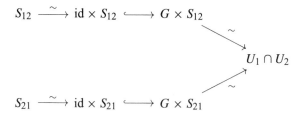

Then the composition

$$S_{12} \longleftrightarrow U_1 \cap U_2 \xrightarrow{\;\sim\;} G \times S_{21} \xrightarrow{\text{pr}_2} S_{21}$$

is smooth. Finally, we need to show that $\pi : M \to M/G$ is a smooth fiber map. For $q \in M$ with $p := \pi(q) \in M/G$, choose a G-invariant neighborhood U of the G-orbit through q of the form $\eta : G \times S_\epsilon \xrightarrow{\sim} U$. Then $V = U/G \simeq S_\epsilon$ is the corresponding neighborhood of $p \in M/G$:

$$
\begin{array}{ccccc}
M \supseteq U & \xrightarrow{\;\eta^{-1}\;} & G \times S_\epsilon & \xrightarrow{\;\sim\;} & G \times V \\
{\scriptstyle \pi}\downarrow & & & & \downarrow{\scriptstyle \text{pr}_2} \\
M/G \supseteq V & \xrightarrow{\qquad\text{id}\qquad} & & & V
\end{array}
$$

since the projection pr_2 is smooth. It is then easy to check that the transition maps for the bundle defined by π are smooth. We leave this as an exercise. $\square$

3.9.2   The Marsden–Weinstein Theorem

Theorem 3.9.12 (Marsden–Weinstein[MW74]) *Let* (M, ω, G, μ) *be a Hamiltonian G-space for a compact Lie group G. Let* $i : \mu^{-1}(0) \hookrightarrow M$ *be the inclusion map. Assume that G acts freely on $\mu^{-1}(0)$. Then*

- *the orbit space* $M_{\text{red}} := \mu^{-1}(0)/G$ *is a manifold,*
- $\pi : \mu^{-1}(0) \to M_{\text{red}}$ *is a principal G-bundle, and*
- *there is a symplectic form* ω_{red} *on* M_{red} *satisfying* $i^*\omega = \pi^*\omega_{\text{red}}$.

Definition 3.9.13 (Reduction) The pair $(M_{\text{red}}, \omega_{\text{red}})$ is called the *reduction* of (M, ω) with respect to G and μ.

Remark 3.9.14 The reduction is sometimes also called *reduced space, symplectic quotient* or the *Marsden–Weinstein quotient*.

Lemma 3.9.15 *Let* (V, ω) *be a symplectic vector space. Suppose that* I *is an isotropic subspace of* V. *Then* ω *induces a canonical symplectic form* Ω *on* I^ω/I, *where* I^ω *denotes the symplectic orthocomplement of* I.

Exercise 3.9.16 Prove Lemma 3.9.15.

Proof of Theorem 3.9.12 Note that since G acts freely on $\mu^{-1}(0)$, we get that $d_q \mu$ is surjective for all $q \in \mu^{-1}(0)$ since

$$\operatorname{im} d_q \mu = \operatorname{Ann}(\mathfrak{g}_q) := \{\xi \in \mathfrak{g}^* \mid \langle \xi, X \rangle = 0, \ \forall X \in \mathfrak{g}_q\},$$

where Ann denotes the *annihilator* and $\mathfrak{g}_q$ the Lie algebra of the stabilizer G_q which in this case is trivial and hence $\operatorname{im} d_q \mu = \mathfrak{g}^*$. Thus, 0 is a *regular value* and therefore $\mu^{-1}(0)$ is a closed submanifold of codimension dim G. The first part of Theorem 3.9.12 is just an application of Theorem 3.9.7 to the free action of G on $\mu^{-1}(0)$.

Lemma 3.9.15 gives a canonical symplectic structure on the quotient $T_q \mu^{-1}(0)/T_q \mathcal{O}_q$ since, by the fact that G acts freely on $\mu^{-1}(0)$, we have that $\mathcal{O}_q \cong G$ and thus

$$T_q \mu^{-1}(0) = \ker d_q \mu = (T_q \mathcal{O}_q)^{\omega_q}.$$

One can indeed check that $T_q \mathcal{O}_q$ is an isotropic subspace of $T_q M$. The point $[q] \in M_{\text{red}} = \mu^{-1}(0)/G$ has tangent space $T_{[q]} M_{\text{red}} \cong T_q \mu^{-1}(0)/T_q \mathcal{O}_q$. Thus Lemma 3.9.15 defines a non-degenerate 2-form ω_{red} on M_{red}. This is well-defined because ω is G-invariant.

By construction, we have $i^*\omega = \pi^*\omega_{\text{red}}$, where

$$
\begin{array}{ccc}
\mu^{-1}(0) & \xrightarrow{\ i\ } & M \\
\pi \downarrow & & \\
M_{\text{red}} & &
\end{array}
$$

Hence, $\pi^* d\omega_{\text{red}} = d\pi^*\omega_{\text{red}} = di^*\omega = i^*d\omega = 0$. The closedness of ω_{red} follows from the injectivity of π^*. $\square$

Remark 3.9.17 The Marsden–Weinstein quotient is often also denoted by $M//G$.

Example 3.9.18 Consider the action of S^1 on $\mathbb{C}^n$ given by $u \cdot (z_1, \ldots, z_n) = (u z_1, \ldots, u z_n)$. Moreover, consider $\mathbb{C}^n$ as a symplectic manifold with its standard symplectic structure (see Example 3.1.29) and use the Killing form (see Definition 2.7.42) on $T_1 S^1$ to identify $T_1^* S^1 \cong T_1 S^1 \cong \mathbb{R}$. The moment map for this action is, as we have seen in Example 3.8.12, given by

$$\mu(z_1, \ldots, z_n) = \frac{1}{2} \sum_{1 \leq j \leq n} |z_j|^2.$$

Note that we have $\mu^{-1}(0) = \{0\}$ and $\mu^{-1}(0)/G = \{\text{pt}\}$. If we consider the level set for the point 1, we get

$$\mu^{-1}(1) = S^{2n-1}$$

and the Marsden–Weinstein quotient is given by the complex projective space:

$$\mathbb{C}^n //S^1 = \mu^{-1}(1)/S^1 = S^{2n-1}/S^1 = \mathbb{C}P^n.$$

Exercise 3.9.19 Consider the action of $G := U_m$ on the space of $n \times m$ matrices $\mathbb{C}^{n \times m}$ given by $g \cdot A = Ag^{-1}$ with symplectic form given by the imaginary part of the Hermitian inner product on $\mathbb{C}^{n \times m}$. Show that for $A \in \mathbb{C}^{n \times m}$ and $B \in \mathfrak{u}_m$ the map

$$\mu(A) \cdot B = \frac{i}{2} \operatorname{Tr}(ABA^*)$$

is a moment map.

Example 3.9.20 Consider the same action and setting as in Exercise 3.9.19. By using the Killing form on $\mathfrak{u}_m$, we can identify $\mathfrak{u}_m^* \cong \mathfrak{u}_m$ and thus view the moment map as in Exercise 3.9.19 as a map

$$\mu : \mathbb{C}^{n \times m} \to \mathfrak{u}_m,$$

$$A \mapsto \mu(A) = \frac{i}{2} A^* A.$$

If we consider the level set at $\frac{i}{2} I_m$, where I_m denotes the $m \times m$ identity matrix, we can see that this point fixed by the adjoint action of the unitary group U_m on $\mathfrak{u}_m$. We get

$$\mu^{-1}\left(\frac{i}{2} I_m\right) = \{A \in \mathbb{C}^{n \times m} \mid A^* A = I_m\}.$$

The Marsden–Weinstein quotient $\mathbb{C}^{n \times m} //U_m = \mu^{-1}\left(\frac{i}{2} I_m\right)/U_m$ is then given by the *Grassmannian* $\operatorname{Gr}(m, n)$ consisting of m-planes in $\mathbb{C}^n$.

3.9.3 Noether's Theorem

Let (M, ω, G, μ) be a Hamiltonian G-space.

Theorem 3.9.21 (Noether[Noe18]) *A function* $f : M \to \mathbb{R}$ *is* G-*invariant if and only if* μ *is constant along the trajectories of the Hamiltonian vector field of* f.

Proof Let X_f be the Hamiltonian vector field of f. Moreover, let $X \in \mathfrak{g}$ and

$$\mu^X = \langle \mu, X \rangle : M \to \mathbb{R}.$$

Then we have

$$L_{X_f} \mu^X = \iota_{X_f} \mathrm{d}\mu^X = \iota_{X_f} \iota_{X^\#} \omega = -\iota_{X^\#} \iota_{X_f} \omega = -\iota_{X^\#} \mathrm{d}f = -L_{X^\#} f = 0,$$

(3.26)

because f is G-invariant. □

Definition 3.9.22 (Integral of Motion/Symmetry) A G-invariant function $f : M \to \mathbb{R}$ is called an *integral of motion* of the Hamiltonian G-space (M, ω, G, μ). If μ is constant along the trajectories of a Hamiltonian vector field X_f, then the corresponding 1-parameter group of diffeomorphisms $(\exp(tX)_f)_{t \in \mathbb{R}}$ is called a *symmetry* of (M, ω, G, μ).

Remark 3.9.23 The *Noether principle* asserts that there is a one-to-one correspondence between symmetries and integrals of motion.

3.10 Kähler Manifolds and Complex Geometry

Kähler manifolds are complex manifolds of dimension n together with a Hermitian metric such that the associated 2-form ω is closed and hence directly yields a symplectic structure, thus they form an important class of symplectic manifolds.

3.10.1 Complex Structures

Definition 3.10.1 (Complex Structure) A *complex structure* on a real vector space V is a linear map $J : V \to V$ such that $J^2 = -\mathrm{id}_V$.

Definition 3.10.2 (Complexification of a Vector Space) Let V be a real vector space. The *complexification* of V is given by the choice of a complex structure J such that

$$(a + ib)v := av + bJv, \quad v \in V, \, a, b \in \mathbb{R}.$$

We will denote the complexification of a real vector space V by $V^c := V \otimes_{\mathbb{R}} \mathbb{C}$. We can then define *complex conjugation* on V^c as

$$z = x + iy \mapsto \bar{z} = x - iy, \quad x, y \in V.$$

Remark 3.10.3 For a complex vector space V, one can always define a complex structure J by defining $Jv := iv$ for $v \in V$.

Example 3.10.4 Consider the complex vector space $\mathbb{C}^n$ with coordinates $z_1, \ldots, z_n$ such that $z_j = x_j + iy_j$ for $x_j, y_j \in \mathbb{R}$ and $j = 1, \ldots, n$. We identify $\mathbb{C}^n$ with $\mathbb{R}^{2n}$ by $(z_1, \ldots, z_n) \mapsto (x_1, \ldots, x_n, y_1, \ldots, y_n)$. The canonical complex structure of $\mathbb{R}^{2n}$ is then given by

$$J_0 = \begin{pmatrix} 0 & I_n \\ -I_n & 0 \end{pmatrix},$$

where I_n denotes the identity matrix of size n.

If we have a real vector space V with a complex structure J, we can induce a complex structure on its dual V^* by setting

$$\langle Jv, w \rangle = \langle vJw \rangle, \quad v \in V, \, w \in V^*.$$

3.10.2   Kähler Manifolds

Definition 3.10.5 (Hermitian Form) A *Hermitian form* h on a complex vector space V is a sesquilinear form $h : V \times V \to \mathbb{C}$, i.e. we have $h(v, w) = \overline{h(w, v)}$ for all $v, w \in V$. Denote the set of all Hermitian forms on V by $\mathrm{Herm}(V)$.

Definition 3.10.6 (Kähler Space) A real symplectic vector space (V, Ω) is called a *Kähler space* if there is a compatible complex structure J such that $\Omega(v, Jv)$ is positive.

Remark 3.10.7 Note that for any symmetric bilinear form $g : V \times V \to \mathbb{R}$ on some real vector space V is invariant under multiplication with i, i.e. we have $g(v, w) = g(w, v) = g(iv, iw)$ for any $v, w \in V$. Denote the set of such forms by $\mathrm{Bil}(V)^{\mathrm{i}}$ The same is true for anti-symmetric bilinear forms on V. Denote the set of such forms Ω by $\mathrm{Bil}_-(V)^{\mathrm{i}}$

Remark 3.10.8 Let V be a real vector space and let W be a complex vector space. Then there is a chain of isomorphisms

$$\mathrm{Herm}(W) \xrightarrow{\sim} \mathrm{Bil}(V)^{\mathrm{i}} \xrightarrow{\sim} \mathrm{Bil}_-(V)^{\mathrm{i}}.$$

These isomorphisms are given by

$$g = \Re(h), \tag{3.27}$$

$$\Omega = -\Im(h), \tag{3.28}$$

$$h(a, b) = g(a, b) + ig(a, ib) = \Omega(ia, b) + i\Omega(a, b), \quad a, b \in W. \tag{3.29}$$

This shows that h is positive definite if and only if g is.

Let now M be a complex n-manifold. Then the tangent space at some point $q \in M$ is given by $T_q M \cong \mathbb{C}^n$ and thus isomorphic to $\mathbb{R}^{2n}$ as a real vector space. Regarded as a real vector space, there is a natural complex structure J_q by Exercise 3.10.4. This complex structure varies smoothly with respect to $q \in M$. It corresponds to the choice of local coordinates $z_j = x_j + iy_j$ for $j = 1, \ldots, n$. The basis of the tangent space $T_q M$ regarded as a complex vector space is given by

$$\left.\frac{\partial}{\partial z_j}\right|_q = \frac{1}{2}\left(\left.\frac{\partial}{\partial x_j}\right|_q - i\left.\frac{\partial}{\partial y_j}\right|_q\right), \quad j = 1, \ldots, n$$

and regarded as a real vector space it is given by

$$\left.\frac{\partial}{\partial x_j}\right|_q, \quad \left.\frac{\partial}{\partial y_j}\right|_q, \quad j = 1, \ldots, n.$$

Thus, the complex structure J_q acts as

$$J_q\left(\left.\frac{\partial}{\partial x_j}\right|_q\right) = -\left.\frac{\partial}{\partial y_j}\right|_q, \quad j = 1, \ldots, n \tag{3.30}$$

$$J_q\left(\left.\frac{\partial}{\partial y_j}\right|_q\right) = \left.\frac{\partial}{\partial x_j}\right|_q, \quad j = 1, \ldots, n. \tag{3.31}$$

Definition 3.10.9 (Kähler Manifold I) A complex n-manifold M endowed with a symplectic form ω, when regarded as a real $2n$-dimensional manifold, is called a *Kähler manifold*, if for each point $q \in M$ the real vector space $(T_q M, \omega_q, J_q)$ is a Kähler space.

The following definition if equivalent to Definition 3.10.9.

Definition 3.10.10 (Kähler Manifold II) Consider a complex n-manifold M endowed with a Hermitian metric g. Then M is a *Kähler manifold* if the skew-symmetric bilinear form

$$\omega(\ ,\ ) := g(J\ ,\ )$$

is a symplectic form.

Remark 3.10.11 In the setting of Definition 3.10.10, we mean by Hermitian metric a Riemannian metric g such that for each point $q \in M$ the bilinear form g_q is a J_q-invariant inner product on the $2n$-dimensional real vector space $T_q M$ and J_q is compatible with g_q in the sense that

$$g_q(J_q v, J_q w) = g_q(v, w), \quad \forall v, w \in T_q M.$$

If the differential form ω associated to g_q is closed, we call g a Kähler metric.

3.10.2.1   Mumford's Criterion

In order to see that a complex n-manifold M is indeed Kähler, we need to make sure that the differential form $\omega(\ ,\ ) := g(J\ ,\ )$ is closed. We want to look at a convenient method in order to figure this out. Let $\mathrm{Diff}(M)$ be the group of diffeomorphisms on M. We can observe that the action

$$\mathrm{Diff}(M) \times M \to M,$$
$$(g, q) \mapsto \phi_g(q) =: gq$$

actually leaves the complex structure and the metric h invariant. Let us define $G :=$ $\mathrm{Diff}(M)$. We can see that ϕ_q induces for each g in the isotropy group $G_q = \{g \in G \mid gq = q\}$ of q a map

$$((\phi_g)_*)_q : T_q M \to T_q M,$$

and hence a representation ρ_q of G_q in $T_q M$. Therefore, we have a homomorphism

$$\rho_q : G_q \to \mathrm{Aut}_{\mathbb{C}}(T_q M).$$

Theorem 3.10.12 (Mumford) *If $J_q \in \rho_q(G_q)$ for all $q \in M$, then $\mathrm{d}\omega = 0$.*

Proof Recall that G leaves the complex structure and the metric invariant. This means that G also leaves the forms ω and $\mathrm{d}\omega$ invariant. Thus, for all $g \in G_q$ and $u, v, w \in T_q M$, we get

$$\mathrm{d}\omega_q(\rho_q(g)u, \rho_q(g)v, \rho_q(g)w) = \mathrm{d}\omega_q(u, v, w).$$

If we set $\rho_q(g) = J_q$ and apply the above formula twice, we get

$$\mathrm{d}\omega_q(u, v, w) = \mathrm{d}\omega_q(J_q u, J_q v, J_q w)$$
$$= \mathrm{d}\omega_q(J_q^2 u, J_q^2 v, J_q^2 w)$$
$$= \mathrm{d}\omega_q(-u, -v, -w)$$

$$= -\mathrm{d}\omega_q(u, v, w)$$
$$= 0.$$

$\square$

Exercise 3.10.13 Use Theorem 3.10.12 to prove that the complex projective space $\mathbb{C}P^n$ is a Kähler manifold endowed with the *Fubini–Study Kähler form*

$$\omega_{\mathrm{FS}} = \frac{\mathrm{i}}{2}\partial\bar{\partial}\log(1 + |z|^2).$$

Moreover, by using the fact that $H^2(\mathbb{C}P^n) \cong \mathbb{R}$ and that $H_2(\mathbb{C}P^n, \mathbb{R})$ is generated by $[\mathbb{C}P^1]$, conclude that the class of ω is defined by

$$[\omega][\mathbb{C}P^1] = \int_{\mathbb{C}P^n} \omega_{\mathrm{FS}}.$$

Note that this is an instance of Example 3.9.18, which one can consider as a special case of the Marsden–Weinstein theorem (Theorem 3.9.12).

3.10.2.2 Kähler Forms

Assign to the complex coordinates $z_j = x_j + \mathrm{i}y_j$ for all $j = 1, \ldots, n$ the complex 1-forms

$$\mathrm{d}z_j = \mathrm{d}x_j + \mathrm{i}\mathrm{d}y_j, \qquad \mathrm{d}\bar{z} = \mathrm{d}x_j - \mathrm{i}\mathrm{d}y_j,$$

and the Cauchy–Riemann operators

$$\frac{\partial}{\partial z_j} = \frac{1}{2}\left(\frac{\partial}{\partial x_j} - \mathrm{i}\frac{\partial}{\partial y_j}\right), \tag{3.32}$$

$$\frac{\partial}{\partial \bar{z}_j} = \frac{1}{2}\left(\frac{\partial}{\partial x_j} + \mathrm{i}\frac{\partial}{\partial y_j}\right). \tag{3.33}$$

We then define a $(1, 1)$-form on a complex n-manifold M, where the first entry stands for the holomorphic and the second entry for the anti-holomorphic form-part, to be

$$\omega = \sum_{1 \le j < k \le n} \omega_{jk}\mathrm{d}z_j \wedge \mathrm{d}\bar{z}_k,$$

where the ω_{jk} are complex-valued functions. The space of general (p, q)-forms on M is denoted by $\Omega^{p,q}(M)$ where p denotes the homogeneous degree of the holomorphic part, i.e. of the $\mathrm{d}z$-terms, and q denotes the homogeneous degree of the anti-holomorphic part, i.e. of the $\mathrm{d}\bar{z}$-terms.

Definition 3.10.14 (Dolbeault Differential) The differential $\mathrm{d} = \partial + \bar{\partial}$ on $\Omega^{\bullet,\bullet}(M)$, where

$$\partial := \sum_{1 \le i \le n} \frac{\partial}{\partial z_i} \mathrm{d}z_i, \qquad \bar{\partial} := \sum_{1 \le i \le n} \frac{\partial}{\partial \bar{z}_i} \mathrm{d}\bar{z}_i,$$

is called the *Dolbeault differential*. Moreover, we call ∂ (holomorphic differential) and $\bar{\partial}$ (anti-holomorphic differential) the *Dolbeault operators*.

Remark 3.10.15 (Dolbeault Cohomology) Similarly as in the de Rham case for real manifolds, the Dolbeault operators have the property $\partial^2 = \bar{\partial}^2 = 0$ and thus give rise to a cohomology theory for M called the *Dolbeault cohomology*. The cohomology groups are defined as

$$H^{p,q}(M.\mathbb{C}) := \frac{\ker\left(\bar{\partial}\colon \Omega^{p,q}(M) \to \Omega^{p,q+1}(M)\right)}{\operatorname{im}\left(\bar{\partial}\colon \Omega^{p,q-1}(M) \to \Omega^{p,q}(M)\right)}.$$

Definition 3.10.16 (Hermitian Form) A differential form on a complex manifold M is called *Hermitian* if the tangent bundle TM carries a Hermitian metric h, i.e. for each point $q \in M$ the complex vector space $T_q M$ is equipped with a Hermitian inner product h_q such that the map $q \mapsto h_q$ is smooth.

We can define positive-definite Hermitian matrix $H = (H_{jk})$ in complex coordinates $z_1, \ldots, z_n$ on M as

$$H_{jk} := \left\langle \frac{\partial}{\partial z_j}, \frac{\partial}{\partial z_k} \right\rangle, \quad j, k = 1, \ldots, n.$$

Definition 3.10.17 (Kähler Form) The *Kähler form* is the $(1, 1)$-form on M defined through the Hermitian matrix H as

$$\Omega = \frac{i}{2} \sum_{1 \le j < k \le n} H_{jk} \mathrm{d}z_j \wedge \mathrm{d}\bar{z}_k.$$

Remark 3.10.18 The Kähler form Ω is a *real* form, i.e. we have

$$\bar{\Omega} = -\frac{i}{2} \sum_{1 \le j < k \le n} \bar{H}_{jk} \mathrm{d}\bar{z}_j \wedge \mathrm{d}z_k = \frac{i}{2} \sum_{1 \le j < k \le n} H_{kj} \mathrm{d}z_k \wedge \mathrm{d}\bar{z}_j = \Omega.$$

Using the Kähler form, we can give another definition of a Kähler manifold.

Definition 3.10.19 (Kähler Manifold III) A complex manifold M is a *Kähler manifold* if M is Hermitian and the corresponding Kähler form Ω is closed.

Theorem 3.10.20 *Let M be a Hermitian manifold with Kähler form* Ω. *Then M is a Kähler manifold if and only if locally there exists a real-valued smooth function f such that*

$$\Omega = i\partial\bar{\partial} f.$$

Exercise 3.10.21 Prove Theorem 3.10.20.

3.11   Hodge Theory for Complex Manifolds

3.11.1   Hodge Dual and Hodge Laplacian

Let M be a complex n-manifold and denote by $\Omega^{p,q}(M) := \bigwedge^p \Omega^{1,0}(M) \otimes \bigwedge^q \Omega^{0,1}(M)$ the space of (p,q)-forms on M. Note that we have a decomposition

$$\Omega^k(M) \otimes \mathbb{C} = \bigoplus_{p+q=k} \Omega^{p,q}(M).$$

An element of $\Omega^{p,q}(M)$ is of the form

$$\omega = \sum_{\substack{1 \le j_1 < \cdots < j_p \le n \\ 1 \le k_1 < \cdots < k_q \le n}} \omega^{j_1 \cdots j_p}_{k_1 \cdots k_q} dz_{j_1} \wedge \cdots \wedge dz_{j_p} \wedge d\bar{z}^{k_1} \wedge \cdots \wedge d\bar{z}^{k_q}.$$

Exercise 3.11.1 Show that for any closed Hermitian n-manifold M we have

$$\overline{\Omega^{p,q}(M)} \cong \Omega^{q,p}(M).$$

We can construct a complex Hodge star by taking the Hermitian metric h on M and consider its Riemannian metric $g = \Re(h)$ on M viewed as a real $2n$-manifold. The complex Hodge star operator is then defined by

$$*: \Omega^{p,q}(M) \to \Omega^{n-q,n-p}(M),$$

$$\alpha \wedge \beta \mapsto \alpha \wedge *\beta = g(\alpha, \bar{\beta})d\mathrm{Vol}.$$

Moreover, we define

$$\partial^*: \Omega^{p,q}(M) \to \Omega^{p+1,q}(M),$$

$$\omega \mapsto \partial^*\omega = (-1)^k *^{-1} \bar{\partial} * \omega,$$

and

$$\bar{\partial}^* \colon \Omega^{p,q}(M) \to \Omega^{p,q+1}(M),$$

$$\omega \mapsto \bar{\partial}^*\omega = (-1)^k *^{-1} \partial * \omega.$$

Then we can define two Hodge Laplacians by

$$\Delta_\partial := \partial\partial^* + \partial^*\partial, \tag{3.34}$$

$$\Delta_{\bar{\partial}} := \bar{\partial}\bar{\partial}^* + \bar{\partial}^*\bar{\partial}. \tag{3.35}$$

One can actually check that when defining $\Delta = \Delta_\partial + \Delta_{\bar{\partial}}$, it will not give the correct answer for complex manifolds. This is the reason why we need to work with Kähler manifolds.

Proposition 3.11.2 *If M is a Kähler manifold, then $\Delta = \Delta_\partial + \Delta_{\bar{\partial}}$.*

Proof In particular, one can consider the map $\omega \wedge (\ ) \colon \Omega^{p,q}(M) \to \Omega^{p+1,q+1}(M)$ given by $\alpha \mapsto \omega \wedge \alpha$ where ω is the symplectic form. Then it is not hard to see that $[(\omega \wedge (\ ))^*, \partial] = i\bar{\partial}^*$ and $[(\omega \wedge (\ ))^*, \bar{\partial}] = i\partial^*$. Thus we get $\bar{\partial}\partial^* + \partial^*\bar{\partial} + \partial\bar{\partial}^* + \bar{\partial}^*\partial = 0$ and thus $\Delta_\partial = \Delta_{\bar{\partial}}$. Computing everything, we get $\Delta = 2\Delta_\partial = 2\Delta_{\bar{\partial}}$ and thus $\Delta = \Delta_\partial + \Delta_{\bar{\partial}}$. $\qquad\square$

3.11.2   Hodge Decomposition and Hodge Diamond

We can now define the space of harmonic forms with respect to the Hodge Laplacian $\Delta \colon \Omega^{p,q}(M) \to \Omega^{p+q}(M)$ by

$$\mathrm{Harm}^{p,q}_\Delta(M, \mathbb{C}) := \ker(\Delta) \cap \Omega^{p,q}(M).$$

Remark 3.11.3 We denote by $h^{p,q} := \dim_{\mathbb{C}} \mathrm{Harm}^{p,q}_\Delta(M, \mathbb{C})$ the dimensions of the harmonic spaces.

Theorem 3.11.4 (Hodge) *Let M be a closed Kähler manifold. Then there is a canonical isomorphism*

$$H^k(M) \otimes \mathbb{C} \cong \bigoplus_{p+q=k} \mathrm{Harm}^{p,q}_\Delta(M, \mathbb{C}).$$

Moreover, since Δ is real and $\Delta = \Delta_\partial + \Delta_{\bar{\partial}}$, we get that

$$\mathrm{Harm}^{p,q}_\Delta(M, \mathbb{C}) \cong \overline{\mathrm{Harm}^{p,q}_\Delta(M, \mathbb{C})}, \tag{3.36}$$

$$\mathrm{Harm}^{p,q}_\Delta(M, \mathbb{C}) \cong \overline{\mathrm{Harm}^{n-q,n-p}_\Delta(M, \mathbb{C})}. \tag{3.37}$$

Remark 3.11.5 (Hodge Diamond) Note that Theorem 3.11.4 implies that

$$h^{p,q} = h^{q,p} = h^{n-p,n-q}.$$

This can be visualized in the following scheme:

$$h^{0,0}$$

$$h^{1,0} \qquad\qquad h^{0,1}$$

$$h^{2,0} \qquad\qquad h^{1,1} \qquad\qquad h^{0,2}$$

$$\cdot^{\cdot^{\cdot}} \qquad\qquad \vdots \qquad\qquad \cdot^{\cdot^{\cdot}}$$

$$h^{n,0} \qquad \cdots \qquad\qquad\qquad \cdots \qquad h^{0,n}$$

$$\cdot^{\cdot^{\cdot}} \qquad\qquad \vdots \qquad\qquad \cdot^{\cdot^{\cdot}}$$

$$h^{n,n-2} \qquad\qquad h^{n-1,n-1} \qquad\qquad h^{n-2,n}$$

$$h^{n,n-1} \qquad\qquad h^{n-1,n}$$

$$h^{n,n}$$

3.11.2.1 Application: Cohomology of Complex Torus

Define the *complex 2-dimensional torus* to be $T_{\mathbb{C}}^2 := \mathbb{C}/\mathbb{Z}^2$ with local coordinates $z = x + iy$. Define the Hermitian inner product for complex vector fields by

$$h\left(Z_1 \frac{\partial}{\partial z} + Z_2 \frac{\partial}{\partial \bar{z}}, Z_1' \frac{\partial}{\partial z} + Z_2' \frac{\partial}{\partial \bar{z}} \right) := Z_1 \overline{Z_1'} + Z_2 \overline{Z_2'}.$$

The Riemannian metric is given by the real part of this Hermitian inner product

$$g\left(Z_1 \frac{\partial}{\partial z} + Z_2 \frac{\partial}{\partial \bar{z}}, Z_1' \frac{\partial}{\partial z} + Z_2' \frac{\partial}{\partial \bar{z}} \right) = Z_1 Z_1' + Z_2 Z_2'.$$

The complex structure is defined as

$$J \begin{pmatrix} \partial/\partial x \\ \partial/\partial y \end{pmatrix} = \begin{pmatrix} 0 & -1 \\ 1 & 0 \end{pmatrix} \begin{pmatrix} \partial/\partial x \\ \partial/\partial y \end{pmatrix}.$$

The symplectic form is given by

$$\omega \left(Z_1 \frac{\partial}{\partial z} + Z_2 \frac{\partial}{\partial \bar{z}}, Z_1' \frac{\partial}{\partial z} + Z_2' \frac{\partial}{\partial \bar{z}} \right) = Z_1 Z_2' - Z_2 Z_1'.$$

Note that we have

$$*idz = id\bar{z}, \quad *d\bar{z} = -idz, \quad *(dz \wedge d\bar{z}) = -2i.$$

For $f \in \Omega^0(T_{\mathbb{C}}^2)$, we can compute Δf by

$$\Delta f = *d * f = - \left(\frac{\partial^2 f}{\partial x^2} + \frac{\partial^2 f}{\partial y^2} \right).$$

It is easy to see that the only elements in $\mathrm{Harm}_{\Delta}^{0,0}(T_{\mathbb{C}}^2)$ are constant functions since the coordinates z and $\bar{z}$ are not well-defined on the whole of $T_{\mathbb{C}}^2$. Thus, we get

$$h^{0,0}(T_{\mathbb{C}}^2) := \dim \mathrm{Harm}_{\Delta}^{0,0}(T_{\mathbb{C}}^2) = 1.$$

Exercise 3.11.6 Show that

$$\Delta (f_1 dz + f_2 d\bar{z}) = (\Delta f_1)dz + (\Delta f_2)d\bar{z},$$
$$\Delta(f dz \wedge d\bar{z}) = (\Delta f)dz \wedge d\bar{z}.$$

Similarly, as before we get

$$h^{0,0}(T_{\mathbb{C}}^2) = h^{0,1}(T_{\mathbb{C}}^2) = h^{1,1}(T_{\mathbb{C}}^2) = 1.$$

In general, when we have $T_{\mathbb{C}}^{2n} := \mathbb{C}^n / \mathbb{Z}^{2n}$, we get that $h^{p,q}(T_{\mathbb{C}}^{2n}) = \binom{n}{p}\binom{n}{q}$ and thus

$$\dim H^k(T_{\mathbb{C}}^{2n}, \mathbb{C}) = \sum_{p+q=k} \dim \mathrm{Harm}_{\Delta}^{p,q}(T_{\mathbb{C}}^{2n}) = \sum_{p+q=k} \binom{n}{p}\binom{n}{q} = \binom{2n}{k}.$$

Chapter 4
Poisson Geometry

Poisson geometry appears as a combination of methods of differential geometry, symplectic geometry and non-commutative geometry. Motivated by the dynamical structure induced from the setting of classical mechanics, it connects to the notion of symplectic geometry and provides the actual mathematical structure to talk about deformation quantization. In fact, as we will see, the phase space can be regarded as a *Poisson manifold* with *Poisson structure* coming from the canonical symplectic form since every symplectic manifold naturally induces a Poisson manifold. In this chapter we will start by introducing Poisson manifolds and discuss their local behaviour by considering a certain application of the Darboux theorem together with their induced foliation into symplectic leaves. Then we will discuss the notion of morphisms between such manifolds and discuss its relation to Hamiltonian vector fields as we have seen before. A particularly important construction is the one of *Poisson cohomology*. We will see that Poisson manifolds do naturally define a cohomology theory for which the first few cohomology group have important geometric interpretation also in prospect to deformation theory. In particular, we will see that they form obstructions to certain structure. Then we will move to the notion of *symplectic groupoids*, introduce *Lie groupoids* and *Lie algebroids*, and discuss certain integrability conditions for Poisson manifolds. We will briefly give a generalization which forms a combination of the notion of symplectic structure with the one of a Poisson structure, called a *Dirac structure*. Finally, we will cover the notion of *Morita equivalence*, which can be regarded as another construction for morphisms between Poisson manifolds, by using the concept of symplectic groupoids as discussed before. This chapter is mainly based on [Mor58, Xu91, Xu04, GL92, JSW02, Sch98, Wei83, Wei71, Wei77, Wei81, GW92, BC05, Con95, Cat04, GG01, DZ05, Cou90a, Cou90b, BW04, GRS05, Wal07, CF03, CF04, Fer00, Fer02].

N. Moshayedi, *Kontsevich's Deformation Quantization and Quantum Field Theory*, Lecture Notes in Mathematics 2311, https://doi.org/10.1007/978-3-031-05122-7_4

4.1   Poisson Manifolds

4.1.1   *Poisson Structures and the Schouten–Nijenhuis Bracket*

Let M denote a smooth manifold.

Definition 4.1.1 (Poisson Structure) A *Poisson structure* on M is an $\mathbb{R}$-bilinear Lie bracket $\{\ ,\ \}$ on $C^\infty(M)$ satisfying the Leibniz rule

$$\{f, gh\} = \{f, g\}h + g\{f, h\}, \quad \forall f, g, h \in C^\infty(M).$$

Definition 4.1.2 (Casimir Function) A function $f \in C^\infty(M)$ is called *Casimir* if the Hamiltonian vector field $X_f = \{f,\ \}$ of f vanishes.

Remark 4.1.3 By the Leibniz rule of the Poisson bracket $\{\ ,\ \}$, there exists a bivector field $\pi \in \mathfrak{X}^2(M) := \Gamma(\bigwedge^2 TM)$ such that

$$\{f, g\} = \pi(df, dg).$$

Definition 4.1.4 (Schouten–Nijenhuis Bracket) The *Schouten–Nijenhuis bracket* is a unique extension of the Lie bracket of vector fields to a graded bracket of multivector fields. It is defined by

$$[X_1 \wedge \cdots \wedge X_m, Y_1 \wedge \cdots \wedge Y_n]_{\mathrm{SN}} := \sum_{\substack{1 \leq i \leq m \\ 1 \leq j \leq m}} (-1)^{i+j} [X_i, Y_j] X_1 \wedge \cdots \wedge X_{i-1} \wedge X_{i+1} \wedge \cdots \wedge X_m \wedge$$

$$\wedge Y_1 \wedge \cdots \wedge Y_{j-1} \wedge Y_{j+1} \wedge \cdots \wedge Y_n \tag{4.1}$$

for vector fields $X_1, \ldots, X_m, Y_1, \ldots, Y_n$ and by

$$[f, X_1 \wedge \cdots \wedge X_m]_{\mathrm{SN}} := -\iota_{X_1 \wedge \cdots \wedge X_m} df$$

for a function f.

Exercise 4.1.5 (Jacobi Identity) Show that the Schouten–Nijenhuis bracket satisfies the graded Jacobi identity:

$$(-1)^{(|X|-1)(|Z|-1)} [X, [Y, Z]_{\mathrm{SN}}]_{\mathrm{SN}} + (-1)^{(|Y|-1)(|X|-1)} [Y, [Z, X]_{\mathrm{SN}}]_{\mathrm{SN}} +$$

$$+ (-1)^{(|Z|-1)(|Y|-1)} [Z, [X, Y]_{\mathrm{SN}}]_{\mathrm{SN}} = 0,$$

where $|\ |$ denotes the degree operation, i.e. $|X| = k$ if $X \in \mathfrak{X}^k(M)$.

Exercise 4.1.6 Show that the Jacobi identity for $\{\ ,\ \}$ is equivalent to the condition

$$[\pi, \pi]_{\text{SN}} = 0.$$

See Sect. 5.2.4.1 for the computation.

In local coordinates $(x_1, \ldots, x_n)$ we can determine the components of the bivector field π as

$$\pi^{ij}(x) = \{x_i, x_j\}.$$

Definition 4.1.7 (Poisson Manifold) A pair (M, π), where π is a Poisson bivector field on M, is called a *Poisson manifold*.

Definition 4.1.8 (Symplectic Poisson Structure) If the bivector field π is invertible at each point x, it is called *non-degenerate* or *symplectic*.

Remark 4.1.9 Note that, if π is symplectic, then we can define a symplectic form on M. Indeed, we can locally define the matrices

$$(\omega_{ij}) = (-\pi^{ij})^{-1},$$

which defines globally a 2-form $\omega \in \Omega^2(M)$. Moreover, the condition $[\pi, \pi] = 0$ implies that $d\omega = 0$.

Exercise 4.1.10 Show that a linear bivector field π on some vector space V satisfies the Jacobi identity if and only if the constants c_k^{ij} with $\pi^{ij} = x^k c_k^{ij}$ are the structure constants of a Lie algebra.

4.1.2 Examples of Poisson Structures

Example 4.1.11 (Constant Poisson Structure) Let $M = \mathbb{R}^n$ and suppose that $\pi^{ij}(x)$ is constant. Then we can find local coordinates on M

$$(q_1, \ldots, q_k, p_1, \ldots, p_k, e_1, \ldots, e_\ell), \quad 2k + \ell = n,$$

such that

$$\pi = \sum_{1 \le i \le k} \frac{\partial}{\partial q_i} \wedge \frac{\partial}{\partial p_i}.$$

Expressing it in terms of a bracket, we get

$$\{f, g\} = \sum_{1 \leq i \leq k} \left(\frac{\partial f}{\partial q_i} \frac{\partial g}{\partial p_i} - \frac{\partial f}{\partial p_i} \frac{\partial g}{\partial q_i} \right).$$

Note that this agrees with the usual Poisson bracket on $C^\infty(T^*\mathbb{R}^n)$ in Hamiltonian mechanics (see Sect. 3.7.3). Note that all coordinates e_j for $0 \leq j \leq \ell$ are Casimirs with respect to $\{\ ,\ \}$.

Example 4.1.12 (Poisson Structures on $\mathbb{R}^2$) Let $M = \mathbb{R}^2$ and consider a smooth function $f \in C^\infty(M)$. Such a function f induces a Poisson structure on M by

$$\{x_1, x_2\} := f(x_1, x_2).$$

Moreover, any Poisson structure on $\mathbb{R}^2$ is of this form.

Example 4.1.13 (Lie–Poisson Structure) Let M be a finite-dimensional vector space V with coordinates $(x_1, \ldots, x_n)$. We can define a *linear* Poisson structure by

$$\{x_i, x_j\} := \sum_{1 \leq k \leq n} c_{ij}^k x_k,$$

where c_{ij}^k are determined constants with $c_{ij}^k = -c_{ji}^k$. Such a Poisson structure is usually called *Lie–Poisson structure*, since the Jacobi identity of $\{\ ,\ \}$ implies that the c_{ij}^k are the structure constants of a Lie algebra $\mathfrak{g}$, which might be identified naturally with V^*. Thus, we might also identify $V \cong \mathfrak{g}^*$. Conversely, any Lie algebra $\mathfrak{g}$ with structure constants c_{ij}^k defines through $\{\ ,\ \}$ a linear Poisson structure on $\mathfrak{g}^*$. A Lie–Poisson structure is sometimes also called *Kirillov–Kostant* Poisson structure. In terms of a bivector field π on $\mathfrak{g}^*$ it is described as

$$\pi|_\xi \left(\mathrm{d}f|_\xi, \mathrm{d}g|_\xi \right) = \xi \left([\mathrm{d}f|_\xi, \mathrm{d}g|_\xi] \right),$$

where $\xi \in \mathfrak{g}^*$ and $\mathrm{d}f|_\xi \in T_\xi^*\mathfrak{g}^* \cong \mathfrak{g}^{**} \cong \mathfrak{g}$.

Remark 4.1.14 Deformation quantization of a Lie–Poisson structure on $\mathfrak{g}^*$ leads to the definition of the *universal enveloping algebra* $U(\mathfrak{g})$ (see Sect. 5.1.4). The elements in the center of $U(\mathfrak{g})$ are usually called *Casimir elements*. They exactly correspond to elements in the center of the Poisson algebra of functions on $\mathfrak{g}^*$, and hence, by extension, the Casimir functions for the center of any Poisson algebra.

Exercise 4.1.15 Let $X_1, \ldots, X_n \in \mathfrak{X}(M)$ be vector fields on some manifold M such that $[X_i, X_j] = 0$ for $1 \leq i, j \leq n$. Let (π^{ij}) be some real anti-symmetric $n \times n$ matrix. Show that

$$\pi = \frac{1}{2} \pi^{ij} X_i \wedge X_j$$

is a Poisson structure on M.

Example 4.1.16 (Poisson Quotients) Let G be a Lie group and consider its cotan-gent bundle T^*G which can be trivialized to $G \times \mathfrak{g}^*$. Note that left-multiplication L acts on T^*G by the differential $\mathrm{d}L$.

(1) Determine the group action $\mathrm{d}L$ in the trivialization $G \times \mathfrak{g}^*$ and its fundamental vector fields.
(2) Show that the *Poisson quotient* T^*G/G is canonically isomorphic to $\mathfrak{g}^*$, where we consider the induced Poisson structure on $\mathfrak{g}^*$ with the opposite sign.

4.2 Symplectic Leaves and Local Structure of Poisson Manifolds

4.2.1 Local and Regular Poisson Structures

Let π be a symplectic Poisson structure on M. Then Darboux's theorem asserts that, around each point of M, one can find coordinates $(q_1, \ldots, q_k, p_1, \ldots, p_k)$ such that

$$\pi = \sum_{1 \leq i \leq k} \frac{\partial}{\partial q_i} \wedge \frac{\partial}{\partial p_i}.$$

The symplectic form is then given by

$$\omega = \sum_{1 \leq i \leq k} \mathrm{d}q_i \wedge \mathrm{d}p_i.$$

Definition 4.2.1 (Presymplectic Form) A closed 2-form is called *presymplectic*.

Note that each 2-form ω on M corresponds to a bundle map

$$\omega^\flat : TM \to T^*M,$$
$$v \mapsto \omega^\flat(v) := \omega(v, \ ). \tag{4.2}$$

We can define a similar bundle map for a bivector field $\pi \in \mathfrak{X}^2(M)$ by

$$\pi^\sharp : T^*M \to TM,$$
$$\alpha \mapsto \pi^\sharp(\alpha) := \pi(\ , \alpha). \tag{4.3}$$

such that $\iota_{\pi^\sharp(\alpha)}\beta = \beta(\pi^\sharp(\alpha)) = \pi(\beta, \alpha)$. The matrix representing $\pi^\sharp$ in the basis $(\mathrm{d}x_i)$ and $(\frac{\partial}{\partial x_i})$ for local coordinates $(x_1, \ldots, x_n)$ of M, is (up to a sign) given by

$$\pi^{ij}(x) = \{x_1, x_j\}.$$

Thus, bivector fields (or 2-forms) are non-degenerate if and only if the associated bundle maps are invertible.

In general, the image of $\pi^\sharp \colon T^*M \to TM$ defined as in (4.3) induces an integrable singular distribution on M. In fact, M is a disjoint union of *leaves* $\mathcal{O}$ satisfying

$$T_x\mathcal{O} = \pi^\sharp(T_x^*M), \quad \forall x \in M.$$

Definition 4.2.2 (Regular Poisson Structure) A Poisson structure π is called *regular* if $\pi^\sharp$ has locally constant rank.

Remark 4.2.3 If the considered Poisson structure is regular, then it defines a *foliation* in the ordinary sense. Note that one can always find an open dense subset of M where this is in fact the case. We call this the *regular part* of M.

Locally, by using Darboux's theorem, if π has constant rank k around a given point, there exist coordinates $(q_1, \ldots, q_k, p_1, \ldots, p_k, e_1 \ldots, e_\ell)$ such that

$$\{q_i, p_j\} = \delta_{ij}, \quad \{q_i, q_j\} = \{p_i, p_j\} = \{q_i, e_j\} = \{p_i, e_j\} = 0.$$

4.2.2 Local Splitting and Symplectic Foliation

Theorem 4.2.4 (Local Splitting Theorem) *Around any point x_0 of a Poisson manifold (M, π) there exist local coordinates*

$$(q_1, \ldots, q_k, p_1, \ldots, p_k, e_1, \ldots, e_\ell), \quad (q, p, e)(x_0) = (0, 0, 0)$$

such that

$$\pi = \sum_{1 \leq i \leq k} \frac{\partial}{\partial q_i} \wedge \frac{\partial}{\partial p_i} + \frac{1}{2} \sum_{1 \leq i,j \leq \ell} \eta^{ij}(e) \frac{\partial}{\partial e_i} \wedge \frac{\partial}{\partial e_j} \tag{4.4}$$

with $\eta^{ij}(0) = 0$.

Remark 4.2.5 We have a *symplectic factor* in the splitting (4.4) associated to the coordinates (q_i, p_i) and a factor where all Poisson brackets vanish at $e = 0$ associated to the coordinates (e_j). The latter factor is often called the *totally degenerate factor*. We can identify the symplectic factor with an open subset of the leaf $\mathcal{O}_{x_0}$ through x_0. If we consider the foliation given by the collection of all leaves

$$M = \bigsqcup_{x \in M} \mathcal{O}_x$$

we see that π canonically defines a singular foliation of M by *symplectic leaves*, since π induces a symplectic structure on each leaf. One usually refers to the totally degenerate factor, which is locally well-defined up to isomorphism, as the *transverse structure* to π along a given leaf.

Exercise 4.2.6 (Coadjoint Orbits) Consider a manifold $M = \mathfrak{g}^*$ given by the dual of the Lie algebra $\mathfrak{g} = \mathrm{Lie}(G)$ of some connected n-dimensional Lie group G. Moreover, consider the coadjoint representation Ad^* of G on $\mathfrak{g}^*$. For each $\xi \in \mathfrak{g}^*$, this action defines a *coadjoint orbit* $\mathcal{O}_\xi := \{\mathrm{Ad}_g^*(\xi) \mid g \in G\} \subseteq \mathfrak{g}^*$ and the isotropy groups $G_\xi \subseteq G$. One can show that $\mathcal{O}_\xi \cong G/G_\xi$ is an immersed submanifold of $\mathfrak{g}^*$ with tangent space

$$T_{\xi'}\mathcal{O}_\xi := \{X^\#(\xi') \mid X \in \mathfrak{g}\},$$

where $X^\#$ denotes the fundamental vector field on $\mathfrak{g}^*$. If the Lie group G is compact, one can observe that the coadjoint orbits are closed submanifolds. We want to identify the coadjoint orbits as the symplectic leaves of the Poisson manifold $\mathfrak{g}^*$.

(1) Show that the coadjoint representation $\mathrm{Ad}^* \colon G \to \mathrm{GL}(\mathfrak{g}^*)$ is compatible with the Poisson structure.
(2) Show that the fundamental vector fields $X^\#_{\mathfrak{g}^*} = X_{J(X^\#)}$ are Hamiltonian for the Hamiltonians given by the functionals $J(X^\#) \colon \xi \mapsto \xi(X^\#)$.
(3) Show explicitly that the Hamiltonian vector fields $X_{J(X^\#)}$ define an involutive subset of vector fields on $\mathfrak{g}^*$.
(4) Show that the coadjoint orbits are exactly the symplectic leaves of the Poisson manifold $\mathfrak{g}^*$. Note that these orbits are maximally integral manifolds.
(5) Consider the induced symplectic form $\omega_{\mathcal{O}_\xi}$ on $\mathcal{O}_\xi$ and show explicitly that $\omega_{\mathcal{O}_\xi}$ is symplectic. Moreover, consider the relation

$$\omega_{\mathcal{O}_\xi}\Big|_{\xi'}\left(X_{J(X^\#)}\Big|_{\xi'}, X_{J(X)}\Big|_{\xi'}\right) = \xi'([X, Y]),$$

with $\xi' \in \mathcal{O}_\xi$ and $X, Y \in \mathfrak{g}$, and show explicitly that $\omega_{\mathcal{O}_\xi}$ is well-defined, non-degenerate and symplectic.
(6) Deduce that the coadjoint orbits $\mathcal{O}_\xi$ are symplectic manifolds with a transitive symplectic G-action.

Example 4.2.7 (Symplectic Leaves of Lie–Poisson Structures) Consider a Lie algebra $\mathfrak{g}$ with dual $\mathfrak{g}^*$ equipped with its Lie–Poisson structure as in Example 4.1.13. The symplectic leaves are then the coadjoint orbits for any connected Lie group with Lie algebra $\mathfrak{g}$. Since $\{0\}$ is always an orbit, a Lie–Poisson structure is not regular unless $\mathfrak{g}$ is commutative.

Exercise 4.2.8 Describe the symplectic leaves in the duals of the Lie algebras $\mathfrak{su}_2$ and $\mathfrak{sl}_2(\mathbb{R})$.

Remark 4.2.9 (Linearization Problem) Linearizing the functions η^{ij} as in Theorem 4.2.4 at x_0, we can write

$$\{e_i, e_j\} = \sum_{1 \leq k \leq \ell} c_{ij}^k e_k + O(e^2).$$

Thus, it turns out that c_{ij}^k defines a Lie–Poisson structure on the normal space to the symplectic leaf at x_0. The *linearization problem* consists of determining whether one can choose suitable *transverse* coordinates $(e_1, \ldots, e_\ell)$ with respect to which $O(e^2)$ vanishes. If the Lie algebra structure on the conormal bundle to a symplectic leaf determined by linearization of π at a point x_0 is semi-simple and of *compact type*, then π is linearizable around x_0 through a smooth change of coordinates.

4.3 Poisson Morphisms and Completeness

Definition 4.3.1 (Poisson Map I) Let (M_1, π_1) and (M_2, π_2) be Poisson manifolds. A smooth map $\psi: M_1 \to M_2$ is a *Poisson map* (or *Poisson morphism*) if $\psi^*: C^\infty(M_2) \to C^\infty(M_1)$ is a homomorphism of Poisson algebras, i.e.

$$\psi^*\{f, g\}_2 = \{\psi^*f, \psi^*g\}_1, \quad \forall f, g \in C^\infty(M_2).$$

Equivalently, we can reformulate Definition 4.3.1 in terms of Poisson bivector fields and Hamiltonian vector fields.

Definition 4.3.2 (Poisson Map II) Let (M_1, π_1) and (M_2, π_2) be Poisson manifolds. A smooth map $\psi: M_1 \to M_2$ is a *Poisson map* (or *Poisson morphism*) if and only if either of the following two equivalent conditions hold:

(1) $\psi_*\pi_1 = \pi_2$, i.e. π_1 and π_2 are ψ-related.
(2) $X_f = \psi_*(X_{\psi^*f})$ for all $f \in C^\infty(M_2)$.

Remark 4.3.3 Condition (2) of Definition 4.3.2 shows that trajectories of X_{ψ^*f} project to those of X_f if ψ is a Poisson map. However, X_f being complete does not imply that X_{ψ^*f} is complete. Hence, one defines a Poisson map $\psi: M_1 \to M_2$ to be *complete* if for all $f \in C^\infty(M_2)$ such that X_f is complete, then X_{ψ^*f} is complete.

Example 4.3.4 (Complete Functions) Consider $\mathbb{R}$ as a Poisson manifold endowed with the zero Poisson structure. Then any map $f: M \to \mathbb{R}$ is a Poisson map, which is *complete* if and only if X_f is a complete vector field.

Exercise 4.3.5 Find the Poisson manifolds for which the set of complete functions is closed under addition.

Example 4.3.6 (Open Subsets of Symplectic Manifolds) Let (M, π) be a symplectic manifold, and let $U \subseteq M$ be an open subset. Then the inclusion $U \hookrightarrow M$ is complete if and only if U is closed. More, generally, the image of a complete Poisson map is a union of symplectic leaves.

Exercise 4.3.7 Show that the inclusion of every symplectic leaf in a Poisson manifold is a complete Poisson map.

Exercise 4.3.8 Let M_1 be a Poisson manifold and let M_2 be symplectic. Show that then any Poisson map $\psi : M_1 \to M_2$ is a submersion. Furthermore, if M_2 is connected and ψ is complete, then ψ is surjective (assuming that M_1 is non-empty).

Remark 4.3.9 Exercise 4.3.8 gives a first hint to the fact that complete Poisson maps with symplectic target must be *fibrations*. In fact, if M_1 is symplectic and $\dim M_1 = \dim M_2$, then a complete Poisson map $\psi : M_1 \to M_2$ is a covering map. In general, a complete Poisson map $\psi : M_1 \to M_2$, where M_2 is symplectic, is a locally trivial symplectic fibration with a flat Ehresmann connection (see Definition 2.10.6): the horizontal lift in $T_x M_1$ of a vector $X \in T_{\psi(x)} M_2$ is defined as

$$\pi_1^\sharp((\mathrm{d}_x \psi)^*(\pi_2^\sharp)^{-1}(X)).$$

The horizontal subspaces define a foliation whose leaves are coverings of M_2, and M_1 and ψ are completely determined, up to isomorphism, by the *holonomy*

$$\pi_1(M_2, x) \to \mathrm{Aut}(\psi^{-1}(x)), \tag{4.5}$$

where $\pi_1(M_2, x)$ in (4.5) denotes the *fundamental group* of M_2 with base point x.

Exercise 4.3.10 (Poisson Trace) Let (M, π) be a Poisson manifold and denote by $C_c^\infty(M)$ the space of smooth functions on M with compact support. A compactly supported linear functional $\mu : C_c^\infty(M) \to \mathbb{R}$ is called *Poisson trace* if $\mu(\{f, g\}) = 0$ for all $f, g \in C_c^\infty(M)$. Show that if $M = \mathbb{R}^{2n}$ is symplectic with standard symplectic form ω_0, a linear functional $\mu : C_c^\infty(M) \to \mathbb{R}$ is a Poisson trace if and only if μ is a scalar multiple of an integral of the form

$$\mu(f) = c \int_{\mathbb{R}^{2n}} f(x) \mathrm{d}^{2n} x, \quad c \in \mathbb{R}.$$

In particular, μ is a distribution.

4.4   Poisson Cohomology

4.4.1   Definition and Existence

Poisson manifolds do naturally admit a special type of cohomology theory which was first considered by Lichnerowicz in [Lic77]. In order to understand its existence, let us consider the following lemma:

Lemma 4.4.1 *Let π be a Poisson structure on some manifold M. Then for any multivector field $X \in \mathfrak{X}^\bullet(M)$, we have*

$$[\pi, [\pi, X]_{SN}]_{SN} = 0.$$

Proof Using the Jacobi identity of the Schouten–Nijenhuis bracket as in Exercise 4.1.5, we get

$$(-1)^{k-1}[\pi, [\pi, X]_{SN}]_{SN} - [\pi, [X, \pi]_{SN}]_{SN} + (-1)^{k-1}[X, [\pi, \pi]_{SN}]_{SN} = 0, \ \forall X \in \mathfrak{X}^k(M).$$

Note that $[X, \pi]_{SN} = -(-1)^{k-1}[\pi, X]_{SN}$ and thus $[\pi, [\pi, X]_{SN}]_{SN} = -\frac{1}{2}[X, [\pi, \pi]_{SN}]_{SN}$. Using the fact that π is Poisson, we know by Exercise 4.1.6 that $[\pi, \pi]_{SN} = 0$. Hence, we have $[\pi, [\pi, X]_{SN}]_{SN} = 0$ which proves the claim. □

Let us define an $\mathbb{R}$-linear map

$$d_\pi : \mathfrak{X}^\bullet(M) \to \mathfrak{X}^\bullet(M),$$

$$X \mapsto d_\pi(X) := [\pi, X]_{SN}.$$

Now using Lemma 4.4.1, we can deduce that $d_\pi \circ d_\pi = 0$ and thus d_π indeed defines a differential of degree $+1$ on the space of multivector fields $\mathfrak{X}^\bullet(M)$. This defines a cochain complex $(\mathfrak{X}^\bullet(M), d_\pi)$ and we have

$$\cdots \to \mathfrak{X}^{k-1}(M) \xrightarrow{d_\pi} \mathfrak{X}^k(M) \xrightarrow{d_\pi} \mathfrak{X}^{k+1}(M) \to \cdots$$

This is often called the *Lichnerowicz complex*.

Definition 4.4.2 (Poisson Cohomology) The cohomology induced by the Lichnerowicz complex is called *Poisson cohomology*. Given a Poisson manifold (M, π), the Poisson cohomology groups are defined by

$$H^k_\pi(M) := \frac{\ker(d_\pi : \mathfrak{X}^k(M) \to \mathfrak{X}^{k+1}(M))}{\operatorname{im}(d_\pi : \mathfrak{X}^{k-1}(M) \to \mathfrak{X}^k(M))}.$$

Remark 4.4.3 It is important to mention that the Poisson cohomology groups $H^\bullet_\pi(M)$ are usually infinite-dimensional space, whereas de Rham cohomology groups $H^\bullet(M)$ are often very well finite-dimensional. If e.g. $\pi = 0$, we have $H_\pi^\bullet(M) = \bigoplus_{k \geq 0} H_\pi^k(M) = \mathfrak{X}^\bullet(M)$.

4.4.2   Interpretation

4.4.2.1   Zeroth Cohomology Group

Let us take a look at the Poisson cohomology groups in different orders. The zeroth Poisson cohomology group $H_\pi^0(M)$ is then given by $f \in C^\infty(M)$ such that $X_f := -[\pi, f] = 0$. Namely, it is given by the space of *Casimir functions* of π, i.e. the first integrals of the associated symplectic foliation (see Definition 4.1.2).

4.4.2.2   First Cohomology Group

The first Poisson cohomology group $H_\pi^1(M)$ is given by the quotient of the space of vector fields $X \in \mathfrak{X}^\bullet(M)$ such that $[\pi, X]_{SN} = 0$ by the space of Hamiltonian vector fields, i.e. vector fields of the form $-[\pi, f] = X_f$ for some function $f \in C^\infty(M)$. In particular, this means that the first Poisson cohomology group can be interpreted as the space of outer infinitesimal automorphism groups of π. This can be seen by regarding the space of vector fields $X \in \mathfrak{X}^\bullet(M)$ with the property $[\pi, X]_{SN} = 0$ as infinitesimal automorphisms of the Poisson structure and the space of Hamiltonian vector fields as above as the inner infinitesimal automorphisms of the Poisson structure.

4.4.2.3   Second Cohomology Group

The second Poisson cohomology group $H_\pi^2(M)$ is given by the quotient of the space of bivector fields $X \in \mathfrak{X}^2(M)$ such that $[\pi, X]_{SN} = 0$ by the space of bivector fields of the form $X = [\pi, Y]_{SN}$. Let ε be a formal parameter. Then, if $[\pi, X]_{SN} = 0$, we get that $\pi + \varepsilon X$ satisfies the Jacobi identity up to terms of order ε^2, i.e. we have

$$[\pi, \varepsilon X, \pi + \varepsilon X]_{SN} = \varepsilon^2 [X, X]_{SN} = 0 \quad \mod \varepsilon^2.$$

Therefore, we can think of $\pi + \varepsilon X$ as some infinitesimal deformation of the Poisson structure π. However, up to terms of order ε^2, we know that $\pi + \varepsilon [\pi, Y]_{SN} = (\Phi_\varepsilon^Y)_* \pi$ where Φ_ε^Y denotes the flow of Y at time ε. This means that $\pi + \varepsilon [\pi, Y]_{SN}$ is a trivial infinitesimal deformation of π up to some infinitesimal diffeomorphism. Hence, the second Poisson cohomology group $H_\pi^2(M)$ arises as the quotient of the space of all possible infinitesimal deformations of π by the space of trivial deformations, namely, $H_\pi^2(M)$ can be interpreted as the moduli space of formal infinitesimal deformations of π.

4.4.2.4 Third Cohomology Group

The third Poisson cohomology group $H_\pi^3(M)$ can be interpreted as the space of obstructions to formal deformations. Let us assume we have an infinitesimal deformation $\pi + \varepsilon X$, i.e. $[\pi, X]_{SN} = 0$. Then, as we have seen, $\pi + \varepsilon X$ only satisfies the Jacobi identity up to terms of order ε^2. Thus, in order that the Jacobi identity is satisfied up to terms of order ε^2, one has to add a term $\varepsilon^2 X'$ such that

$$[\pi + \varepsilon X + \varepsilon^2 X', \pi + \varepsilon X + \varepsilon^2 X']_{SN} = 0 \quad \mathrm{mod} \ \varepsilon^3. \tag{4.6}$$

Hence, we have to solve the equation $2[\pi, X']_{SN} = -[X, X]_{SN}$, which is solvable if and only if the cohomology class of $[X, X]_{SN}$ is trivial in $H_\pi^3(M)$. Moreover, assuming (4.6) is satisfied, in order to find a term $\varepsilon^3 X''$ such that

$$[\pi + \varepsilon X + \varepsilon^2 X' + \varepsilon^3 X'', \pi + \varepsilon X + \varepsilon^2 X' + \varepsilon^3 X'']_{SN} = 0 \quad \mathrm{mod} \ \varepsilon^4,$$

we need to make sure that the cohomology class of $[X, X']_{SN}$ is trivial in $H_\pi^3(M)$, and so on.

Definition 4.4.4 (Exact Poisson Structure) If the cohomology class of π in $H_\pi^2(M)$ vanishes, i.e. if there is some vector field X such that $\pi = [\pi, X]_{SN}$, we call π an *exact Poisson structure*.

Exercise 4.4.5 (Relation to Lie Algebra Cohomology) Prove that the Lichnerow-icz complex associated to a Poisson manifold (M, π) can be identified with a subcomplex of the Chevalley–Eilenberg complex (see Definition 2.7.48) of the (infinite-dimensional) Lie algebra $C^\infty(M)$ with coefficients in $C^\infty(M)$ consisting of cochains given by multi-derivations. The construction here is with respect to the adjoint action induced by the Poisson structure.

Remark 4.4.6 If U is an Ad^*-invariant open subset of $\mathfrak{g}^*$, the dual of a Lie algebra $\mathfrak{g}$ of a connected Lie group G, then

$$H_\pi^\bullet(U) \cong H^\bullet(\mathfrak{g}, C^\infty(U)),$$

where the action of $\mathfrak{g}$ on $C^\infty(U)$ is given by the coadjoint action and there is a natural isomorphism on the level of cochain complexes. In particular, if G is compact and semisimple, the above statement together with Remark 2.7.56 gives

$$H_\pi^\bullet(U) = H^\bullet(\mathfrak{g}, \mathbb{R}) \otimes (C^\infty(U))^G = \bigoplus_{\substack{k \geq 0 \\ k \neq 1,2}} H^k(\mathfrak{g}, \mathbb{R}) \otimes (C^\infty(U))^G.$$

4.5 Symplectic Groupoids and Integration of Poisson Manifolds

When we consider a Poisson manifold (M, π) we are often interested in the following question: Is there a Lie group integrating the Lie algebra $(C^\infty(M), \{\ ,\ \})$ where $\{\ ,\ \}$ is the Poisson bracket induced by π. This problem turns out to be quite hard to answer especially because of something called *flux conjecture* in symplectic geometry. One can simplify this problem by considering finite-dimensional objects, called *Lie groupoids*, instead of infinite-dimensional Lie groups.

4.5.1 Lie Algebroids and Lie Groupoids

Definition 4.5.1 (Lie Algebroid) A *Lie algebroid* is a vector bundle $E \to M$ together with a Lie bracket $[\ ,\ ]$ on the space of sections $\Gamma(E)$ and a bundle map $\rho\colon E \to TM$ giving rise to a Lie algebra morphism $\rho\colon \Gamma(E) \to \Gamma(TM)$ such that the analogue of the Leibniz rule is satisfied:

$$[\alpha, f\beta] = f[\alpha, \beta] + \rho(\beta(f)\alpha), \quad \alpha, \beta \in \Gamma(E),\ f \in C^\infty(M).$$

Remark 4.5.2 The map ρ is usually called the *anchor map*.

Example 4.5.3 If $M = \{\mathrm{pt}\}$ is a point, a Lie algebroid is the same as a Lie algebra.

Example 4.5.4 For any manifold M, the tangent bundle TM, together with the Lie bracket given as the commutator of vector fields, is a Lie algebroid where the anchor map is the identity.

Example 4.5.5 For any foliation $\mathcal{F}$ of some manifold M, we can construct a Lie algebroid $T_{\mathcal{F}}M$ which is defined as the tangent bundle of the foliation.

Example 4.5.6 (Action Lie Algebroid) Let $\mathfrak{k}$ be a Lie algebra acting on some manifold M. We can then define the *action Lie algebroid*

$$E = \mathfrak{k} \times M,$$

with anchor map given by the action and the Lie bracket on constant sections is determined through the Leibniz rule, i.e. for $X, Y \in \Gamma(E)$, we have

$$[X, Y] = [X, Y]_{\mathfrak{k}} + L_{\rho(X)}Y - L_{\rho(Y)}X.$$

Example 4.5.7 Consider a 2-form $\omega \in \Omega^2(M)$ on some manifold M. Moreover, let $E = TM \times \mathbb{R}$ together with the bracket of $\Gamma(E) = \mathfrak{X}(M) \oplus C^\infty(M)$ given by

$$[X + f, Y + g] = [X, Y] + X(g) - Y(f) + \omega(X, Y),$$

and the anchor ρ given by the projection $TM \times \mathbb{R} \to TM$. One can see that E is a Lie algebroid if and only if ω is closed. In fact, this Lie algebroid corresponds to a principal $\mathbb{R}/H$-bundle, with H being a discrete subgroup of $\mathbb{R}$, if and only if for any function $f : S^2 \to M$ the integral

$$\int_{S^2} f^* \omega$$

is in H. Thus, if the set of all such integrals is dense in $\mathbb{R}$ for all maps $f : S^2 \to \mathbb{R}$, then there is no such principal bundle for any $H \subset \mathbb{R}$.

The global counterpart to Lie algebroids are called *Lie groupoids*.

Definition 4.5.8 (Lie Groupoid) A *Lie groupoid* consists of a manifold G together with two submersions $s, t : G \to M$, called *source* and *target* maps, onto the base manifold M, an embedding $M \hookrightarrow G$, $x \mapsto e_x$, the identity section, and a smooth multiplication map $G \times_M G \to G$, $(g, h) \mapsto gh$ defined on pairs with $s(g) = t(h)$. Elements of G are called *arrows* and elements of M are called *objects*.

Remark 4.5.9 We often denote a Lie groupoid as $G \rightrightarrows M$, where the two arrows denote the source and target maps.

Example 4.5.10 A Lie group G is a Lie groupoid with unique unit to a point $G \rightrightarrows \{pt\}$. In particular, for any Lie groupoid $G \rightrightarrows M$, the inclusion of the isotropy groups will define Lie subgroupoids, i.e. for $p \in M$

$$
\begin{array}{ccc}
G_p & \rightrightarrows & \{p\} \\
\downarrow & & \downarrow \\
G & \rightrightarrows & M
\end{array}
$$

Example 4.5.11 Any manifold M can be regarded as a trivial Lie groupoid $M \rightrightarrows M$ where all elements are units. The multiplication is trivial. In particular, $p = p_1 \circ p_2$ if and only if $p = p_1 = p_2$. Any Lie groupoid $G \rightrightarrows M$, the units of the base M induce a Lie subgroupoid

$$
\begin{array}{ccc}
M & \rightrightarrows & M \\
\downarrow & & \downarrow \\
G & \rightrightarrows & M
\end{array}
$$

Example 4.5.12 (Pair Groupoid) Let M be a manifold. Then we can construct the *pair groupoid*

$$M \times M \rightrightarrows M,$$

that has a unique arrow between any points p', $p \in M$. Composition is then given by

$$(p', p) = (p'_1, p_1) \circ (p'_2, p_2) \Leftrightarrow p'_1 = p, \quad p_1 = p'_2, \quad p_2 = p.$$

If we have a Lie groupoid $G \rightrightarrows M$, we can see that the source and target maps combine to a Lie groupoid morphism

$$
\begin{array}{ccc}
G & \rightrightarrows & M \\
{\scriptstyle (t,s)}\downarrow & & \downarrow \\
M \times M & \rightrightarrows & M
\end{array}
$$

We call the groupoid morphism (t, s) sometimes the *(groupoid) anchor*. It plays a similar role as the anchor map for Lie algebroids.

Example 4.5.13 (Fundamental Groupoid) The *fundamental groupoid* associated to a manifold M, denoted by

$$\Pi(M) \rightrightarrows M,$$

consists of homotopy classes $[\gamma]$ of continuous paths $\gamma : [0, 1] \rightarrow M$ relative to fixed endpoints. Source and target maps are given by $s([\gamma]) = \gamma(0), t([\gamma]) = \gamma(1)$ and the groupoid multiplication is given by the *concatenation* of paths.

Example 4.5.14 (Action Groupoid) Let G be a Lie group and consider an action of G on a manifold M. We can then consider the *action groupoid* $G \rightrightarrows M$ which can be defined as the subgroupoid of the direct product of the groupoids $G \rightrightarrows \{pt\}$ and $M \times M \rightrightarrows M$. It consists of all $(g, p', p) \in G \times (M \times M)$ with $p' = gp$.

Remark 4.5.15 To any Lie groupoid one can associate a Lie algebroid, similarly as to any Lie group we can associate a Lie algebra. The converse, however, is not true. This is known as the *integrability problem*. In the case where the base manifold M is Poisson, one can formulate the integrability problem as: is there a Lie groupoid integrating T^*M? The integrability of the Lie algebroid T^*M is based on the integrability of $[\, , \,]$ on 1-forms $\Omega^1(M)$. The integrability problem for Poisson manifolds is more accessible since we can use the symplectic structure on the leaves of the characteristic foliation.

4.5.2 Symplectic Groupoids and Integrability Conditions

Definition 4.5.16 (Symplectic Groupoid) A *symplectic groupoid* is a Lie groupoid G together with a symplectic form ω, such that the graph of the multiplication map $\mathsf{m}\colon G \times_M G \to G$ is Lagrangian, i.e. ω vanishes on

$$\text{graph}(\mathsf{m}) := ((g, h), \mathsf{m}(g, h)) \subset G$$

and $\dim\text{graph}(\mathsf{m}) = \frac{1}{2}\dim G$.

Proposition 4.5.17 *A symplectic groupoid induces a Poisson structure on the base manifold M satisfying the following conditions:*

(1) *The source map $s\colon G \to M$ is Poisson and the target map $t\colon G \to M$ is anti-Poisson.*
(2) *Both, s and t are complete maps.*
(3) *The s-fibers and the t-fibers are symplectic orthogonal.*
(4) *The base manifold M, considered as the zero section, is a Lagrangian submanifold of G.*
(5) *The Lie algebroid of G is canonically isomorphic to T^*M.*

Remark 4.5.18 The integrability problem can then be rephrased as: is there a symplectic groupoid integrating M?

Theorem 4.5.19 (Mackenzie–Xu[MX00]) *If M is a Poisson manifold such that T^*M is an integrable Lie algebroid, then M is an integrable Poisson manifold.*

Definition 4.5.20 (Cotangent Paths) A *cotangent path* in a Poisson manifold M is a path $a\colon [0, 1] \to T^*M$, satisfying

$$\frac{\mathrm{d}}{\mathrm{d}t}\pi(a(t)) = \pi^\sharp(a(t)).$$

Definition 4.5.21 (Cotangent Homotopy) A *cotangent homotopy* is a family $a_\varepsilon = a_\varepsilon(t)$ of cotangent paths with the property that the base paths $\gamma_\varepsilon(t)$ have fixed end points and the variation $\delta a_\varepsilon = 0$. Two cotangents paths a_0 and a_1 are said to be *homotopic*, denoted by $a_0 \sim a_1$, if there is a cotangent homotopy joining them.

Definition 4.5.22 (Weinstein Groupoid for T^*M) The *Weinstein groupoid* for the cotangent bundle T^*M is defined as the topological groupoid over M whose set of arrows is given by the quotient

$$G(M) := \text{cotangent paths/cotangent homotopies.}$$

Remark 4.5.23 Two special properties of the cotangent bundle T^*M are important which makes the integrability problem accessible. First, the anchor and the bracket are both induced from the Poisson structure π on M, namely, $\rho = \pi^\sharp\colon T^*M \to TM$. Second, one has the duality between T^*M and TM. This is important to realize the Weinstein groupoid $G(M)$ as a symplectic quotient (symplectic reduction) and the obtained symplectic form on $G(M)$ makes the Weinstein groupoid to a symplectic groupoid. This actually implies Theorem 4.5.19.

Definition 4.5.24 (Monodromy Groups of Poisson Manifolds) Let M be a Poisson manifold and let L denote the symplectic leaf through x. The *monodromy group* $\mathcal{N}_x$ at some point $x \in M$ is the set of vectors in the center of the conormal space $N_x^* L$ at $x \in L$ with the property that the constant cotangent path $a(t) = v$ is cotangent homotopic to the zero cotangent path.

Theorem 4.5.25 (Crainic–Fernandes[CF04]) *For a Poisson manifold M, the following are equivalent:*

(1) M is integrable by a symplectic Lie groupoid.
(2) The algebroid T^*M is integrable.
(3) The Weinstein groupoid $G(M)$ is a smooth manifold.
(4) The monodromy groups $\mathcal{N}_x$ with $x \in M$, are locally uniformly discrete.

Remark 4.5.26 The last condition of Theorem 4.5.25 does not have any reference to algebroids or groupoids and thus describes an integrability condition which is computable in explicit examples. The monodromy groups, in particular, define invariants of Poisson manifolds. It can be actually expressed in terms of variations of symplectic areas of spheres along transverse directions to the symplectic leaves.

4.6 Dirac Manifolds

4.6.1 Courant Algebroids

Using the bundle maps $\omega^\flat$ and $\pi^\sharp$ as in (4.2) and (4.3), we can describe closed 2-forms and Poisson bivector fields as subbundles of $TM \oplus T^*M$. Indeed, we can just consider the graphs

$$L_\omega := \mathrm{graph}(\omega^\flat),$$

$$L_\pi := \mathrm{graph}(\pi^\sharp).$$

Let us introduce the following canonical structure on $TM \oplus T^*M$:

(1) The symmetric bilinear form

$$\langle \ , \ \rangle_+ : TM \oplus T^*M \times TM \oplus T^*M \to \mathbb{R},$$

$$((X, \alpha), (Y, \beta)) \mapsto \langle (X, \alpha), (Y, \beta) \rangle_+ := \alpha(Y) + \beta(X).$$

(2) The bracket

$$[\![\ , \ ]\!] : \Gamma(TM \oplus T^*M) \times \Gamma(TM \oplus T^*M) \to \Gamma(TM \oplus T^*M),$$

$$((X, \alpha), (Y, \beta)) \mapsto [\![(X, \alpha), (Y, \beta)]\!] := ([X, Y], L_X \beta - \iota_Y d\alpha).$$

Definition 4.6.1 (Courant Algebroid) A *Courant algebroid* is a vector bundle E equipped with a non-degenerate symmetric bilinear form $\langle \ , \ \rangle \colon E \times E \to \mathbb{R}$, a bilinear bracket $[\ , \ ] \colon \Gamma(E) \times \Gamma(E) \to \Gamma(E)$ and a bundle map $\rho \colon E \to TM$ satisfying the following properties:

(1) $[X, [Y, Z]] = [[X, Y], Z] + [Y, [X, Z]]$ (left Jacobi identity),
(2) $\rho([X, Y]) = [\rho(X), \rho(Y)]$,
(3) $[X, fY] = f[X, Y] + \rho(X)(f) \cdot Y$, (Leibniz rule)
(4) $[X, X] = \frac{1}{2} \mathsf{D}\langle X, X \rangle$, where

$$\mathsf{D} := \rho^* \mathrm{d} \colon C^\infty(P) \xrightarrow{\mathrm{d}} \Omega^1(P) \xrightarrow{\rho^*} E^* \cong E.$$

(5) $\rho(X)\langle Y, Z \rangle = \langle [X, Y], Z \rangle + \langle Y, [X, Z] \rangle$, (self-adjoint)

Definition 4.6.2 (Split Courant Algebroid) The bundle $E := TM \oplus T^*M$ together with the bilinear form $\langle \ , \ \rangle_+$ and the bracket $[\![\ , \ ]\!]$ is an example of a *Courant algebroid* which we call the *split Courant algebroid*.

4.6.2 Dirac Structures

Proposition 4.6.3 *A subbundle $L \subset TM \oplus T^*M$ is of the form L_π (resp. L_ω) for a bivector field π (resp. 2-form ω) if and only if*

(1) $TM \cap L = \{0\}$ (resp. $L \cap T^*M = \{0\}$) at all points of M,
(2) L is maximal isotropic with respect to $\langle \ , \ \rangle_+$,

Moreover, $[\pi, \pi] = 0$ (resp. $\mathrm{d}\omega = 0$) if and only if

(3) $\Gamma(L)$ is closed under the Courant bracket $[\![\ , \ ]\!]$, i.e.

$$[\![\Gamma(L), \Gamma(L)]\!] \subset \Gamma(L).$$

Definition 4.6.4 (Dirac Structure) A *Dirac structure* on M is a subbundle $L \subset TM \oplus T^*M$ which is maximal isotropic with respect to $\langle \ , \ \rangle_+$ and whose sections are closed under the Courant bracket $[\![\ , \ ]\!]$.

Definition 4.6.5 (Dirac Manifold) A tuple (M, L), where L is a Dirac structure on M is called a *Dirac manifold*.

Remark 4.6.6 Dirac structures are exactly those which satisfy conditions (2) and (3) of Proposition 4.6.3, but do not necessarily appear as the graph of some bivector field or 2-form.

Definition 4.6.7 (Almost Dirac Structure) If L only satisfies condition (2) of Proposition 4.6.3, it is called an *almost Dirac structure*.

Remark 4.6.8 Usually, condition (3) of Proposition 4.6.3 is referred to as the *integrability condition* of a Dirac structure.

Example 4.6.9 (Regular Foliations) Let $F \subseteq TM$ be a subbundle, and let $\mathrm{Ann}(F) \subset T^*M$ be its annihilator. Then $L = F \oplus \mathrm{Ann}(F)$ is an almost Dirac structure. It is a Dirac structure if and only if F satisfies the *Frobenius condition* (see Theorem 2.9.15)

$$[\Gamma(F), \Gamma(F)] \subset \Gamma(F).$$

Hence, *regular foliations* are examples of Dirac structures.

Example 4.6.10 (Linear Dirac Structure) If V is a finite-dimensional real vector space, then a *linear Dirac structure* on V is a subspace $L \subset V \oplus V^*$ which is maximal isotropic with respect to the symmetric pairing $\langle\ ,\ \rangle_+$.

Let L be a linear Dirac structure on V. Let $\mathrm{pr}_1 : V \oplus V^* \to V$ and $\mathrm{pr}_2 : V \oplus V^* \to V^*$ be the canonical projections, and consider the subspace

$$W := \mathrm{pr}_1(L) \subset V.$$

Then L induces a skew-symmetric bilinear form θ on W defined by

$$\theta(X, Y) := \alpha(Y), \qquad (4.7)$$

where $X, Y \in W$ and $\alpha \in V^*$ such that $(X, \alpha) \in L$.

Exercise 4.6.11 Show that θ is well-defined, i.e. (4.7) is independent of the choice of α.

Conversely, any pair (W, θ), where $W \subseteq V$ is a subspace and θ is a skew-symmetric bilinear form W, defines a linear Dirac structure by

$$L := \{(X, \alpha) \mid X \in W,\ \alpha \in V^* \text{ such that } \alpha|_W = \iota_X\theta\}.$$

Exercise 4.6.12 Check that this L is a linear Dirac structure on V with associated subspace W and bilinear form θ.

Note that Example 4.6.10 induces a simple way in which linear Dirac structures can be restricted to subspaces.

4.6.3 Dirac Structures for Constrained Manifolds

Example 4.6.13 (Restriction of Dirac Structures to Subspaces) Let L be a linear Dirac structure on V, let $U \subseteq V$ be a subspace and consider the pair (W, θ)

associated to L. Then U inherits the linear Dirac structure L_U from L defined by

$$W_U := W \cap U, \qquad \theta_U := i^*\theta,$$

where $i : U \hookrightarrow V$ denotes the inclusion.

Exercise 4.6.14 Show that there is a canonical isomorphism

$$L_U \cong \frac{L \cap (U \oplus V^*)}{L \cap \mathrm{Ann}(U)}.$$

Remark 4.6.15 Let (M, L) be a Dirac manifold, and let $i : N \hookrightarrow M$ be a submanifold. The constructions of Example 4.6.13, when applied to $T_x N \subseteq T_x M$ for all $x \in M$, define a maximal isotropic *subbundle* $L_N \subset TN \oplus T^*N$. However, L_N might not be a continuous family of subspaces. When L_N *is* a continuous family, it is a smooth bundle which then directly satisfies the integrability condition (condition (3) of Proposition 4.6.3). Hence, L_N defines a Dirac structure.

Example 4.6.16 (Moment Level Sets) Let $\mu : M \to \mathfrak{g}^*$ be the moment map for a Hamiltonian action of a Lie group G on a Poisson manifold (M, π). Let $\xi \in \mathfrak{g}^*$ be a regular value for μ and let G_ξ be the isotropy group at ξ with respect to the coadjoint action, and consider

$$Q = \mu^{-1}(\xi) \hookrightarrow M.$$

At each point $x \in Q$, we have a linear Dirac structure on $T_x Q$ given by

$$(L_Q)_x := \frac{L_x \cap (T_x Q \oplus T_x^* M)}{L_x \cap \mathrm{Ann}(T_x Q)}.$$

One can verify that L_Q defines a smooth bundle by verifying that $L_x \cap \mathrm{Ann}(T_x Q)$ has constant dimension. In fact, one can show that $L_x \cap \mathrm{Ann}(T_x Q)$ has constant dimension if and only if the stabilizer groups of the G_ξ-action on Q have constant dimension, which happens if the G_ξ-orbits on Q have constant dimension. In this case, L_Q is a Dirac structure on Q.

4.7   Morita Equivalence for Poisson Manifolds

Morita equivalence was developed by Morita in [Mor58] to prove that two rings have equivalent categories of left-modules if and only if there is an equivalence bimodule for the corresponding rings. This was generalized to the concept of C^*-algebras and referred to as *strong Morita equivalence* by Riefel in [Rie74]. We want to take a look at an equivalence relation for integrable Poisson manifolds as constructed by Xu in [Xu91].

4.7.1   Morita Equivalence of Symplectic Groupoids

Definition 4.7.1 (Symplectic Left-Module)   For a symplectic groupoid $G \rightrightarrows G_0$, we define a *symplectic left G-module* as a symplectic manifold M together with a smooth map $J : M \to G_0$ such that J has a symplectic left G-action. The map J is usually called the *momentum mapping* of the module M.

Remark 4.7.2   The symplectic left-modules of a symplectic groupoid $G \rightrightarrows G_0$ form a *category* $\zeta(G)$ in the following sense: the objects of $\zeta(G)$ are given by symplectic left G-modules and the morphisms are given by canonical relations satisfying certain compatible conditions with the groupoid actions. Composition of morphisms is defined as the set-theoretic composition of relations. However, since the composition of morphisms may not be a morphism again, this is actually not a real category.

Definition 4.7.3 (Morita Equivalence of Symplectic Groupoids)   Two symplectic groupoids $G \rightrightarrows G_0$ and $H \rightrightarrows H_0$ are called *Morita equivalent* if there is a symplectic manifold M and some surjective submersion $\rho : M \to G_0$ and $\sigma : M \to H_0$ such that

(1) G has a free and proper left-action on M,
(2) H has a free and proper right-action on M,
(3) the two actions commute with each other,
(4) ρ induces a diffeomorphism from the quotient space M/H onto G_0,
(5) σ induces a diffeomorphism from the quotient space $G \setminus M$ onto H_0,
(6) the graph of the G- and H-actions

$$\Gamma := \{g, x, h, gxh) \mid g \in G, \ h \in H, \ x \in M \text{ such that } gxh \text{ is defined}\} \subset G \times M \times H \times \overline{M},$$

is a Lagrangian submanifold, where $\overline{M}$ denotes the symplectic manifold M with opposite symplectic structure.

Remark 4.7.4   The triple (M, ρ, σ) is called an *equivalence bimodule* between the symplectic groupoids $G \rightrightarrows G_0$ and $H \rightrightarrows H_0$.

Theorem 4.7.5   *Morita equivalent symplectic groupoids have equivalent categories of symplectic left-modules.*

4.7.2   Morita Equivalence of Poisson Manifolds

Definition 4.7.6 (Morita Equivalence of Poisson Manifolds)   Two Poisson manifolds (M_1, π_1) and (M_2, π_2) are called *Morita equivalent* if there is a symplectic manifold M together with a Poisson morphism $\rho : M \to M_1$ and an anti-Poisson morphism $\sigma : M \to M_2$ such that

$$M_1 \xleftarrow{\ \rho\ } M \xrightarrow{\ \sigma\ } M_2$$

is a complete dual pair with connected and simply connected fibers. The symplectic manifold M is then called an *equivalence bimodule*.

Remark 4.7.7 A dual pair $M_1 \xleftarrow{\rho} M \xrightarrow{\sigma} M_2$ is said to be *complete* if ρ and σ are complete in the sense that the pullback of any complete Hamiltonian is still a complete Hamiltonian on M. See also [Wei83] for more about dual pairs.

Example 4.7.8 Let S be a simply connected symplectic manifold and N a connected manifold with zero Poisson structure. Then the product $S \times N$ is Morita equivalent to N. In particular, if we take $M = S \times T^*N$ together with $\rho = (\text{id}, \text{pr}) \colon M \to S \times N$ and $\sigma = \text{pr} \colon M \to N$, where pr is the natural projection from T^*N onto N, we can check that $S \times N \xleftarrow{\rho} M \xrightarrow{\sigma} N$ is a Morita equivalence.

Example 4.7.9 Let (S, ω) be a symplectic manifold. Then, the rescaling $(S, a\omega)$ is Morita equivalent to $(S, b\omega)$ whenever $a, b \in \mathbb{R} \setminus \{0\}$. This can be seen as follows: Let M be the fundamental groupoid $\Pi_1(S)$ equipped with the symplectic structure $\widetilde{\omega} = a\rho^*\omega - b\sigma^*\omega$ with ρ and σ being the source and target maps of the groupoid $\Pi_1(S)$, respectively. We can observe that $\rho \colon \Pi_1(S) \to (S, a\omega)$ is a Poisson morphism, while $\sigma \colon \Pi_1(S) \to (S, b\omega)$ is an anti-Poisson morphism. Moreover, for any tangent vectors $v_1, v_2 \in T_q M$ such that v_1 is tangent to the ρ-fiber through $q \in M$ and v_2 is tangent to the σ-fiber through $q \in M$, we have

$$\begin{aligned}
\widetilde{\omega}(v_1, v_2) &= (a\rho^*\omega)(v_1, v_2) - (b\sigma^*\omega)(v_1, v_2) \\
&= a\omega(T\rho(v_1), T\rho(v_2)) - b\omega(T\sigma(v_1), T\sigma(v_2)) \qquad (4.8) \\
&= 0.
\end{aligned}$$

Thus, ρ-fibers and σ-fibers are $\widetilde{\omega}$-orthogonal. Hence, we get immediately that $(S, a\omega)$ and $(S, b\omega)$ are Morita equivalent as Poisson manifolds.

Proposition 4.7.10 (Weinstein) *Let S_1 and S_2 be two symplectic manifolds. Then S_1 is Morita equivalent to S_2 if and only if $\pi_1(S_1) \cong \pi_1(S_2)$.*

Proof Assume that S_1 and S_2 are Morita equivalent with equivalence bimodule (M, ρ, σ). Then $\rho \colon M \to S_1$ is a fibration with simply connected fibers. Thus, from the exact sequence

$$0 = \pi_1(\rho\text{-fiber}) \to \pi_1(M) \to \pi_1(S_1) \to \pi_0(\rho\text{-fiber}) = 0,$$

that $\pi_1(M) \cong \pi_1(S_1)$ and similarly we get that $\pi_1(M) \cong \pi_1(S_2)$. On the other hand, let $\widetilde{S}_1$ and $\widetilde{S}_2$ be the universal coverings of S_1 and S_2, respectively and $G = \pi_1(S_1) \cong \pi_1(S_2)$. Let then $M := (\widetilde{S}_1 \times \widetilde{S}_2)/G$ equipped with the obvious symplectic structure defined similarly as the one in (4.8), where G acts on $\widetilde{S}_1 \times \widetilde{S}_2$ diagonally. Define $\rho: M \to S_1$ and $\sigma: M \to S_2$ to be the natural projections. Then one can check that $S_1 \xleftarrow{\rho} M \xrightarrow{\sigma} S_2$ is indeed an equivalence bimodule. $\square$

4.7.3 Symplectic Realization of Morita Equivalent Poisson Manifolds

Definition 4.7.11 (Symplectic Realization) A *symplectic realization* of a Poisson manifold N is a pair (M, ρ) where M is a symplectic manifold and $\rho: M \to N$ is a Poisson morphism.

Definition 4.7.12 (Complete Symplectic Realization) A symplectic realization $\rho: M \to N$ is called *complete* if ρ is complete as a Poisson morphism.

Definition 4.7.13 (Full Symplectic Realization) A symplectic realization $\rho: M \to N$ is called *full* if ρ is a submersion.

Theorem 4.7.14 (Xu[Xu91]) *Let* M_1 *and* M_2 *be integrable Poisson manifolds.* M_1 *and* M_2 *are Morita equivalent if and only if their source-simply connected symplectic groupoids are Morita equivalent.*

Corollary 4.7.15 *The Morita equivalence of Poisson manifolds gives rise to an equivalence relation in the category of all integrable Poisson manifolds.*

Theorem 4.7.16 (Xu[Xu91]) *Morita equivalent Poisson manifolds have equivalent* categories *of complete symplectic realizations.*

Chapter 5
Deformation Quantization

Deformation quantization was proposed at first by Dirac in [Dir30] in order to capture the algebraic non-commutativity structure of quantum observables, given by certain self-adjoint operators on some Hilbert space, in a theory which relates the classical observables, given by the commutative algebra of smooth functions on the phase space, through a quantization procedure. As opposed to other quantization methods, as already mentioned, deformation quantization lies its focus on the algebraic structure of the system. As we have seen, the algebraic structure encoding the dynamics of the classical system is described by a Poisson bracket on the algebra of smooth functions. This Poisson bracket is needed in order to *deform* the classical point-wise product on the algebra of smooth functions to some non-commutative *star product* by using a small parameter. In order to be consistent with the physical world, we need our theory to reproduce the classical setting when passing to the limit where this small parameter approaches zero. One first builds such a deformation theory upon the setting of a Poisson bracket induced by a symplectic structure, which was already done in the first half of the twentieth century by e.g. Moyal [Moy49] and Weyl [Wey35] and developed further in the second half of the twentieth century by e.g. Bayen et al. [Bay+78a, Bay+78b], de Wilde and Lecomte [DL83] and Fedosov [Fed94]. The more general Poisson case was done by Kontsevich in his famous paper [Kon03]. In this chapter we will consider the mathematical framework needed to understand Kontsevich's construction of deformation quantization and more generally his proof of the *formality theorem*. We will start by introducing the general concepts of formal power series and star products and answer the existence problem first for the local symplectic case and then for the global case by using the methods provided by Fedosov [Fed94, Fed96]. Then we will prove Kontsevich's formality theorem by using the concept of *strong homotopy Lie algebras* (or also L_∞-algebras). After proving the more general formality theorem, we can deduce the existence of a deformation quantization of general Poisson manifolds as a special case. We will then discuss the construction of the explicit star product provided by Kontsevich, using the notion of *graphs*. Finally,

N. Moshayedi, *Kontsevich's Deformation Quantization and Quantum Field Theory*, Lecture Notes in Mathematics 2311, https://doi.org/10.1007/978-3-031-05122-7_5

we will also discuss another point of view of the formality theorem by using the notion of *operads*, provided by Tamarkin [Tam03] and Kontsevich [Kon99], which leads to the idea of a higher version of deformation quantization from the point of view of Lurie's higher algebra construction [Lur17]. This chapter is mainly based on [Moy49, Fed94, Fed96, Kon99, Kon03, Tam03, CI05, GRS05, Cat+05, Wal07, Bay+78a, Bay+78b, Wey31, Wey35, CFS92, Con85, LS93, LM95, Del95, GR99, Gut05, Ale+16].

5.1 Star Products

5.1.1 Formal Power Series

In the framework of quantization, we are often interested in power series in a small parameter, for which convergence does not play any role.

Definition 5.1.1 (Ring of Formal Power Series) The *ring of formal power series* in a variable X with coefficients in a commutative ring k, denoted by $k[[X]]$, is defined as follows: Elements of $k[[X]]$ are given by infinite power series of the form

$$f(X) = a_0 + a_1 X + a_2 X^2 + \cdots + a_n X^n + \cdots, \quad a_n \in k, \ \forall n \geq 0.$$

We can define addition and multiplication similarly as for the ring of polynomials with coefficients in k, denoted by $k[X]$. Commutativity of $k[[X]]$ follows from the fact that k is commutative. Clearly, we have $k[X] \subset k[[X]]$. Moreover, $k[[X]]$ is not a field since there are elements, for example the element $X \in k[[X]]$, which are not invertible in $k[[X]]$.

Proposition 5.1.2 *Let k be a commutative ring and let $f(X) = \sum_{i \geq 0} a_i X^i \in k[[X]]$ be a formal power series. Then $f(X)$ is invertible in $k[[X]]$ if and only if a_0 is invertible in k.*

Proof We need to show that there exists a formal power series $g(X) := \sum_{j \geq 0} b_j X^j \in k[[X]]$ such that $f(X)g(X) = 1$. Note that, by the Cauchy formula, we have

$$f(X)g(X) = \left(\sum_{i \geq 0} a_i X^i \right) \left(\sum_{j \geq 0} b_j X^j \right) \tag{5.1}$$

$$= \sum_{i \geq 0} \sum_{j \geq 0} a_i b_j X^{i+j} = \sum_{\ell \geq 0} \left(\sum_{i \geq 0} a_i b_{\ell-i} \right) X^\ell. \tag{5.2}$$

Thus, if we compare the coefficients of X^k on both sides of $f(X)g(X) = 1$, we can observe that $g(X)$ satisfies the equation if and only if $a_0 b_0 = 1$ and $\sum_{0 \leq i \leq \ell} a_i b_{\ell-i} = 0$ for all $\ell \geq 1$. Hence, if a_0 is invertible in k, we get that $b_0 :=$

a_0^{-1} exists. So we can rewrite the remaining equations as $a_0 b_\ell = -\sum_{1 \leq i \leq \ell} a_i b_{\ell-i}$
or

$$b_\ell = -b_0 \sum_{1 \leq i \leq \ell} a_i b_{\ell-i}.$$

Example 5.1.3 (Binomial Series) Consider the ring $\mathbb{Q}[Y]$. There, we can define the polynomials

$$\binom{Y}{n} := \frac{Y(Y-1)\cdots(Y-n+1)}{n!}, \qquad \forall n \geq 0.$$

We define the *binomial series* as

$$(1+X)^Y := \sum_{n \geq 0} \binom{Y}{n} X^n.$$

If we consider the ring $\mathbb{Q}[Y, Z][\![X]\!]$, we get the following identity:

$$(1+X)^{Y+Z} = \sum_{n \geq 0} \binom{Y+Z}{n} X^n \tag{5.3}$$

$$= \sum_{n \geq 0} \sum_{0 \leq j \leq n} \binom{Y}{j} \binom{Z}{n-j} X^n \tag{5.4}$$

$$= \left(\sum_{j \leq 0} \binom{Y}{j} X^j \right) \left(\sum_{\ell \geq 0} \binom{Z}{\ell} X^\ell \right) \tag{5.5}$$

$$= (1+X)^Y (1+X)^Z. \tag{5.6}$$

We have used the *Vandermonde convolution* formula here. If we consider a complex number $z \in \mathbb{C}$ instead of Y in $(1+X)^Y$, we get a formal power series $(1+X)^z$ in $\mathbb{C}[\![X]\!]$.

Exercise 5.1.4 Show that in $k[Y][\![X]\!]$ we have

$$(1-X)^{-Y} = \sum_{n \geq 0} \binom{Y+n-1}{n} X^n.$$

Moreover, show that in $k[Y, Z][\![X]\!]$ we have

$$\frac{1}{(1-X)^{Y+Z}} = \frac{1}{(1-X)^Y (1-X)^Z}.$$

Definition 5.1.5 (Ring of Formal Laurent Series) The *ring of formal Laurent series* in X with coefficients in k, denoted by $k((X))$, is defined as follows: Elements of $k((X))$ are of the form

$$f(X) = a_r X^r + a_{r+1} X^{r+1} + a_{r+2} X^{r+2} + \cdots , \quad r \in \mathbb{Z}, \ a_n \in k, \ n \geq r.$$

A formal Laurent series is basically a generalization of a formal power series, where finitely many negative exponents are allowed. We define addition and multiplication similarly as for $k[[X]]$ and $k((X))$ is commutative since k is commutative. Moreover, note that $k[[X]] \subset k((X))$

Exercise 5.1.6 Check explicitly that the coefficients of the product of two formal Laurent series are polynomial functions of the coefficients of the factors. Deduce that multiplication on $k((X))$ is indeed well-defined.

5.1.1.1 Formal Derivatives and Formal Integrals

Definition 5.1.7 (Index of a Formal Laurent Series) Let $f(X) \in k((X))$ be a formal Laurent series such that $f(X) \neq 0$. Then there is an integer n such that $[X^n] f(X) \neq 0$. The smallest such integer is called the *index* of $f(X)$. We denote the index by $i(f)$. Moreover, we define the index of zero to be $+\infty$.

Let $f(X) = \sum_{n \geq 0} a_n X^n \in k((X))$. We define the *formal derivative* of $f(X)$ as

$$f'(X) := \frac{\mathrm{d}}{\mathrm{d}X} f(X) := \sum_{n \geq i(f)} n a_n X^{n-1}.$$

We can define the *formal integral* of $f(X)$ only for the case when $\mathbb{Q} \subseteq k$ and $a_{-1} = 0$. In that case, we get

$$\int f(X)\mathrm{d}X := \sum_{\substack{n \geq i(f) \\ n \neq -1}} a_n \frac{X^{n+1}}{n+1}.$$

5.1.2 Formal Deformations

Let $\mathcal{A}$ be a commutative associative algebra with unit over some commutative base ring k and let $\hbar$ denote a formal parameter.[1]

---

[1] In the introduction we have denoted the formal parameter by ε. We want to use $\hbar$ instead to emphasize the reduced Planck constant from the physics literature, which takes the role of a (small)

Definition 5.1.8 (Formal Deformation) A *formal deformation* of $\mathcal{A}$ is the algebra $\mathcal{A}[[\hbar]]$ of formal power series over the ring $k[[\hbar]]$ of formal power series.

Remark 5.1.9 Elements of the deformed algebra $\mathcal{A}[[\hbar]]$ are of the form

$$C = \sum_{i \geq 0} c_i \hbar^i, \quad c_i \in \mathcal{A}$$

and the product between such elements is given by the Cauchy formula

$$\left(\sum_{i \geq 0} a_i \hbar^i \right) \bullet_\hbar \left(\sum_{j \geq 0} b_j \hbar^j \right) = \sum_{r \geq 0} \left(\sum_{\ell \geq 0} a_{r-\ell} b_\ell \right) \hbar^r.$$

Definition 5.1.10 (Star Product (General)) A *star product* is a $k[[\hbar]]$-linear associative product $\star$ on $\mathcal{A}[[\hbar]]$ which deforms the trivial extension $\bullet_\hbar \colon \mathcal{A}[[\hbar]] \otimes_{k[[\hbar]]} \mathcal{A}[[\hbar]] \to \mathcal{A}[[\hbar]]$ in the sense that for any two $v, w \in \mathcal{A}[[\hbar]]$ we have

$$v \star w = v \bullet_\hbar w \quad \mod \hbar.$$

Remark 5.1.11 In fact, the star product is a formal non-commutative deformation of the usual point-wise product for the case where $\mathcal{A} = C^\infty(M)$ for some Poisson manifold (M, π) and $k = \mathbb{R}$.

Definition 5.1.12 (Star Product) A *star product* on a Poisson manifold (M, π) is an $\mathbb{R}[[\hbar]]$-bilinear map

$$\star \colon C^\infty(M)[[\hbar]] \times C^\infty(M)[[\hbar]] \to C^\infty(M)[[\hbar]],$$
$$(f, g) \mapsto f \star g, \tag{5.7}$$

such that

(1) $f \star g = fg + \sum_{i \geq 1} B_i(f, g)\hbar^i$,
(2) $(f \star g) \star h = f \star (g \star h), \quad \forall f, g, h \in C^\infty(M)$,
(3) $1 \star f = f \star 1 = f, \quad \forall f \in C^\infty(M)$.

Remark 5.1.13 The B_i are bidifferential operators on $C^\infty(M)$ of *globally bounded order*. We can write

$$B_i(f, g) = \sum_{K,L} \beta_i^{KL} \partial_K f \partial_L g,$$

---

deformation parameter in this theory. More precisely, the deformation parameter is usually given by $\frac{i\hbar}{2}$.

where we sum over all multi-indices $K = (k_1, \ldots, k_m)$ and $L = (\ell_1, \ldots, \ell_n)$ for some lengths $m, n \in \mathbb{N}$. The β_i^{KL} are smooth functions which are non-zero only for finitely many choices of the multi-indices K and L.

Remark 5.1.14 A star product on $\mathcal{A}[\![\hbar]\!]$ is sometimes also called a *formal deformation* of $\mathcal{A} \subset \mathcal{A}[\![\hbar]\!]$. We will sometimes also denote a star product simply as $\star = \sum_{k \geq 0} \hbar^k B_k$.

Remark 5.1.15 It can happen that the operators B_k, defining a star product $\star$, are not differential operators of order 2 (bidifferential operators) but just $k[\![\hbar]\!]$-bilinear maps when defined on a general formally deformed $k[\![\hbar]\!]$-algebra $\mathcal{A}[\![\hbar]\!]$ (see Definition 5.1.10). A star product on a Poisson manifold as in Definition 5.1.12, i.e. where the B_k are indeed bidifferential operators, is sometimes also called a *differential star product*.

Definition 5.1.16 (Natural Star Product) A star product $\star = \sum_{k \geq 0} \hbar^k B_k$ is called *natural* if each B_k is a bidifferential operator of order at most k in each argument.

Let us now look at some important examples of star products.

5.1.3 The Moyal Product

Consider the standard symplectic manifold $(\mathbb{R}^{2n}, \omega_0)$ endowed with the canonical symplectic form ω_0 regarded as a Poisson manifold with Poisson structure induced by ω_0. Moreover, choose Darboux coordinates $(q, p) = (q_1, \ldots, q_n, p_1, \ldots, p_n)$ on $\mathbb{R}^{2n}$.

Definition 5.1.17 (Moyal Product[Moy49]) The *Moyal product* is the star product on $C^\infty(\mathbb{R}^{2n})$ defined by

$$f \star g := f(q, p) \exp\left(\frac{i\hbar}{2} \left(\overleftarrow{\partial}_q \overrightarrow{\partial}_p - \overleftarrow{\partial}_p \overrightarrow{\partial}_q \right) \right) g(q, p), \qquad f, g \in C^\infty(\mathbb{R}^{2n}), \tag{5.8}$$

where $\overleftarrow{\partial}$ denotes the derivative on f and $\overrightarrow{\partial}$ the derivative on g.

Remark 5.1.18 Note that in (5.8) the formal parameter is actually replaced by $\hbar \to \frac{i\hbar}{2}$, which is typical in the physics literature, especially in quantum field theory.

If we consider $\mathbb{R}^d$ regarded as a Poisson manifold endowed with a constant Poisson structure π, we can define a star product similarly to (5.8) by

$$f \star g(x) = \exp\left(\frac{i\hbar}{2} \pi^{ij} \frac{\partial}{\partial x^i} \frac{\partial}{\partial y^j} \right) f(x) g(y) \Big|_{y=x}, \qquad f, g \in C^\infty(\mathbb{R}^d). \tag{5.9}$$

Proposition 5.1.19 *The star product defined in (5.9) is associative for any choice of* π^{ij}.

Proof

$$((f \star g) \star h)(x) = \exp\left(\frac{i\hbar}{2}\pi^{ij}\frac{\partial}{\partial x^i}\frac{\partial}{\partial z^j}\right)(f \star g)(x)h(z)\Big|_{x=z}$$

$$= \exp\left(\frac{i\hbar}{2}\pi^{ij}\left(\frac{\partial}{\partial x^i} + \frac{\partial}{\partial y^i}\right)\frac{\partial}{\partial z^j}\right)\exp\left(\frac{i\hbar}{2}\pi^{k\ell}\frac{\partial}{\partial x^k}\frac{\partial}{\partial y^\ell}\right)f(x)g(y)h(z)\Big|_{x=y=z}$$

$$= \exp\left(\frac{i\hbar}{2}\left(\pi^{ij}\frac{\partial}{\partial x^i}\frac{\partial}{\partial z^j} + \pi^{k\ell}\frac{\partial}{\partial y^k}\frac{\partial}{\partial z^\ell} + \pi^{mn}\frac{\partial}{\partial x^m}\frac{\partial}{\partial y^n}\right)\right)f(x)g(y)h(z)\Big|_{x=y=z}$$

$$= \exp\left(\frac{i\hbar}{2}\pi^{ij}\frac{\partial}{\partial x^i}\left(\frac{\partial}{\partial y^j} + \frac{\partial}{\partial z^j}\right)\right)\exp\left(\frac{i\hbar}{2}\pi^{ij}\frac{\partial}{\partial y^k}\frac{\partial}{\partial z^\ell}\right)f(x)g(y)h(z)\Big|_{x=y=z}$$

$$= (f \star (g \star h))(x).$$

5.1.4 The Canonical Star Product on $\mathfrak{g}^*$

Let $\mathfrak{g}$ be a Lie algebra and consider its dual $\mathfrak{g}^*$ together with the induced Lie–Poisson structure. Moreover, consider the algebra of polynomials on $\mathfrak{g}^*$ given by $\mathrm{Sym}(\mathfrak{g})$. Denote by $U(\mathfrak{g})$ the *universal enveloping algebra* of $\mathfrak{g}$ defined as

$$U(\mathfrak{g}) := T(\mathfrak{g})/I,$$

where $T(\mathfrak{g})$ denotes the tensor algebra of $\mathfrak{g}$ and $I \subset T(\mathfrak{g})$ denotes the ideal generated by elements of the form $X \otimes Y - Y \otimes X - [X, Y]$.

Theorem 5.1.20 (Poincaré–Birkhoff–Witt) *A Lie algebra* $\mathfrak{g}$ *embeds into its universal enveloping algebra* $U(\mathfrak{g})$ *and the associated graded algebra of* $U(\mathfrak{g})$ *is isomorphic to the symmetric algebra* $\mathrm{Sym}(\mathfrak{g})$ *through the map*

$$\sigma : \mathrm{Sym}(\mathfrak{g}) \xrightarrow{\sim} U(\mathfrak{g}),$$

$$X_1 \otimes \cdots \otimes X_k \mapsto \frac{1}{k!}\sum_{\tau \in S_k} X_{\tau(1)} \circ \cdots \circ X_{\tau(k)}. \tag{5.10}$$

Using Theorem 5.1.20, we get

$$U(\mathfrak{g}) = \bigoplus_{j \geq 0} \sigma(\mathrm{Sym}^j(\mathfrak{g})).$$

Let $P \in \mathrm{Sym}^p(\mathfrak{g})$ and $Q \in \mathrm{Sym}^q(\mathfrak{g})$. Then we can define a star product on $\mathfrak{g}^*$ by

$$P \star Q = \sum_{n \geq 0} \hbar^n \sigma^{-1}((\sigma(P) \circ \sigma(Q))_{p+q-n}).$$

This star product can be characterized by the multiplication in $U(\mathfrak{g})$ as

$$X \star X_1 \otimes \cdots \otimes X_k = X \otimes X_1 \otimes \cdots \otimes X_k$$

$$+ \sum_{j=1}^{k} \frac{(-1)^j}{j!} \hbar^j B_j [[[X, X_{r_1}], \ldots], X_{r_j}] \otimes X_1 \otimes \cdots \otimes \widehat{X}_{r_1}$$

$$\otimes \cdots \otimes \widehat{X}_{r_j} \otimes \cdots \otimes X_k, \tag{5.11}$$

where B_j are *Bernoulli numbers*.

Exercise 5.1.21 Consider a homogeneous star product $\star = \sum_{k \geq 0} \hbar^k B_k$ on the cotangent bundle $M = T^*N$ of some manifold N endowed with its standard symplectic form ω_0. Moreover, let $D \colon C^\infty(T^*N) \to C^\infty(T^*N)$ be a homogeneous differential operator such that $[L_X, D] = -kD$ for $k \geq 1$. Show that then

$$\frac{1}{n!} \int_{T^*N} D(f) \omega_0^n = 0, \quad \forall f \in C_c^\infty(T^*N).$$

Let now $f \in \mathfrak{X}^j(T^*N)$ and $g \in C^\infty(T^*N)$. Show that the map $g \mapsto B_k(f, g)$ as well as the map $g \mapsto B_k(g, f)$ induces a homogeneous differential operator with $[L_X, D] = -(j - k)D$. Show then that

$$\frac{1}{n!} \int_{T^*N} f \star g \omega_0^n = \sum_{k \geq 0} \frac{\hbar^k}{n!} \int_{T^*N} B_k(f, g) \omega_0^n,$$

and similarly for $g \star f$. Show also that for $f \in \mathfrak{X}^j(T^*N)$ with $j = 0, 1$ and $g \in C^\infty(T^*N)$ we get

$$\frac{1}{n!} \int_{T^*N} (f \star g - g \star f) \omega_0^n = 0.$$

5.1.5   Equivalent Star Products

Exercise 5.1.22 Check that the B_i of a star product $\star$ are *strict* bidifferential operators, i.e. there is no term in order zero and

$$B_i(1, f) = B_i(f, 1) = 0, \quad \forall i \in \mathbb{N}.$$

Moreover, check that the skew-symmetric part B_1^- of the first bidifferential operator, defined as

$$B_1^-(f, g) := \frac{1}{2} \left(B_1(f, g) - B_1(g, f) \right),$$

satisfies:

- $B_1^-(f, g) = -B_1^-(g, f)$,
- $B_1^-(f, gh) = g B_1^-(f, h) + B_1^-(f, g)h$,
- $B_1^-(B_1^-(f, g), h) + B_1^-(B_1^-(g, h), f) + B_1^-(B_1^-(h, f), g) = 0$.

Deduce that $\{f, g\} := B_1^-(f, g)$ is a Poisson structure.

Remark 5.1.23 Exercise 5.1.22 shows that given a star product $\star$, we can always deduce a Poisson structure by the formula

$$\{f, g\} := \lim_{\hbar \to 0} \frac{f \star g - g \star f}{\hbar}. \tag{5.12}$$

On the other hand we want to consider the following problem: Given a Poisson manifold (M, π), can we define an associative, but possibly non-commutative, product $\star$ on the algebra of smooth functions, which is the point-wise product and such that (5.12) is satisfied.

Definition 5.1.24 (Equivalent Star Products) Two star products $\star$ and $\star'$ on $C^\infty(M)$ are said to be *equivalent* if and only if there is a linear operator $D \colon C^\infty(M)[[\hbar]] \to C^\infty(M)[[\hbar]]$ of the form

$$Df := f + \sum_{i \geq 1} D_i f \hbar^i,$$

such that

$$f \star' g = D^{-1}(Df \star Dg), \tag{5.13}$$

where D^{-1} denotes the inverse in the sense of formal power series.

Remark 5.1.25 Note that by the definition of a star product, the D_i have to be differential operators vanishing on constants.

Lemma 5.1.26 *For any equivalence class of star products, there is a representative whose first term B_1 in the $\hbar$-expansion is skew-symmetric.*

Proof Given any star product

$$f \star g = fg + B_1(f, g)\hbar + B_2(f, g)\hbar^2 + \cdots$$

one can define an equivalent star product $\star'$ as in (5.13) by using a formal differential operator

$$D = \mathrm{id} + D_1\hbar + D_2\hbar^2 + \cdots$$

The skew-symmetric condition for B_1' implies that

$$D_1(fg) = D_1 fg + f D_1 g + \underbrace{\frac{1}{2}\left(B_1(f,g) + B_1(g,f)\right)}_{=:B_1^+(f,g)}, \tag{5.14}$$

which can be checked for polynomials and further be completed to hold for any smooth functions on M. If we choose D_1 to vanish on linear functions, Eq. (5.14) describes uniquely how D_1 acts on quadratic terms, given by the symmetric part B_1^+ of B_1. Namely, we get

$$D_1(x^i x^j) = B_1^+(x^i, x^j) := \frac{1}{2}\left(B_1(x^i, x^j) + B_1(x^j, x^i)\right),$$

where (x^i) are local coordinates on M. By the associativity of $\star$, we get a well-defined operator since it does not depend on the way how we order the factors.

Exercise 5.1.27 Check that $D_1((x^i x^j)x^\ell) = D_1(x^i(x^j x^\ell))$.

Remark 5.1.28 A natural problem appearing in this setting concerns *classification* of star products. In fact, it was shown that equivalence classes of star products on a symplectic manifold (M, ω) are in one-to-one correspondence with elements in $H^2(M)[\![\hbar]\!]$ [Del95, GR99]. The general construction was formulated by Kontsevich [Kon03]. He constructed an explicit formula for a star product on $\mathbb{R}^d$ endowed with any Poisson structure. However, we want to start with a much more general statement, formulated and also proved by Kontsevich [Kon03], which provides a solution to the existence of deformation quantization for a special case as we will see. This general result is called *formality*.

5.1.6   Fedosov's Globalization Approach

The Moyal product as in (5.8) is only defined locally on $\mathbb{R}^{2n}$. In [Fed94] Fedosov showed how to obtain a star product on any symplectic manifold (M, ω) by using a symplectic connection.

5.1.6.1   Weyl Bundles and the Operators δ and δ^*

Let (M, ω) be a $2n$-dimensional symplectic manifold.

Definition 5.1.29 (Weyl Algebra) To the tangent space $T_x M$ at some point $x \in M$, we can associate the formal *Weyl algebra* W_x which is the associative algebra over $\mathbb{C}$ with unit whose elements are of the form

$$a(y, \hbar) = \sum_{2k+\ell \geq 0} \hbar^k a_{k,i_1,\ldots,i_\ell} \, y^{i_1} \cdots y^{i_\ell},$$

where $y = (y^1, \ldots, y^{2n}) \in T_x M$ is some tangent vector and $a_{k,i_1,\ldots,i_\ell}$ are covariant tensors.

The product of two elements $a, b \in W_x$ is given by

$$a \circ b = \exp\left(-\frac{i\hbar}{2}\omega^{ij}\frac{\partial}{\partial y^i}\frac{\partial}{\partial z^j}\right) a(y, \hbar) b(z, \hbar)|_{z=y}$$

$$= \sum_{k \geq 0}\left(-\frac{i\hbar}{2}\right)^k \frac{1}{k!}\omega^{i_1 j_1}\cdots\omega^{i_k j_k}\frac{\partial^k a}{\partial y^{i_1}\cdots\partial y^{i_k}}\frac{\partial^k b}{\partial y^{j_1}\cdots\partial y^{j_k}}.$$

Exercise 5.1.30 Check that $\circ$ does not depend on any choice of basis of $T_x M$ and is associative.

Definition 5.1.31 (Weyl Bundle) The *Weyl bundle* is defined as the union of all Weyl algebras

$$W := \bigsqcup_{x \in M} W_x.$$

Remark 5.1.32 Sections of W are of the form

$$a(x, y, \hbar) = \sum_{2k+\ell \geq 0} \hbar^k a_{k,i_1,\ldots,i_\ell}(x) y^{i_1} \cdots y^{i_\ell}, \tag{5.15}$$

where $a_{k,i_1,\ldots,i_\ell}$ are symmetric covariant tensor fields on M. Moreover, $(\Gamma(W), \circ)$ forms an associative algebra. Note that the *center* of $\Gamma(W)$ consists of sections which do not contain any y-coordinates. Hence, the center of $\Gamma(W)$ can be identified as the linear space Z defined as follows: elements of Z are of the form

$$a = a(x, \hbar) = \sum_{k \geq 0} \hbar^k a_k(x),$$

where $a_k(x) \in C^\infty(M)$. Moreover, for $a, b \in Z$, we have an associative product

$$a \star b = \sum_{k \geq 0} \hbar^k c_k,$$

with

- c_k being polynomials in a_k and b_k and their derivatives,
- $c_0(x) = a_0(x)b_0(x)$,
- $[a, b] = a \star b - b \star a = -i\hbar\{a_0, b_0\} + O(\hbar^2)$.

Thus, $(Z, \star)$ defines an associative algebra which is usually called the algebra of *quantum observables*.

Definition 5.1.33 (Differential Form with Values in W) A *differential q-form on M with values in W* is given by

$$a = \sum \hbar^k a_{k, i_1, \ldots, i_p, j_1, \ldots, j_q}(x) y^{i_1} \cdots y^{i_p} \mathrm{d}x^{j_1} \wedge \cdots \wedge \mathrm{d}x^{j_q}, \tag{5.16}$$

where the coefficients are covariant tensor fields which are symmetric with respect to the $i_1, \ldots, i_p$ indices and anti-symmetric with respect to the $j_1, \ldots, j_q$ indices.

Remark 5.1.34 These differential forms form an algebra

$$\Omega^\bullet(M, W) := \Omega^\bullet(M) \otimes \Gamma(W) = \bigoplus_{q=0}^{2n} \Omega^q(M) \otimes \Gamma(W),$$

where the multiplication is given by means of the exterior product of 1-forms $\mathrm{d}x^i$ and the Weyl product $\circ$ of polynomials in y^i. Note that $\mathrm{d}x^i$ commutes with y^i. We will also denote the product of two forms a and b by $a \circ b$.

Let $a \in \Omega^{q_1}(M, W)$ and $b \in \Omega^{q_2}(M, W)$. Then we can define a graded commutator as

$$[a, b] = a \circ b - (-1)^{q_1} b \circ a.$$

By definition, a form $a \in \Omega^\bullet(M, W)$ is said to be *central*, if the commutator $[a, b]$ vanishes for any $b \in \Omega^\bullet(M, W)$. The central forms are given by $\Omega^\bullet(M, Z)$. We define two projections of $a = a(x, y, \mathrm{d}x, \hbar)$ onto the center: $a_0 := a(x, 0, \mathrm{d}x, \hbar)$ and $a_{00} := a(x, 0, 0, \hbar)$.

Definition 5.1.35 (Symbol) For $a = a(x, y, \mathrm{d}x, \hbar) \in \Gamma(W)$, we write $\sigma(a)$ instead of a_0 and call $\sigma(a)$ the *symbol* of the section a.

Define the following two operators

$$\delta a = dx^k \wedge \frac{\partial a}{\partial y^k}, \qquad \delta^* a = y^k \iota_{\frac{\partial}{\partial x^k}} a. \tag{5.17}$$

Note hat there is a filtration in the algebra W: $W \supset W_1 \supset W_2 \supset \cdots$ with respect to the total degree $2k + \ell$ of terms as in (5.15). Hence, there is also a filtration $\Omega^\bullet(M, W) \supset \Omega^\bullet(M, W_1) \supset \Omega^\bullet(M, W_2) \supset \cdots$ with respect to the total degree $2k + p$ corresponding to the variables $\hbar$ and y^i of terms as in (5.16). The operator $\delta : \Omega^q(M, W_p) \to \Omega^{q+1}(M, W_{p-1})$ reduces the filtration by -1 and is similar to the exterior derivative. The operator $\delta^* : \Omega^q(M, W_p) \to \Omega^{q-1}(M, W_{p+1})$ will raise the filtration by $+1$. In particular, we have

$$\delta\left(y^{i_1} \cdots y^{i_p} dx^{j_1} \wedge \cdots \wedge dx^{j_q}\right) = dx^{i_1} \wedge \cdots \wedge dx^{i_p} \wedge dx^{j_1} \wedge \cdots \wedge dx^{j_q} \tag{5.18}$$

$$\delta^*\left(y^{i_1} \cdots y^{i_p} dx^{j_1} \wedge \cdots \wedge dx^{j_q}\right) = -(-1)^{3+\cdots+q} y^{i_1} \cdots y^{i_p} y^{j_1} \cdots y^{j_q}. \tag{5.19}$$

Lemma 5.1.36 (Fedosov[Fed94]) *The operators δ and δ^* do not depend on the choice of local coordinates and have the following properties:*

- $\delta^2 = (\delta^*)^2 = 0,$
- *for monomials of the form as in (5.18) and (5.19), we have*

$$\delta\delta^* + \delta^*\delta = (p+q)\mathrm{id}.$$

Exercise 5.1.37 Check Lemma 5.1.36 by a direct computation.

Remark 5.1.38 We can check that δ is in fact an anti-derivation, i.e., for $a \in \Omega^{q_1}(M, W)$ and $b \in \Omega^{q_2}(M, W)$, we get

$$\delta(a \circ b) = (\delta a) \circ b + (-1)^{q_1} a \circ (\delta b).$$

Note also that this is not true for δ^*.

Define now the operator δ^{-1} acting on monomials by $\delta^{-1} = \frac{1}{p+q}\delta^*$ for $p+q > 0$ and $\delta^{-1} = 0$ for $p+q = 0$. By Lemma 5.1.36 we get that any form $a \in \Omega^\bullet(M, W)$ can be written as

$$a = \delta\delta^{-1}a + \delta^{-1}\delta a + a_{00},$$

which is similar to the Hodge decomposition as in Sect. 2.6.1.

5.1.6.2 Fedosov's Connection

Definition 5.1.39 (Symplectic Connection) A *symplectic connection* is a torsion-free connection preserving the tensor ω_{ij}, i.e. $\delta_i \omega_{jk} = 0$, where δ_i is a covariant derivative with respect to $\frac{\partial}{\partial x^i}$.

Remark 5.1.40 In local Darboux coordinates, the coefficients $\Gamma_{ijk} = \omega_{im} \Gamma^m_{jk}$ of the symplectic connection are completely symmetric with respect to indices ijk. It can be shown that a symplectic connection does always exist, but is not unique.

Let ∂ be a symplectic connection on M. The corresponding connection in W can be defined as an operator

$$\partial : \Omega^q(M, W) \to \Omega^q(M, W),$$
$$a \mapsto \partial a := dx^i \wedge \partial_i a. \tag{5.20}$$

Remark 5.1.41 By Definition 5.1.39, we get

(1) $\partial(a \circ b) = (\partial a) \circ b + (-1)^{q_1} a \circ (\partial b)$ for all $a \in \Omega^{q_1}(M, W)$,
(2) For any q-form $\varphi \in \Omega^q(M)$ we have

$$\partial(\varphi \wedge a) = d\varphi \wedge a + (-1)^q \varphi \wedge \partial a.$$

If we choose Darboux coordinates, we can write the connection ∂ in the form

$$\partial a = da + \left[\frac{i}{\hbar} \Gamma, a \right],$$

where $\Gamma = \frac{1}{2} \Gamma_{ijk} y^i y^j dx^k$ is some local 1-form with values in W and $d := dx^i \wedge \frac{\partial}{\partial x^i}$ is the de Rham differential with respect to x. Using this, we can rewrite the operator δ defined in 5.17 as

$$\delta a = -\left[\frac{i}{\hbar} \omega_{ij} y^i dx^j, a \right].$$

Lemma 5.1.42 (Fedosov[Fed94]) *Le ∂ be a symplectic connection. Then*

$$\partial \delta a + \delta \partial a = 0, \tag{5.21}$$

$$\partial^2 a = \left[\frac{i}{\hbar} R, a \right], \tag{5.22}$$

where $R = \frac{1}{4} R_{ijk\ell} y^i y^j dx^k \wedge dx^\ell$, with $R_{ijk\ell} = \omega_{im} R^m_{jk\ell}$ being the curvature tensor of the symplectic connection.

Remark 5.1.43 We can also consider more general connections D in W, namely, connections of the form

$$Da = \partial a + \left[\frac{i}{\hbar}\gamma, a\right] = da + \left[\frac{i}{\hbar}(\Gamma + \gamma), a\right], \tag{5.23}$$

where $\gamma \in \Omega^1(M, W)$ is a globally defined 1-form on M with values in W.

Definition 5.1.44 (Curvature) Let D be a connection on W of the form (5.23) such that $\gamma_0 = 0$. We define the *curvature* of D to be the 2-form

$$\frac{i}{\hbar}\Omega = \frac{i}{\hbar}\left(R + \partial\gamma + \frac{i}{\hbar}\gamma^2\right). \tag{5.24}$$

Lemma 5.1.45 (Fedosov[Fed94]) *For any $a \in \Omega^\bullet(M, W)$, we have*

$$D^2 a = \left[\frac{i}{\hbar}\Omega, a\right].$$

Exercise 5.1.46 Prove Lemma 5.1.45.

Definition 5.1.47 (Abelian Connection) We say that a connection D on W is *abelian*, if for any $a \in \Omega^\bullet(M, W)$, we have

$$D^2 a = \left[\frac{i}{\hbar}\Omega, a\right].$$

5.1.6.3   Fedosov's Main Theorems and the Global Star Product

Theorem 5.1.48 (Fedosov[Fed94]) *There exists an abelian connection*

$$D = -\delta + \partial + \left[\frac{i}{\hbar}r, \ \right] = \partial + \left[\frac{i}{\hbar}\left(\omega_{ij}\,y^i\mathrm{d}x^j + r\right), \ \right],$$

where ∂ is a fixed symplectic connection and $r \in \Omega^1(M, W_3)$ is a globally defined 1-form satisfying the Weyl normalization condition $r_0 = 0$. *In particular, the abelian property is satisfied whenever*

$$\delta r = R + \partial r + \frac{i}{\hbar}r^2,$$

and then the curvature is given by $\Omega = -\omega \in \Omega^2(M)$.

Remark 5.1.49 In fact, the form r can be constructed explicitly as

$$r = \frac{1}{8} R_{ijk\ell} y^i y^j y^k dx^\ell + \frac{1}{20} \partial_m R_{ijk\ell} y^i y^j y^k y^m dx^\ell + \cdots$$

where ∂_m is the covariant derivative with respect to the vector field $\frac{\partial}{\partial x^m}$.

Let $H_D^0(\Gamma(W)) \subset \Gamma(W)$ be the subalgebra consisting of *flat sections*, i.e.

$$H_D^0(\Gamma(W)) := \{a \in \Gamma(W) \mid Da = 0\}.$$

Theorem 5.1.50 (Fedosov[Fed94]) *For any $a_0 \in Z$ there is a unique section $a \in H_D^0(\Gamma(W))$ such that $\sigma(a) = a_0$.*

Using Theorem 5.1.50, we can easily see that for each $a(y, \hbar) \in W_{x_0}$ with fixed $x_0 \in M$, there is a flat section $a(x, y, \hbar) \in H_D^0(\Gamma(W))$ such that

$$a(x_0, y, \hbar) = a(y, \hbar).$$

Hence, the centralizer of $H_D^0(\Gamma(W)) \subset \Gamma(W)$ coincides with the center Z of W. In other words, if a section $b \in \Gamma(W)$ commutes with any flat section $a \in H_D^0(\Gamma(W))$, then $b \in Z$. Similarly, the centralizer of $H_D^0(\Gamma(W)) \subset \Omega^\bullet(M, W)$ is $\Omega^\bullet(M, Z)$. The section $a \in H_D^0(\Gamma(W))$ can be explicitly constructed by its symbol $\sigma(a) = a_0$:

$$a = a_0 + \partial_i a_0 y^i + \frac{1}{2} \partial_i \partial_j a_0 y^i y^j + \frac{1}{6} \partial_i \partial_j \partial_k a_0 y^i y^j y^k - \frac{1}{24} R_{ijk\ell} \omega^{\ell m} a_0 y^i y^j y^k + \cdots$$

When the curvature tensor is zero, we would get the explicit expression

$$a = \sum_{k \geq 0} \frac{1}{k!} (\partial_{i_1} \cdots \partial_{i_k} a_0) y^{i_1} \cdots y^{i_k}.$$

One can see that, having a fixed abelian connection D, flat sections form a subalgebra $H_D^0(\Gamma(W))$ with respect to the fiberwise Weyl multiplication $\circ$ in the algebra $\Gamma(W)$. Theorem 5.1.50 states that the map $\sigma \colon H_D^0(\Gamma(W)) \to Z$ is a bijection.

The star product in Z is then constructed as follows: Using the bijections σ and σ^{-1}, the associative product $\circ$ in $\Gamma(W)$ can be transferred to the set Z, i.e., for $a, b \in Z$ we have

$$a \star b = \sigma \left(\sigma^{-1}(a) \circ \sigma^{-1}(b) \right).$$

It is easy to check that this indeed defines a star product on M, or more precisely on $C^\infty(M)$.

Remark 5.1.51 Another approach to globalize the Moyal product was given by de Wilde and Lecomte [DL83] using Darboux's theorem. However, it turns out that Fedosov's approach is more important in the sense that it provides a natural extension to the general Poisson case (see Sect. 5.4).

5.1.7   Symmetries of Star Products

In the physics literature, when having a quantum theory, a *symmetry* is an automorphism of some algebra of observables. Let us define what a symmetry is in the context of deformation quantization.

Definition 5.1.52 (Symmetry)   A *symmetry of a star product* $\star = \sum_{k \geq 0} \hbar^k B_k$ is an automorphism σ of $C^\infty(M)[[\hbar]]$ with product given by $\star$, i.e. $\sigma(u \star v) = \sigma(u) \star \sigma(v)$ and $\sigma(1) = 1$, where σ is defined as a formal power series

$$\sigma(u) = \sum_{r \geq 0} \hbar^k \sigma_k(u)$$

on linear maps $\sigma_k \colon C^\infty(M) \to C^\infty(M)$. The group of such automorphisms is denoted by $\mathrm{Aut}_{\mathbb{R}[[\hbar]]}(\star)$.

Lemma 5.1.53   *Let (M, π) be a Poisson manifold with star product $\star$ and let $\sigma \in \mathrm{Aut}_{\mathbb{R}[[\hbar]]}(\star)$ be an automorphism of $\star$. Then we can write*

$$\sigma(u) = L(u \circ \varphi^{-1}),$$

where φ is a Poisson diffeomorphism of (M, π) and $L = \mathrm{id} + \sum_{k \geq 1} \hbar^k L_k$ is a formal power series of linear maps. In particular, all the L_k are differential operators.

Let σ_t be a 1-parameter group of symmetries of a star product $\star$. The generator for this star product is given by a derivation D of $\star$, i.e. $D = \sum_{k \geq 0} \hbar^k D_k$ where D_0 is a vector field with the property $L_{D_0} \pi = 0$. Moreover, if $\star$ is *natural*, then each D_k for $k \geq 0$ is a differential operator of order at most $k + 1$. We will denote the algebra of $\hbar$-linear derivations of the star product $\star$ by $\mathrm{Der}(M, \star)$.

Definition 5.1.54 (Action of Lie Group on Star Product)   Let $\star$ be a star product on some Poisson manifold (M, π) and let G be a Lie group. The *action of G on $\star$* is defined to be a homomorphism

$$\sigma \colon G \to \mathrm{Aut}_{\mathbb{R}[[\hbar]]}(\star).$$

In particular, we get that $\sigma_g = (\varphi_g)^{-1} + O(\hbar)$, for any $g \in G$, and there is an induced Poisson action φ of G on (M, π).

Note that when passing to the infinitesimal automorphisms, they will provide homomorphisms of Lie algebras $D: \mathfrak{g} \to \mathrm{Der}(M, \star)$. For $X \in \mathfrak{g}$, we can construct $D_X := X^\sharp + \sum_{k \geq 1} \hbar^k D_X^k$, where $X^\sharp$ denotes the fundamental vector field on M defined by φ, i.e. we have

$$X^\sharp(x) := \frac{\mathrm{d}}{\mathrm{d}t}\Big|_{t=0} \varphi\left(\exp(-tX)x\right).$$

Definition 5.1.55 (Action of Lie Algebra on Star Product) A homomorphism $D: \mathfrak{g} \to \mathrm{Der}(M, \star)$ is called an *action of the Lie algebra $\mathfrak{g}$ on $\star$*.

Definition 5.1.56 (Invariant Star Product) A star product $\star = \sum_{k \geq 0} \hbar^k B_k$ on some Poisson manifold (M, π) is called *invariant* under a diffeomorphism φ of M, if $u \mapsto u \circ \varphi$ defines a symmetry of $\star$.

5.1.8 ⋆-Hamiltonians and Quantum Moment Maps

Definition 5.1.57 (Hamiltonian Derivation) A derivation $D \in \mathrm{Der}(M, \star)$ for some Poisson manifold (M, π) is called *Hamiltonian* (or *essentially inner*) if $D = \frac{1}{\hbar} \mathrm{ad}_\star f$ for some $f \in C^\infty(M)[[\hbar]]$. The space of Hamiltonian derivations of a star product $\star$ will be denoted by $\mathrm{Ham}(M, \star)$.

Remark 5.1.58 Note that $\mathrm{Ham}(M, \star) \subset \mathrm{Der}(M, \star)$ is a linear subspace. In particular, we should think about it as the *quantum analogue* for Hamiltonian vector fields.

Definition 5.1.59 (Almost ⋆-Hamiltonian) An action of a Lie algebra $\mathfrak{g}$ on some star product $\star$ is called *almost ⋆-Hamiltonian* if each D_X is a Hamiltonian derivation and the choice of a function λ_X satisfying

$$D_X = \frac{1}{\hbar} \mathrm{ad}_\star \lambda_X, \quad X \in \mathfrak{g}$$

is called a *(quantum) Hamiltonian*.

Remark 5.1.60 Note that an almost ⋆-Hamiltonian action is equivalent to a linear map

$$\begin{aligned}
\lambda: \mathfrak{g} &\to C^\infty(M)[[\hbar]], \\
X &\mapsto \lambda_X := \lambda(X),
\end{aligned} \tag{5.25}$$

such that

$$\mathrm{ad}_\star \frac{1}{\hbar}[\lambda_X, \lambda_Y]_\star = \mathrm{ad}_\star \lambda_{[X,Y]}.$$

Definition 5.1.61 ($\star$-Hamiltonian) An action of a Lie algebra $\mathfrak{g}$ on some star product $\star$ is called $\star$-*Hamiltonian* if λ_X can be chosen such that the map (5.25) is a homomorphism of Lie algebras where we consider the Lie bracket $\frac{1}{\hbar}[\ ,\ ]_\star$ on $C^\infty(M)[[\hbar]]$ defined as $\frac{1}{\hbar}[f, g]_\star := \frac{1}{\hbar}(f \star g - g \star f)$.

Remark 5.1.62 A homomorphism as in (5.25) in the setting of Definition 5.1.61 is also called a *generalized moment map*.

Remark 5.1.63 Note that the homomorphism condition for any linear map

$$\mu: \mathfrak{g} \to C^\infty(M)[[\hbar]],$$

$$X \mapsto \mu(X) = \mu^0(X) + O(\hbar),$$

can be translated into the condition

$$\frac{1}{\hbar}\Big(\mu(X) \star \mu(Y) - \mu(Y) \star \mu(X)\Big) = \mu([X, Y])$$

and thus

$$\{\mu^0(X), \mu^0(Y)\} = \mu^0([X, Y]).$$

One can check that such a μ defines an action of the Lie algebra $\mathfrak{g}$ on the star product $\star$ by

$$D_X = \frac{1}{\hbar} \operatorname{ad}_\star \mu(X) = \{\mu^0(X), \ \} + O(\hbar),$$

such that the inifnitesimal action of $\mathfrak{g}$ on M is actually Hamiltonian.

Definition 5.1.64 (Covariant Star Product) The star product $\star$ associated to a generalized moment map $\mu^0: \mathfrak{g} \to C^\infty(M)$, i.e. when we have

$$\frac{1}{\hbar}\Big(\mu^0(X) \star \mu^0(Y) - \mu^0(Y) \star \mu^0(X)\Big) = \mu^0([X, Y]),$$

is called *covariant* under $\mathfrak{g}$.

Definition 5.1.65 (Quantum Moment Map) A generalized moment map $\mu: \mathfrak{g} \to C^\infty(M)[[\hbar]]$ such that D_X has no terms in $\hbar$ of degree > 0, hence $D_X = X^\sharp$, is called a *quantum moment map*.

Proposition 5.1.66 *Let G be a Lie group of symmetries of a star product $\star$ on some symplectic manifold (M, ω) and let $d\sigma: \mathfrak{g} \to \operatorname{Der}(M, \star)$ be the induced infinitesimal action. Then, if $H^1(M) = 0$ or $\mathfrak{g}$ is involutive, i.e. $[\mathfrak{g}, \mathfrak{g}] \subset \mathfrak{g}$, the action is almost $\star$-Hamiltonian.*

Remark 5.1.67 Note that the action is indeed almost $\star$-Hamiltonian if $d\sigma(\mathfrak{g}) \subset \operatorname{Ham}(M, \star)$, which happens exactly when either $H^1(M) = 0$ or $[\mathfrak{g}, \mathfrak{g}] \subset \mathfrak{g}$.

5.2 Formality

5.2.1 Some Formal Setup

For a Poisson manifold (M, π), one can define a bracket

$$\{f, g\}_\hbar := \sum_{m \geq 0} \hbar^m \sum_{\substack{0 \leq i, j, \ell \leq m \\ i+j+\ell=m}} \pi_i(\mathrm{d}f_j, \mathrm{d}g_\ell), \tag{5.26}$$

where

$$f = \sum_{j \geq 0} f_j \hbar^j, \qquad g = \sum_{\ell \geq 0} g_\ell \hbar^\ell.$$

Definition 5.2.1 (Formal Poisson Structure) We call

$$\pi_\hbar := \pi_0 + \pi_1 \hbar + \pi_2 \hbar^2 + \cdots$$

a formal *Poisson structure* if $\{\ ,\ \}_\hbar$, defined as in (5.26), is a Lie bracket on $C^\infty(M)[[\hbar]]$.

The gauge group in this case is given by *formal diffeomorphisms*, i.e. formal power series of the form

$$\Psi_\hbar := \exp(\hbar X), \tag{5.27}$$

where $X := \sum_{i \geq 0} X_i \hbar^i \in \mathfrak{X}(M)[[\hbar]]$ is a *formal vector field*, i.e. a formal power series where each coefficients $X_i \in \mathfrak{X}(M)$ are vector fields on M.

Definition 5.2.2 (Baker–Campbell–Hausdorff Formula) For two vector fields X and Y on a manifold we have the *Baker–Campbell–Hausdorff* (BCH) formula for solving the equation $\exp(X)\exp(Y) = \exp(Z)$ for a vector field Z:

$$\exp(X) \cdot \exp(Y) = \exp\left(X + \left(\int_0^1 \psi\left(\exp(\mathrm{ad}_X) \exp(t\,\mathrm{ad}_Y) \right) \mathrm{d}t \right) Y \right)$$

$$= \exp\left(X + Y + \frac{1}{2}[X, Y] + \cdots \right) \tag{5.28}$$

$$= \exp\left(\mathrm{BCH}(X, Y) \right),$$

where $\psi(x) := \frac{x \log x}{x-1} = 1 - \sum_{n \geq 0} \frac{(1-x)^n}{n(n+1)}$ and $\mathrm{ad}_X = [X,\ ]$.

Exercise 5.2.3 Check that on the set of formal vector fields $\mathfrak{X}(M)[[\hbar]]$ on M there is a group structure regarding the exponential induced by the BCH formula, i.e. we have a group structure on $\{\exp(\hbar X) \mid X \in \mathfrak{X}(M)[[\hbar]]\}$ by

$$\exp(\hbar X) \cdot \exp(\hbar Y) := \exp\left(\hbar X + \hbar Y + \frac{1}{2}\hbar^2[X, Y] + \cdots\right).$$

The group action is given by

$$\exp(\hbar X)_* \pi_\hbar := \sum_{m \geq 0} \hbar^m \sum_{\substack{0 \leq i,j,\ell \leq m \\ i+j+\ell=m}} (L_{X_i})^j \pi_\ell \qquad (5.29)$$

Remark 5.2.4 The identification between the set of all star products modulo the action given by differential operators D, as in Definition 5.1.24, and the set of formal Poisson structures modulo the (gauge) group action induced by the formal diffeomorphisms as in (5.27) was explained by Kontsevich. This was one of his famous results on deformation quantization.

Remark 5.2.5 (BCH Star Product) Using the BCH formula, we can formulate the star product defined as in 5.1.4 as

$$u \star v(\xi) = \int_{\mathfrak{g} \times \mathfrak{g}} \hat{u}(X) \hat{v}(Y) \exp\left(2\pi i \langle \xi, \mathrm{BCH}(X, Y) \rangle\right) dX dY,$$

where $\hat{u}(X) = \int_{\mathfrak{g}^*} u(\zeta) \exp(-2\pi i \langle \zeta, X \rangle) d\zeta$ and formal parameter $\hbar = 2\pi i$. This star product is often also called the *BCH star product*.

5.2.2 Differential Graded Lie Algebras

Definition 5.2.6 (Graded Lie Algebra) A *graded Lie algebra* (GLA) is a $\mathbb{Z}$-graded vector space $\mathfrak{g} = \bigoplus_{i \in \mathbb{Z}} \mathfrak{g}^i$ endowed with a bilinear (operation)

$$[\ ,\ ] : \mathfrak{g} \otimes \mathfrak{g} \to \mathfrak{g}$$

satisfying:

(1) $[a, b] \in \mathfrak{g}^{\alpha+\beta}$ (homogeneity).
(2) $[a, b] = -(-1)^{\alpha\beta}[b, a]$ (graded skew-symmetry),
(3) $[a, [b, c]] = [[a, b], c] + (-1)^{\alpha\beta}[b, [a, c]]$ (graded Jacobi identity)

for all $a \in \mathfrak{g}^\alpha$, $b \in \mathfrak{g}^\beta$ and $c \in \mathfrak{g}^\gamma$.

Definition 5.2.7 (Shift) Given any graded vector space $\mathfrak{g} = \bigoplus_{i \in \mathbb{Z}} \mathfrak{g}^i$, we can obtain a new graded vector space $\mathfrak{g}[n]$ by shifting each component by n, i.e.

$$\mathfrak{g}[n] := \bigoplus_{i \in \mathbb{Z}} \mathfrak{g}[n]^i, \qquad \mathfrak{g}[n]^i := \mathfrak{g}^{i+n}.$$

Definition 5.2.8 (Differential Graded Lie Algebra) A *differential graded Lie algebra* (DGLA) is a GLA $\mathfrak{g}$ together with a differential d: $\mathfrak{g} \to \mathfrak{g}$, i.e. a linear operator of degree $+1$ (d: $\mathfrak{g}^i \to \mathfrak{g}^{i+1}$) which satisfies the Leibniz rule

$$d[a, b] = [da, b] + (-1)^\alpha [a, db], \quad a \in \mathfrak{g}^\alpha, b \in \mathfrak{g}^\beta$$

and squares to zero (d $\circ$ d $= 0$).

Example 5.2.9 Note that one can turn any Lie algebra into a DGLA concentrated in degree zero by using the trivial zero differential d $= 0$.

Given a DGLA $\mathfrak{g}$, we can consider its cohomology

$$H^i(\mathfrak{g}) := \frac{\ker(d \colon \mathfrak{g}^i \to \mathfrak{g}^{i+1})}{\operatorname{im}(d \colon \mathfrak{g}^{i-1} \to \mathfrak{g}^i)}.$$

Remark 5.2.10 Note that $H^\bullet(\mathfrak{g}) := \bigoplus_i H^i(\mathfrak{g})$ has a natural structure of a graded vector space and further has a GLA structure defined on equivalence classes $[a], [b] \in H^\bullet(\mathfrak{g})$ by

$$[[a], [b]]_{H^\bullet(\mathfrak{g})} := [[a, b]_\mathfrak{g}],$$

where $[\ ,\ ]_\mathfrak{g}$ denotes the Lie bracket on $\mathfrak{g}$. Moreover, extending the GLA $H^\bullet(\mathfrak{g})$ by the zero differential will turn it into a DGLA.

Remark 5.2.11 Note that each morphism $\varphi \colon \mathfrak{g}_1 \to \mathfrak{g}_2$ between two DGLAs induces a morphism $H^\bullet(\varphi) \colon H^\bullet(\mathfrak{g}_1) \to H^\bullet(\mathfrak{g}_2)$ between the cohomologies. In particular, we are interested in *quasi-isomorphisms* between DGLAs.

Definition 5.2.12 (Quasi-Isomorphism) A *quasi-isomorphism* is a morphism between DGLAs which induces an isomorphism on the level of cohomology.

Remark 5.2.13 The notion of quasi-isomorphism naturally holds more generally for any (co)chain complex and the induced (co)homology theory.

Definition 5.2.14 (Quasi-Isomorphic) Two DGLAs $\mathfrak{g}_1$ and $\mathfrak{g}_2$ are called *quasi-isomorphic* if there is a quasi-isomorphism $\varphi \colon \mathfrak{g}_1 \to \mathfrak{g}_2$.

Remark 5.2.15 The existence of a quasi-isomorphism $\varphi \colon \mathfrak{g}_1 \to \mathfrak{g}_2$ does not imply that there is also a quasi-isomorphism *inverse* $\varphi^{-1} \colon \mathfrak{g}_2 \to \mathfrak{g}_1$ and therefore they do not directly define an equivalence relation. We will deal with this issue by considering the category of L_∞-*algebras*.

5.2.3 L_∞-Algebras

The notion of an L_∞-algebra, first introduced in [Sta92], gives a generalization of a DGLA. It has a lot of applications in modern constructions of quantum field theory and homotopy theory. In this section, we will define what an L_∞-algebra is and show how one can extract an L_∞-structure out of any DGLA. Moreover, we briefly explain its representation in terms of higher brackets. We refer to [Sta92, LS93, LM95] for more details.

Definition 5.2.16 (Formal) A DGLA $\mathfrak{g}$ is called *formal* if it is quasi-isomorphic to its cohomology, considered as a DGLA endowed with the zero differential and the induced bracket.

Remark 5.2.17 Kontsevich's *formality theorem* [Kon03] consists of the result that the DGLA of *multidifferential operators* (see later) is formal.

Definition 5.2.18 (Graded Coalgebra) A *graded coalgebra* (GCA) over some ring k is a $\mathbb{Z}$-graded vector space $\mathfrak{h} = \bigoplus_{i \in \mathbb{Z}} \mathfrak{h}^i$ endowed with a *comultiplication*, i.e. a graded linear map

$$\Delta : \mathfrak{h} \to \mathfrak{h} \otimes \mathfrak{h}$$

such that

$$\Delta(\mathfrak{h}^i) \subset \bigoplus_{j+\ell=i} \mathfrak{h}^j \otimes \mathfrak{h}^\ell$$

which also satisfies the *coassociativity* condition

$$(\Delta \otimes \mathrm{id})\Delta(a) = (\mathrm{id} \otimes \Delta)\Delta(a), \quad \forall a \in \mathfrak{h}.$$

Definition 5.2.19 (GCA with Counit) We say that a GCA $\mathfrak{h}$ has a *counit* if there is a morphism

$$\varepsilon : \mathfrak{h} \to k$$

such that $\varepsilon(\mathfrak{h}^i) = 0$ for all $i > 0$ and

$$(\varepsilon \otimes \mathrm{id})\Delta(a) = (\mathrm{id} \otimes \varepsilon)\Delta(a) = a, \quad \forall a \in \mathfrak{h}.$$

Definition 5.2.20 (Cocommutative GCA) We say that a GCA $\mathfrak{h}$ is *cocommutative* if there is a *twisting map* defined on homogeneous elements $x, y \in \mathfrak{h}$ by

$$\mathsf{T} : \mathfrak{h} \otimes \mathfrak{h} \to \mathfrak{h} \otimes \mathfrak{h},$$

$$x \otimes y \mapsto \mathsf{T}(x \otimes y) := (-1)^{|x||y|} y \otimes x,$$

where $|x| \in \mathbb{Z}$ denotes the degree of $x \in \mathfrak{h}$, and extended linearly, such that

$$\mathsf{T} \circ \Delta = \Delta.$$

Example 5.2.21 (Tensor Algebra) Let V be a (graded) vector space over k and consider its tensor algebra $T(V) = \bigoplus_{n=0}^{\infty} V^{\otimes n}$. Denote by 1 the unit of k. Then, $T(V)$ can be endowed with a coalgebra structure. We define a comultiplication $\Delta_{T(V)}$ on homogeneous elements by

$$\Delta_{T(V)}(v_1 \otimes \cdots \otimes v_n) := 1 \otimes (v_1 \otimes \cdots \otimes v_n)$$

$$+ \sum_{j=1}^{n-1} (v_1 \otimes \cdots \otimes v_j) \otimes (v_{j+1} \otimes \cdots \otimes v_n)$$

$$+ (v_1 \otimes \cdots \otimes v_n) \otimes 1.$$

In particular, we have a counit $\varepsilon_{T(V)}$ as the canonical projection $\varepsilon_{T(V)} \colon T(V) \to V^{\otimes 0} := k$. Note that if we consider the *reduced tensor algebra* $\bar{T}(V) := \bigoplus_{n=1}^{\infty} V^{\otimes n}$, then the projection $\bar{\pi} \colon T(V) \to \bar{T}(V)$ and the inclusion $\bar{i} \colon T(V) \hookrightarrow \bar{T}(V)$ induce a comultiplication on $\bar{T}(V)$, thus a coalgebra structure but without counit.

Example 5.2.22 (Symmetric and Exterior Algebra) Let V be a (graded) vector space over k. Then we can also consider the *Symmetric algebra* $\mathrm{Sym}(V)$ and *exterior algebra* $\bigwedge V$. Note that for the graded case we have to take the quotients with respect to the two-sided ideals generated by homogeneous elements of the form $v \otimes w - \mathsf{T}(v \otimes w)$ and $v \otimes w + \mathsf{T}(v \otimes w)$. They can be endowed with a coalgebra structure (without counit if constructed for the reduced tensor algebra). Indeed, e.g. we can define a comultiplication $\Delta_{\mathrm{Sym}(V)}$ on homogeneous elements $v \in V$ by

$$\Delta_{\mathrm{Sym}(V)}(v) := 1 \otimes v + v \otimes 1.$$

Definition 5.2.23 (Coderivation) A *coderivation* of degree ℓ on some GCA $\mathfrak{h}$ is a graded linear map $\delta \colon \mathfrak{h}^i \to \mathfrak{h}^{i+\ell}$ which satisfies the *co-Leibniz identity*

$$\Delta\delta(v) = (\delta \otimes \mathrm{id})\Delta(v) + ((-1)^{\ell|v|}\mathrm{id} \otimes \delta)\Delta(v), \quad \forall v \in \mathfrak{h}^{|v|}.$$

Definition 5.2.24 (Codifferential) A *codifferential* Q on a coalgebra is a coderivation of degree $+1$ which squares to zero ($Q^2 = 0$).

Definition 5.2.25 (L_∞-Algebra) An L_∞-*algebra* is a graded vector space $\mathfrak{g}$ over k endowed with a degree $+1$ coalgebra differential Q on the reduced symmetric space $\overline{\mathrm{Sym}}(\mathfrak{g}[1])$.

Definition 5.2.26 (L_∞-Morphism) An L_∞-morphism $F \colon (\mathfrak{g}, Q) \to (\widetilde{\mathfrak{g}}, \widetilde{Q})$ is a morphism

$$F \colon \overline{\mathrm{Sym}}(\mathfrak{g}[1]) \to \overline{\mathrm{Sym}}(\widetilde{\mathfrak{g}}[1]).$$

of graded coalgebras, which also commutes with the differentials, i.e.

$$FQ = \widetilde{Q}F.$$

Remark 5.2.27 As in the dual case an algebra morphism $f \colon \mathrm{Sym}(\mathcal{A}) \to \mathrm{Sym}(\mathcal{A})$ (resp. a derivation $\delta \colon \mathrm{Sym}(\mathcal{A}) \to \mathrm{Sym}(\mathcal{A})$) is uniquely determined by its restriction to the algebra $\mathcal{A} = \mathrm{Sym}^1(\mathcal{A})$ due to the homomorphism condition $f(a \otimes b) = f(a) \otimes f(b)$ (resp. Leibniz rule), an L_∞-morphism F (resp. a coderivation Q) is uniquely determined by its projection onto the first component F^1 (resp. Q^1).

Let us introduce the generalized notation F^i_j (resp. Q^i_j) for the projection to the i-th component of the target space restricted to the j-th component of the source. We can then rephrase the condition for F (resp. Q) to be an L_∞-morphism (resp. a codifferential). Indeed, using this notation, we get that QQ, FQ, and $\widetilde{Q}F$ are coderivations. It is sufficient to show this on the projection onto the first factor for each term.

Corollary 5.2.28 *A coderivation Q is a codifferential if and only if*

$$\sum_{1 \le i \le n} Q^1_i Q^i_n = 0, \quad \forall n \in \mathbb{N}. \tag{5.30}$$

Corollary 5.2.29 *A morphism F of graded coalgebras is an L_∞-morphism if and only if*

$$\sum_{1 \le i \le n} F^1_i Q^i_n = \sum_{1 \le i \le n} \widetilde{Q}^1_i F^i_n, \quad \forall n \in \mathbb{N}. \tag{5.31}$$

Example 5.2.30 For the case when $n = 1$, we get

$$Q^1_1 Q^1_1 = 0, \qquad F^1_1 Q^1_1 = \widetilde{Q}^1_1 F^1_1.$$

Hence, every coderivation Q induces the structure of a cochain complex of vector spaces on $\mathfrak{g}$ and every L_∞-morphism restricts to a cochain map F^1_1.

Remark 5.2.31 The structure of a DGLA can actually be easily generalized to the L_∞ case. In particular, we can define a quasi-isomorphism of L_∞-algebras to be an L_∞-morphism F such that F^1_1 is a quasi-isomorphism of cochain complexes. Similarly, one can extend the notion of *formality*.

Lemma 5.2.32 *Let $F: (\mathfrak{g}, Q) \to (\tilde{\mathfrak{g}}, \tilde{Q})$ be an L_∞-morphism. If F is a quasi-isomorphism, it admits a* quasi-inverse, *i.e. there is an L_∞-morphism $G: (\tilde{\mathfrak{g}}, \tilde{Q}) \to (\mathfrak{g}, Q)$ that induces the inverse isomorphism for the corresponding cohomologies.*

Remark 5.2.33 Lemma 5.2.32 implies an equivalence relation defined by L_∞-quasi-isomorphisms, i.e. two L_∞-algebras are L_∞-quasi-isomorphic if and only if there is an L_∞-quasi-isomorphism between them. This solves the problem that one has to face regarding DGLAs, where the equivalence relation only holds on the level of quasi-isomorphisms.

Let us see how we can extract an L_∞-structure from any DGLA $(\mathfrak{g}, [\ ,\ ], \mathrm{d})$ $\mathfrak{g}$ which indeed implies that an L_∞-algebra is a generalization of a DGLA. Define Q_1^1 to be a multiple of the differential of the DGLA $\mathfrak{g}$. For $n = 2$, we can write condition (5.30) as

$$Q_1^1 Q_2^1 + Q_2^1 Q_2^2 = 0. \tag{5.32}$$

We identify Q_1^1 with the differential d of $\mathfrak{g}$ (up to some sign) and note that (5.32) suggests that Q_2^1 should be expressed in terms of the bracket $[\ ,\ ]$ of $\mathfrak{g}$ since (5.32) has the same form as the compatibility condition between the bracket and the differential. In particular, we get

$$Q_1^1(a) := (-1)^\alpha \mathrm{d}a, \qquad a \in \mathfrak{g}^\alpha, \tag{5.33}$$

$$Q_2^1(b \otimes c) := (-1)^{\beta(\gamma-1)}[b, c], \qquad b \in \mathfrak{g}^\beta, c \in \mathfrak{g}^\gamma, \tag{5.34}$$

$$Q_n^1 = 0, \qquad \forall n \geq 3. \tag{5.35}$$

For $n = 3$, we get

$$Q_1^1 Q_3^1 + Q_2^1 Q_3^2 + Q_3^1 Q_3^3 = 0. \tag{5.36}$$

If we use the definition of the bracket (5.34) and expand Q_3^2 in terms of Q_2^1 in (5.36), we get

$$(-1)^{(\alpha+\beta)(\gamma-1)}\left[(-1)^{\alpha(\beta-1)}[a, b], c\right]$$
$$+ (-1)^{(\alpha+\gamma)(\beta-1)}(-1)^{(\gamma-1)(\beta-1)}\left[(-1)^{\alpha(\gamma-1)}[a, c], b\right]$$
$$+ (-1)^{(\beta+\gamma)(\alpha-1)}(-1)^{(\beta+\gamma)(\alpha-1)}\left[(-1)^{\beta(\gamma-1)}[b, c], a\right] = 0. \tag{5.37}$$

Remark 5.2.34 Note that, by rearranging signs, (5.37) is the same as the *graded Jacobi identity*.

Similar constructions hold for a DGLA morphism $F : \mathfrak{g} \to \widetilde{\mathfrak{g}}$. It induces an L_∞-morphism $\bar{F}$ that is fully determined by its first component $\bar{F}_1^1 := F$. The non-trivial conditions for $n = 1$ and $n = 2$ on $\bar{F}$, induced by (5.31), are actually given by

$$\bar{F}_1^1 Q_1^1(f) = \widetilde{Q}_1^1 \bar{F}_1^1(f) \Leftrightarrow F(df) = \tilde{d}F(f),$$

$$\bar{F}_1^1 Q_2^1(f \otimes g) + \bar{F}_1^2 Q_2^2(f \otimes g) = \widetilde{Q}_1^1 \bar{F}_1^2(f \otimes g) + \widetilde{Q}_2^1 \bar{F}_2^2(f \otimes g)$$

$$= \Leftrightarrow F([f, g])[F(f), F(g)].$$

Remark 5.2.35 For $n = 3$, where the Q_3^1 do not vanish, one can show that (5.37) would have been satisfied up to *homotopy*, i.e. up to a term of the form

$$d\rho(g, h, k) \pm \rho(dg, h, k) \pm \rho(g, dh, k) \pm \rho(g, h, dk),$$

where $\rho : \bigwedge^3 \mathfrak{g} \to \mathfrak{g}[-1]$. In this case, one says that $\mathfrak{g}$ has the structure of a *homotopy Lie algebra*.

Remark 5.2.36 The concept of Remark 5.2.35 can be generalized by introducing the canonical isomorphism between the symmetric and exterior algebra (usually called *déclage isomorphism*) given by

$$\mathrm{dec}_n : \mathrm{Sym}^n(\mathfrak{g}[1]) \xrightarrow{\sim} \bigwedge^n \mathfrak{g}[n],$$

$$x_1 \otimes \cdots \otimes x_n \mapsto (-1)^{\sum_{1 \le i \le n}(n-i)(|x_i|-1)} x_1 \wedge \cdots \wedge x_n,$$

to define for each n a *multibracket* of degree $2 - n$

$$[\ ,\ldots,\ ]_n : \bigwedge^n \mathfrak{g} \to \mathfrak{g}[2 - n]$$

by starting from the corresponding Q_n^1. In fact, (5.30) induces an infinite family of conditions on these multibrackets. A graded vector space $\mathfrak{g}$ together with such a family of multibrackets is usually called a *strong homotopy Lie algebra* (SHLA) (see also [LM95]).

Definition 5.2.37 (Maurer–Cartan Equation) The *Maurer–Cartan equation* of a DGLA $(\mathfrak{g}, [\ ,\ ], d)$ is given by

$$da + \frac{1}{2}[a, a] = 0, \quad a \in \mathfrak{g}^1. \tag{5.38}$$

Remark 5.2.38 It is easy to show that the set of solutions to the Maurer–Cartan equation (5.38) is preserved under the action of a morphism between DGLAs. Moreover, the group action by the gauge group that is canonically defined through the degree zero part of any formal DGLA also preserves the set of solutions to

(5.38). This can be extended to the formal setting.[2] For a DGLA $\mathfrak{g}$, we will denote by $MC(\mathfrak{g})$ the set of solutions to (5.38) in $\mathfrak{g}$.

5.2.4   The DGLA of Multivector Fields $\mathcal{V}$

Recall from Sect. 2.3.2 that a multivector field of degree $j \geq 1$ on some manifold M is an element

$$X \in \mathfrak{X}^j(M) := \Gamma\left(\bigwedge^j TM\right).$$

In local coordinates, we can write

$$X = \sum_{1 \leq i_1 < \cdots < i_j \leq \dim M} X^{i_1 \cdots i_j} \partial_{i_1} \wedge \cdots \wedge \partial_{i_j}.$$

Hence, if we consider the collection of these vector spaces in all degrees, we naturally get a graded vector space structure

$$\tilde{\mathcal{V}}(M) := \bigoplus_{j \geq 0} \tilde{\mathcal{V}}^j(M), \qquad \tilde{\mathcal{V}}^j(M) := \begin{cases} C^\infty(M), & j = 0 \\ \mathfrak{X}^j(M), & j \geq 1 \end{cases}$$

Using the *Schouten–Nijenhuis bracket* as defined in (4.1), we can endow $\tilde{\mathcal{V}}(M)$ with a GLA structure. We will write $[\ ,\ ]_{SN}$ to indicate that we mean the Schouten–Nijenhuis bracket. We note/recall the following identities for the Schouten–Nijenhuis bracket:

1. $[X, Y]_{SN} = -(-1)^{(|X|+1)(|Y|+1)}[Y, X]_{SN}$,
2. $[X, Y \wedge Z]_{SN} = [X, Y]_{SN} \wedge Z + (-1)^{(|Y|+1)|Z|}Y \wedge [X, Z]_{SN}$,
3. $[X, [Y, Z]_{SN}]_{SN} = [[X, Y]_{SN}, Z]_{SN} + (-1)^{(|X|+1)(|Y|+1)}[Y, [X, Z]_{SN}]_{SN}$.

---

[2] Note that we can set $\mathfrak{g}[\![\hbar]\!] := \mathfrak{g} \otimes k[\![\hbar]\!]$ and show that this is again a DGLA. As we have seen before, the gauge group is then formally defined as $G := \exp(\hbar\mathfrak{g}^0[\![\hbar]\!])$, where $\mathfrak{g}^0[\![\hbar]\!]$ denotes the Lie algebra given as the degree zero part of $\mathfrak{g}[\![\hbar]\!]$. Note that the action of $\hbar\mathfrak{g}^1[\![\hbar]\!]$ is defined by generalizing the adjoint action. Namely, we have

$$\exp(\hbar X)a := \sum_{n \geq 0} \frac{(\mathrm{ad}_X)^n}{n!}(a) - \sum_{n \geq 0} \frac{(\mathrm{ad}_X)^n}{(n+1)!}(\mathrm{d}X) = a + \hbar[X, a] - \hbar\mathrm{d}X + O(\hbar^2),$$

$$\forall X \in \mathfrak{g}^0[\![\hbar]\!], a \in \mathfrak{g}^1[\![\hbar]\!].$$

In order, to obtain the correct signs, we have to shift everything by $+1$. Thus, we define the GLA of multivector fields to be $\mathcal{V}(M) := \widetilde{\mathcal{V}}(M)[1]$. Namely, we have

$$\mathcal{V}(M) := \bigoplus_{j \geq -1} \mathcal{V}^j(M), \qquad \mathcal{V}^j(M) := \widetilde{\mathcal{V}}^{j+1}(M).$$

Finally, using the zero differential, we can turn it into a DGLA $(\mathcal{V}(M), [\ ,\ ]_{\mathrm{SN}}, \mathrm{d} = 0)$.

5.2.4.1 The Case of Poisson Bivector Fields

Consider a manifold M together with a Poisson bivector field $\pi \in \mathcal{V}^1(M)$. Then we can consider its induced Poisson bracket $\{\ ,\ \}$ on $C^\infty(M)$. Recall that, by Exercise 4.1.6, we can translate the Jacobi identity of the Poisson bracket to the condition $[\pi, \pi]_{\mathrm{SN}} = 0$. Indeed, we have

$$\{\{f, g\}, h\} + \{\{g, h\}, f\} + \{\{h, f\}, g\} = 0 \Leftrightarrow$$

$$\pi^{ij} \partial_j \pi^{k\ell} \partial_i f \partial_k g \partial_\ell h + \pi^{ij} \partial_j \pi^{k\ell} \partial_i g \partial_k h \partial_\ell f + \pi^{ij} \partial_j \pi^{k\ell} \partial_i h \partial_k \partial_\ell g = 0 \Leftrightarrow$$

$$\pi^{ij} \partial_j \pi^{k\ell} \partial_i \wedge \partial_k \wedge \partial_\ell = 0 \Leftrightarrow [\pi, \pi]_{\mathrm{SN}} = 0.$$

Note that, since we have endowed $\mathcal{V}(M)$ with the zero differential, we get that Poisson bivector fields are exactly solutions to the Maurer–Cartan equation on the DGLA $\mathcal{V}(M)$:

$$\mathrm{d}\pi + \frac{1}{2}[\pi, \pi]_{\mathrm{SN}} = 0, \qquad \pi \in \mathcal{V}^1(M).$$

In particular, *formal* Poisson structures $\{\ ,\ \}_\hbar$ are associated to formal bivectors $\pi_\hbar \in \hbar \mathcal{V}^1(M)[[\hbar]]$.

5.2.5 The DGLA of Multidifferential Operators $\mathcal{D}$

Note that to any associative algebra $\mathcal{A}$ over a field k, we can assign the complex of multilinear maps given by

$$C^\bullet(\mathcal{A}, \mathcal{A}) := \bigoplus_{j \geq -1} C^j(\mathcal{A}, \mathcal{A}), \qquad C^j(\mathcal{A}, \mathcal{A}) := \mathrm{Hom}_k(\mathcal{A}^{\otimes(j+1)}, \mathcal{A}).$$

We will sometimes just write $\mathcal{C}^j$ instead of $\mathcal{C}^j(\mathcal{A}, \mathcal{A})$ whenever it is clear which algebra $\mathcal{A}$ was considered. Define a family of composition $\{\circ_i\}$ such that for an $(m+1)$-linear operator $\phi \in \mathcal{C}^m$ and an $(n+1)$-linear operator $\psi \in \mathcal{C}^n$ we have

$$(\phi \circ_i \psi)(f_0 \otimes \cdots \otimes f_{m+n}) := \phi(f_0 \otimes \cdots \otimes f_{i-1} \otimes \psi(f_i \otimes \cdots \otimes f_{i+n})$$
$$\otimes f_{i+n+1} \otimes \cdots \otimes f_{m+n}).$$

If we sum over all the different ways of composition, we get

$$\phi \circ \psi := \sum_{0 \le i \le m} (-1)^{in} \phi \circ_i \psi, \tag{5.39}$$

which we can use to endow $\mathcal{C}$ with a GLA structure.

Definition 5.2.39 (Gerstenhaber Bracket) The *Gerstenhaber bracket* on the graded vector space $\mathcal{C}$ is defined by

$$[\ ,\ ]_G \colon \mathcal{C}^m \otimes \mathcal{C}^n \to \mathcal{C}^{m+n},$$
$$(\phi, \psi) \mapsto [\phi, \psi]_G := \phi \circ \psi - (-1)^{mn} \psi \circ \phi. \tag{5.40}$$

Remark 5.2.40 There is a more general notion of a *Gerstenhaber algebra* [Ger63], which describes the connection between *supercommutative rings* [Var04] and *graded Lie superalgebras* [Kac77]. In particular, it consists of a graded commutative algebra endowed with a Poisson bracket of degree -1. This structure plays also a fundamental role in other constructions of theoretical physics, such as in *DeDonder–Weyl theory* [DeD30, Wey35] (a generalization of the Hamiltonian formalism to field theory) and in the *Batalin–Vilkovisky formalism* [BV77, BV81, BV83] (a way of treating perturbative quantization of gauge theories).

Proposition 5.2.41 *The graded vector space $\mathcal{C}$ together with the Gerstenhaber bracket (5.40) is a GLA, which we call* Hochschild GLA.

Proof Clearly, the Gerstenhaber bracket is linear. Note that the sign $(-1)^{mn}$ ensures that it is also (graded) skew-symmetric, since

$$[\phi, \psi]_G = -(-1)^{mn}\left(\psi \circ \phi - (-1)^{mn} \phi \circ \psi\right) = -(-1)^{mn}[\psi, \phi]_G, \ \forall \phi \in \mathcal{C}^m, \psi \in \mathcal{C}^n.$$

Next, we need to make sure that $[\ ,\ ]_G$ satisfies the (graded) Jacobi identity

$$[\phi, [\psi, \chi]_G]_G = [[\phi, \psi]_G, \chi]_G + (-1)^{mn}[\psi, [\phi, \chi]_G]_G, \quad \forall \phi \in \mathcal{C}^m, \psi \in \mathcal{C}^n, \chi \in \mathcal{C}^p. \tag{5.41}$$

Note that the left-hand-side of (5.41) gives

$$\left(\phi \circ \psi - (-1)^{mn}\psi \circ \phi\right) \circ \chi - (-1)^{(m+n)p}\chi \circ \left(\phi \circ \psi - (-1)^{mn}\psi \circ \phi\right)$$

$$= \sum_{\substack{0\le i\le m \\ 0\le k\le m+n}} (-1)^{in+kp}(\phi \circ_i \psi) \circ_k \chi - \sum_{\substack{0\le j\le n \\ 0\le k\le m+n}} (-1)^{m(j+n)+kp}(\psi \circ_j \phi) \circ_k \chi$$

$$- \sum_{\substack{0\le i\le m \\ 0\le k\le p}} (-1)^{(m+n)(k+p)+in}\chi \circ_k (\phi \circ_i \psi)$$

$$+ \sum_{\substack{0\le j\le n \\ 0\le k\le p}} (-1)^{(m+n)(k+p)+m(j+n)}\chi \circ_k (\psi \circ_j \phi). \tag{5.42}$$

We can rewrite the first sum according to the following rule

$$(\phi \circ_i \psi) \circ_k \chi = \begin{cases} (\phi \circ_k \chi) \circ_i \psi, & k < i \\ \phi \circ_i (\psi \circ_{k-i} \chi), & i \le k \le i+n \\ (\phi \circ_{k-n} \chi) \circ_i \psi, & i+n < k \end{cases}$$

as a term of the form

$$\sum_{\substack{0\le i\le m \\ i\le k\le i+n}} (-1)^{in+kp}\phi \circ_i (\psi \circ_{k-i} \chi) = \sum_{\substack{0\le i\le m \\ 0\le k\le n}} (-1)^{(n+p)i+kp}\phi \circ_i (\psi \circ_k \chi). \tag{5.43}$$

Note that the sign of (5.43) is the same as the one in the corresponding term appearing in $(\phi \circ \psi) \circ \chi$ on the left-hand-side of (5.42) plus the terms in which the i-th and k-th composition commute. These then cancel the corresponding terms appearing in the expansion of the second term of the right-hand-side of (5.41). Using the same approach for all the other terms, we get the proof.

Remark 5.2.42 Note that associative multiplications on $\mathcal{A}$ are elements in $\mathcal{C}^1$. In particular, if we denote by $m \in \mathcal{C}^1$ a multiplication map, the associativity condition is given by

$$m(m(f \otimes g) \otimes h) - m(f \otimes m(g \otimes h)) = 0.$$

This is in fact equivalent to the condition that the Gerstenhaber bracket of m with itself vanishes. Indeed, we have

$$[m, m]_G(f \otimes g \otimes h) = \sum_{0 \leq i \leq 1} (-1)^i (m \circ_i m)(f \otimes g \otimes h)$$

$$- (-1)^1 \sum_{0 \leq i \leq 1} (-1)^i (m \circ_i m)(f \otimes g \otimes h)$$

$$= 2\Big(m(m(f \otimes g) \otimes h) - m(f \otimes m(g \otimes h))\Big).$$

For an element ϕ of a (DG) Lie algebra $\mathfrak{g}$ (of degree ℓ), we get that

$$\mathrm{ad}_\phi := [\phi, \ ]$$

is a derivation (of degree ℓ). Indeed, by the Jacobi identity, we have

$$\mathrm{ad}_\phi[\psi, \xi] = [\mathrm{ad}_\phi \psi, \xi] + (-1)^{\ell m}[\psi, \mathrm{ad}_\phi \xi], \quad \forall \psi \in \mathfrak{g}^m, \xi \in \mathfrak{g}^n.$$

Definition 5.2.43 (Hochschild Differential) The *Hochschild differential* is given by

$$\mathrm{d}_m \colon \mathcal{C}^i \to \mathcal{C}^{i+1}$$

$$\psi \mapsto \mathrm{d}_m \psi := [m, \psi]_G. \tag{5.44}$$

Proposition 5.2.44 *The GLA $\mathcal{C}$ endowed with the Hochschild differential d_m is a DGLA.*

Proof The only thing we need to check is that the Hochschild differential indeed squares to zero (i.e. $\mathrm{d}_m \circ \mathrm{d}_m = 0$). This follows immediately from the Jacobi identity and the associativity condition of m expressed in terms of the Gerstenhaber bracket:

$$(\mathrm{d}_m \circ \mathrm{d}_m)\psi = [m, [m, \psi]_G]_G = [[m, m]_G, \psi]_G - [m, [m, \psi]_G]_G$$

$$= -[m, [m, \psi]_G]_G \Leftrightarrow \mathrm{d}_m \circ \mathrm{d}_m = 0. \qquad \square$$

Remark 5.2.45 In fact, we explicitly have

$$(\mathrm{d}_m \psi)(f_0 \otimes \cdots \otimes f_{n+1}) = \sum_{0 \leq i \leq n} (-1)^{i+1} \psi(f_0 \otimes \cdots \otimes f_{i-1}$$

$$\otimes m(f_i \otimes f_{i+1}) \otimes \cdots \otimes f_{n+1})$$

$$+ m(f_0 \otimes \psi(f_1 \otimes \cdots \otimes f_{n+1}))$$

$$+ (-1)^{n+1} m(\psi(f_0 \otimes \cdots \otimes f_n) \otimes f_{n+1}).$$

Let now $\mathcal{A} = C^\infty(M)$ and consider the subalgebra $\widetilde{\mathcal{D}}(M) \subset \mathcal{C}$, which is the (graded) vector space given as the collection

$$\widetilde{\mathcal{D}}^i(M) := \bigoplus_i \widetilde{\mathcal{D}}^i(M)$$

of subspaces $\widetilde{\mathcal{D}}^i(M) \subset \mathcal{C}^i$ which consist of differential operators on $C^\infty(M)$. One can check that $\widetilde{\mathcal{D}}(M)$ is closed with respect to the Gerstenhaber bracket $[\ ,\ ]_G$ and the differential d_m and hence is a DGL subalgebra[3] of $\mathcal{C}$.

Moreover, we want to restrict everything to differential operators which vanish on constant functions. They will in fact form another DGL subalgebra $\mathcal{D}(M) \subset \widetilde{\mathcal{D}}(M)$. Note that we want to have this restriction because of the fact that the coefficients in the expansion of a star product vanish on constant functions, i.e. $B_i(1, f) = 0$ for all $i \in \mathbb{N}$ and $f \in C^\infty(M)$. Note however that the Hochschild differential d_m is not an inner derivation anymore for $\widetilde{\mathcal{D}}(M)$ since the multiplication does not vanish on constants.

Consider an element $B \in \mathcal{D}^1(M)$. Then we want to regard $m + B$ as a deformation of the actual product. The associativity condition for $m + B$ implies

$$[m + B, m + B]_G = 0,$$

and by associativity of m we get

$$[m, B]_G = [B, m]_G = d_m B.$$

This does exactly lead to the Maurer–Cartan equation:

$$d_m B + \frac{1}{2}[B, B]_G = 0. \tag{5.45}$$

Remark 5.2.46 If we consider the *formal* version of $\mathcal{D}(M)$, we can see that the deformed product does exactly describe a star product since $B \in \hbar \mathcal{D}^1(M)[[\hbar]]$. Moreover, the gauge group is given by formal differential operators and the action on the star product is given by (5.13) since the adjoint action, by the definition of the Gerstenhaber bracket, is exactly the composition of D_i with B_j.

---

[3] Note that we will also allow differential operators of degree zero, i.e. operators which do not "derive" anything. Thus, the associative product m is also an element of $\widetilde{\mathcal{D}}^1(M)$.

5.2.6 The Hochschild–Kostant–Rosenberg Map

In [HKR62], Hochschild–Kostant–Rosenberg have shown that the cohomology of the algebra of multidifferential operators and the algebra of multivector fields, which is equal to its cohomology, are isomorphic. They constructed an isomorphism

$$\text{HKR} \colon H^\bullet(\widetilde{\mathcal{D}}(M)) \xrightarrow{\sim} \widetilde{\mathcal{V}}(M) \cong H^\bullet(\widetilde{\mathcal{V}}(M)). \tag{5.46}$$

In particular, this isomorphism is induced by the map

$$U_1^{(0)} \colon \widetilde{\mathcal{V}}(M) \to \widetilde{\mathcal{D}}(M),$$

$$\xi_0 \wedge \cdots \wedge \xi_n \mapsto \left(f_0 \otimes \cdots \otimes f_n \mapsto \frac{1}{(n+1)!} \sum_{\sigma \in S_{n+1}} \text{sign}(\sigma) \xi_{\sigma(0)}(f_0) \cdots \xi_{\sigma(n)}(f_n) \right)$$

$$\tag{5.47}$$

Remark 5.2.47 This is extended to vector fields of order zero by the identity map. However, the map (5.47) fails to preserve the Lie algebra structure. This can be easily seen for order two. Indeed, given homogeneous bivector fields $\chi_1 \wedge \chi_2$ and $\xi_1 \wedge \xi_2$, we get

$$U_1^{(0)}\left([\chi_1 \wedge \chi_2, \xi_1 \wedge \xi_2]_{\text{SN}} \right) \neq \left[U_1^{(0)}(\chi_1 \wedge \chi_2), U_1^{(0)}(\xi_1 \wedge \xi_2) \right]_G.$$

The left-hand-side gives

$$U_1^{(0)}\Big([\chi_1, \xi_1] \wedge \chi_2 \wedge \xi_2 - [\chi_1, \xi_2] \wedge \chi_2 \wedge \xi_1 - [\chi_2, \xi_1] \wedge \chi_1 \wedge \xi_2$$

$$+ [\chi_2, \xi_2] \wedge \chi_1 \wedge \xi_1 \Big)(f \otimes g \otimes h)$$

$$= \frac{1}{6}\Big(\chi_1\big(\xi_1(f)\chi_2(g)\xi_2(h)\big) - \xi_1\big(\chi_1(f)\chi_2(g)\xi_2(h)\big)$$

$$- \chi_1\big(\xi_2(f)\chi_2(g)\xi_1(h)\big) + \xi_2\big(\chi_1(f)\chi_2(g)\xi_1(h)\big) - \chi_2\big(\xi_1(f)\chi_1(g)\xi_2(h)\big)$$

$$+ \xi_1\big(\chi_2(f)\chi_1(g)\xi_2(h)\big)$$

$$+ \chi_2\big(\xi_2(f)\chi_1(g)\xi_1(h)\big) + \xi_2\big(\chi_2(f)\chi_1(g)\xi_1(h)\big) \Big) + \text{permutations}.$$

The right-hand-side is given by

$$\frac{1}{4}\Big(\chi_1\big(\xi_1(f)\xi_2(g)\big)\chi_2(h) + \cdots \Big).$$

One can, however, still show that the difference of the two terms is given by the image of a closed term in the cohomology of $\mathcal{D}(M)$. Thus, there is some hope to control everything and still extend it to a Lie algebra morphism whose first order approximation is given by (5.47). Again, to resolve this problem, we will consider an L_∞-morphism U.

5.2.7 The Dual Point of View

Let V be a vector space and note that we can naturally identify polynomials on V as the *symmetric functions* on the dual space V^*

$$f(v) := \sum_j \frac{1}{j!} f_j(v \otimes \cdots \otimes v), \quad \forall v \in V,$$

where $f_j \in \mathrm{Sym}^j(V^*)$. We want to extend this to the case when V is a graded vector space. In this case, we need to consider the exterior algebra instead. Note that, if we consider the *injective limit* of the $\mathrm{Sym}^j(V^*)$ (resp. $\bigwedge^j V^*$) endowed with the induced topology, we can define the corresponding completion $\widehat{\mathrm{Sym}}(V^*)$ (resp. $\widehat{\bigwedge V^*}$). Now we can define a function in a formal neighborhood of 0 to be given by the formal Taylor expansion in $\hbar$

$$f(\hbar v) := \sum_j \frac{\hbar^j}{j!} f_j(v \otimes \cdots \otimes v), \quad \forall v \in V.$$

Thus, a vector field X on V can be regarded as a derivation on $\widehat{\bigwedge V^*}$ and the Leibniz rule makes sure that X is completely determined by its restriction on V^*. Similarly, if we have an algebra morphism

$$\varphi: \widehat{\bigwedge W^*} \to \widehat{\bigwedge V^*},$$

it induces a map

$$f := \varphi^*: \widehat{\bigwedge V} \to \widehat{\bigwedge W}$$

whose components f_j are completely defined by their projection onto W as the φ_j are defined by their restriction on W^*.

Definition 5.2.48 ((Formal) Pointed Map) A *(formal) pointed map* is an algebra morphism between the reduced exterior algebras

$$\varphi : \overline{\bigwedge W^*} \to \overline{\bigwedge V^*},$$

where the overline means that we are considering the *reduced tensor algebra* $\bar{T}(W^*)$ (resp. $\bar{T}(V^*)$) as in Example 5.2.22.

Definition 5.2.49 (Pointed Vector Field) A *pointed vector field* X on a manifold M is a vector field in $\mathfrak{X}(M)$ which fixes the point zero, i.e.

$$X(f)(0) = 0, \quad \forall f \in C^\infty(M).$$

Definition 5.2.50 (Cohomological Pointed Vector Field) A pointed vector field X is called *cohomological* (or *Q-field*) if and only if it commutes with itself, i.e. $X^2 = \frac{1}{2}[X, X] = 0$.

Definition 5.2.51 (Pointed Q-Manifold) A *pointed Q-manifold* is a (formal) pointed manifold together with a cohomological vector field.

Consider now a Lie algebra $\mathfrak{g}$. Note that the bracket $[\ ,\ ] : \bigwedge^2 \mathfrak{g} \to \mathfrak{g}$ induces a linear map

$$[\ ,\ ]^* : \mathfrak{g}^* \to \bigwedge^2 \mathfrak{g}^*.$$

This can be extended to the whole exterior algebra by a map

$$\delta : \bigwedge \mathfrak{g}^* \to \bigwedge \mathfrak{g}^*[1]$$

such that it satisfies the Leibniz rule and $\delta\big|_{\mathfrak{g}^*} = [\ ,\ ]^*$.

Remark 5.2.52 We can regard the exterior algebra as an *odd* analogue of a manifold on which δ takes the role of a (pointed) vector field. Note also that, since $[\ ,\ ]$ satisfies the Jacobi identity, we have $\delta^2 = 0$ and thus δ defines a cohomological (pointed) vector field.

Consider two Lie algebras $\mathfrak{g}$ and $\mathfrak{h}$ together with the corresponding cohomological vector fields $\delta_\mathfrak{g}$ and $\delta_\mathfrak{h}$ on the respective exterior algebras. Then a Lie algebra morphism $\varphi : \mathfrak{g} \to \mathfrak{h}$ will correspond to a chain map $\varphi^* : \mathfrak{h}^* \to \mathfrak{g}^*$ since

$$\varphi\left([\ ,\ ]_\mathfrak{g} \right) = [\varphi(\ ), \varphi(\ )]_\mathfrak{h} \iff \delta_\mathfrak{g} \circ \varphi^* = \varphi^* \circ \delta_\mathfrak{h}. \tag{5.48}$$

Remark 5.2.53 Equation (5.48) gives a first hint to the correspondence between L_∞-algebras and pointed Q-manifolds. Since a Lie algebra is a particular case of a

DGLA, which again can be endowed with an L_∞-structure, we can see that the map φ satisfies exactly the same condition for the first component of an L_∞-morphism as in (5.31) for $n = 1$.

Let us extend this picture to the case of a graded vector space Z over a field k. Let us decompose the graded space Z into its *even* and *odd* part, i.e.

$$Z = Z_{[0]} \oplus Z_{[1]}$$

where $Z_{[0]}$ denotes the even part, i.e. $Z_{[0]} := \bigoplus_{i \in 2\mathbb{Z}} Z^i$ and $Z_{[1]} := \bigoplus_{i \in 2\mathbb{Z}+1} Z^i$. Here we have denoted by [0], [1] the equivalence classes in $\mathbb{Z}_2$. For a vector space W, denote by ΠW the (odd) space defined by a *parity reversal* on W, which we can also denote by $W[1]$ using the notation of Definition 5.2.7. Let us define $V := Z_{[0]}$ and $\Pi W := Z_{[1]}$. Then $Z = V \oplus \Pi W$ and functions can be identified by elements of

$$\text{Sym}(Z^*) := \text{Sym}(V^*) \otimes \bigwedge W^*.$$

We can express the condition for a vector field $\delta : \text{Sym}(Z^*) \to \text{Sym}(Z^*)[1]$ to be *cohomological* in terms of its coefficients

$$\delta_j : \text{Sym}^j(Z^*) \to \text{Sym}^{j+1}(Z^*)$$

by expanding the equation $\delta^2 = 0$. Hence, we get an infinite family of equations

$$\delta_0 \delta_0 = 0, \tag{5.49}$$

$$\delta_1 \delta_0 + \delta_0 \delta_1 = 0, \tag{5.50}$$

$$\delta_2 \delta_0 + \delta_1 \delta_1 + \delta_0 \delta_2 = 0, \tag{5.51}$$

$$\vdots$$

Let now $m_j := (\delta_j|_{Z^*})^*$ be the dual coefficients and let $\langle \ , \ \rangle : Z^* \otimes Z \to k$. Then we can rewrite these conditions in terms of $m_j : \text{Sym}^j(Z) \to Z$. Note that the first Eq. (5.49), which is now given by $m_0 m_0 = 0$, implies that m_0 is a differential on Z and hence induces a cohomology $H^\bullet_{m_0}(Z)$. For $j = 1$, i.e. for the second Eq. (5.50), we get

$$\langle \delta_1 \delta_0 f, x \otimes y \rangle = \langle \delta_0 f, m_1(x \otimes y) \rangle = \langle f, m_0(m_1(x \otimes y)) \rangle \tag{5.52}$$

and

$$\langle \delta_0 \delta_1 f, x \otimes y \rangle = \langle \delta_1 f, m_0(x) \otimes y \rangle + (-1)^{|x|} \langle \delta_1 f, x \otimes m_0(y) \rangle$$

$$= \langle f, m_1(m_0(x) \otimes y) \rangle + (-1)^{|x|} \langle f, m_1(x \otimes m_0(y)) \rangle. \tag{5.53}$$

If we put (5.52) and (5.53) together, we get that m_0 is a derivation with respect to the multiplication defined by m_1.

Remark 5.2.54 If we consider $Z := \mathfrak{g}[1]$ and consider the exterior algebra by the déclage isomorphism $(\mathrm{Sym}^n(\mathfrak{g}[1]) \xrightarrow{\sim} \bigwedge^n \mathfrak{g}[n])$, we can consider m_1 to be a bilinear skew-symmetric operator on $\mathfrak{g}$.

Note that Eq. (5.51) implies that m_1 is indeed a Lie bracket for which the Jacobi identity holds up to terms containing m_0. Since m_0 is a differential, this means that m_1 is a Lie bracket up to *homotopy*. Hence, considering all equations, we will get a strong homotopy Lie algebra structure on $\mathfrak{g}$.

Remark 5.2.55 Note that this will lead to a one-to-one correspondence between pointed Q-manifolds and SHLAs. Since SHLAs are equivalent to L_∞-algebras, we get a one-to-one correspondence between pointed Q-manifolds and L_∞-algebras.

Definition 5.2.56 (Q-Map) A Q-*map* is a (formal) pointed map between two Q-manifolds Z and $\widetilde{Z}$ which commutes with the Q-fields, i.e. a map

$$\varphi \colon \overline{\mathrm{Sym}}(\widetilde{Z}^*) \to \overline{\mathrm{Sym}}(Z^*)$$

$$\text{such that} \quad \varphi \circ \widetilde{\delta} = \delta \circ \varphi. \tag{5.54}$$

We want to express the condition (5.31) more explicitly in this setting by using Q-maps. Consider the vector field δ and its restriction to $\widetilde{Z}$. We define the coefficients of the dual map as

$$U_j := \left(\varphi_j|_{Z^*}\right)^* \colon \mathrm{Sym}^j(Z) \to \widetilde{Z}.$$

Then we can express the condition (5.54) of a Q-map on the dual coefficients. The first equation gives

$$\langle \varphi(\widetilde{\delta}f), x \rangle = \langle \delta\varphi(f), x \rangle \Rightarrow \langle \widetilde{\delta}_0 f, U_1(x) \rangle = \langle \varphi(f), m_0(x) \rangle$$

$$\Rightarrow \langle f, \widetilde{m}_0(U_1(x)) \rangle = \langle f, U_1(m_0(x)) \rangle.$$

This tells us that the first coefficient U_1 is a chain map with respect to the differential defined by the first coefficient of the Q-structure:

$$H^\bullet(U_1) \colon H^\bullet_{m_0}(Z) \to H^\bullet_{\widetilde{m}_0}(\widetilde{Z}).$$

Remark 5.2.57 Similarly, we can consider the equation for the next coefficient:

$$\widetilde{m}_1(U_1(x) \otimes U_1(y)) + \widetilde{m}_1(U_2(x \otimes y))$$

$$= U_2(m_0(x) \otimes y) + (-1)^{|x|}U_2(x \otimes m_0(y)) + U_1(m_1(x \otimes y)). \tag{5.55}$$

This shows that U_1 preserves the Lie algebra structure induced by m_1 and $\tilde{m}_1$ up to terms containing m_0 and $\tilde{m}_0$, i.e. up to *homotopy*. Recall that this solves the problem that the map $U_1^{(0)}$, defined in (5.47), is a chain map but not a DGLA morphism. Namely, a Q-map U (or equivalently an L_∞-morphism) induces a map U_1 which has the same property as $U_1^{(0)}$.

Definition 5.2.58 (Koszul Sign) Let V be a vector space endowed with a graded commutative product $\otimes$. The *Koszul sign* $\varepsilon(\sigma)$ of a permutation σ is the sign defined by

$$x_1 \otimes \cdots \otimes x_n = \varepsilon(\sigma)x_{\sigma(1)} \otimes \cdots \otimes x_{\sigma(n)}, \quad x_i \in V.$$

Definition 5.2.59 (Shuffle Permutation) An $(\ell, n - \ell)$-*shuffle permutation* is a permutation σ of $(1, \ldots, n)$ such that $\sigma(1) < \cdots < \sigma(\ell)$ and $\sigma(\ell + 1) < \cdots < \sigma(n)$.

Remark 5.2.60 The shuffle permutation associated to a partition $I_1 \sqcup \cdots \sqcup I_j = \{1, \ldots, n\}$ is the permutation that takes at first all the elements indexed by the subset I_1 in the given order, then those of I_2 and so on.

The n-th coefficient of U satisfies the following condition:

$$\tilde{m}_0(U_n(x_1 \otimes \cdots \otimes x_n)) + \frac{1}{2} \sum_{\substack{I \sqcup J = \{1, \ldots, n\} \\ I, J \neq \emptyset}} \varepsilon_x(I, J)\tilde{m}_1(U_{|I|}(x_I) \otimes U_{|J|}(x_J))$$

$$= \sum_{1 \leq j \leq n} \varepsilon_x^j U_n(m_0(x_j) \otimes x_1 \otimes \cdots \otimes \widehat{x}_j \otimes \cdots \otimes x_n)$$

$$+ \frac{1}{2} \sum_{j \neq \ell} \varepsilon_x^{j\ell} U_{n-1}(m_1(x_j \otimes x_\ell) \otimes x_1 \otimes \cdots \otimes \widehat{x}_j \otimes \cdots \otimes \widehat{x}_\ell \otimes \cdots \otimes x_n),$$

$$(5.56)$$

where $\varepsilon_x(I, J)$ is the *Koszul sign* associated to the $(|I|, |J|)$-shuffle permutation for the partition $I \sqcup J = \{1, \ldots, n\}$ and ε_x^j (resp. $\varepsilon_x^{j\ell}$) for the particular case $I = \{j\}$ (resp. $I = \{j, \ell\}$). Moreover, we have set $x_I := \bigotimes_{i \in I} x_i$.

Remark 5.2.61 We will modify (5.56) to an L_∞-morphism where Z is chosen to be the DGLA $\mathcal{V}$ of multivector fields and $\tilde{Z}$ the DGLA $\mathcal{D}$ of multidifferential operators.

5.2.8 Formality of $\mathcal{D}$ and Classification of Star Products on $\mathbb{R}^d$

We want to analyze the relation between the formality of $\mathcal{D}$ and the classification problem of all star products on $\mathbb{R}^d$. Recall that the *associativity* of the star product and the *Jacobi identity* for a bivector field are equivalent to certain Maurer–Cartan equations.

Definition 5.2.62 (Maurer–Cartan Equation on (Formal) L_∞-Algebras) The Maurer–Cartan equation on a (formal) L_∞-algebra $(\mathfrak{g}[[\hbar]], Q)$ is given by

$$Q(\exp(\hbar x)) = 0, \qquad x \in \mathfrak{g}^1[[\hbar]], \tag{5.57}$$

where the exponential sends an element of degree $+1$ to a formal power series in $\hbar\mathfrak{g}[[\hbar]]$.

Remark 5.2.63 Note that, from the dual point of view, condition (5.57) tells us that x is a fixed point of the cohomological vector field δ. This means that for each $f \in \mathrm{Sym}(\mathfrak{g}^*[[\hbar]][1])$ we have

$$\delta f(\hbar x) = 0. \tag{5.58}$$

Moreover, using that $(\delta f)_j = \delta_{j-1} f$, we can expand Eq. (5.58) into a formal Taylor series. In fact, using the pairing

$$\langle \delta_{j-1} f, x \otimes \cdots \otimes x \rangle = \langle f, m_{j-1}(x \otimes \cdots \otimes x) \rangle,$$

we can write the Maurer–Cartan equation (5.57) as

$$\sum_{j \geq 1} \frac{\hbar^j}{j!} m_{j-1}(x \otimes \cdots \otimes x) = \hbar m_0(x) + \frac{\hbar^2}{2} m_1(x \otimes x) + O(\hbar^3) = 0. \tag{5.59}$$

For two DGLAs $\mathfrak{g}$ and $\mathfrak{h}$, an L_∞-morphism $\varphi \colon \mathrm{Sym}(\mathfrak{h}^*[[\hbar]][1]) \to \mathrm{Sym}(\mathfrak{g}^*[[\hbar]][1])$ preserves the Maurer–Cartan equation (5.59), similarly as a morphism of DGLAs preserves the Maurer–Cartan equation (5.38). In particular, if x is a solution of (5.59) on $\mathfrak{g}[[\hbar]]$, we get that

$$U(\hbar x) = \sum_{j \geq 1} \frac{\hbar^j}{j!} U_j(x \otimes \cdots \otimes x)$$

is also a solution of (5.59) on $\mathfrak{h}[[\hbar]]$.

Remark 5.2.64 The action of the gauge group on solutions to the Maurer–Cartan equation (5.38) can be generalized to the case of L_∞-algebras. Namely, if x and x' are equivalent modulo this generalized action, then $U(x)$ and $U(x')$ are still equivalent solutions.

Remark 5.2.65 Remember, we are especially interested in the case where $\mathfrak{g} = \mathcal{V}$ and $\mathfrak{h} = \mathcal{D}$. Thus, we get an L_∞-morphism U which gives as a formula of how to construct an associative star product out of any (formal) Poisson bivector π, given by

$$U(\pi) = \sum_{j \geq 0} \frac{\hbar^j}{j!} U_j(\pi \wedge \cdots \wedge \pi), \qquad (5.60)$$

where we insert the coefficient of degree zero to be the original non-deformed product. Note that if U is a quasi-isomorphism, there is a bijection between (formal) Poisson structures on a manifold M and formal deformations of the point-wise product on $C^\infty(M)$. In particular, as soon as we have a formality map, we have solved the *existence* and *classification* of star products on M.

5.3   Kontsevich's Star Product

We want to construct an explicit expression of the formality map U. The idea is to introduce graphs and to rewrite things in those terms. An easy example is given by the Moyal product $f \star g$ for two smooth functions f and g on $(\mathbb{R}^{2n}, \omega_0)$. We can write $f \star g$ as the sum of graphs as in Fig. 5.1:

Remark 5.3.1 This construction can be generalized by allowing more than two outgoing arrows for gray vertices when considering general multivector fields and allowing incoming arrows for gray vertices (i.e. derivations of the tensor coefficients for the assigned multivector fields). Note that there are no incoming arrows for the Moyal product since the Poisson structure is constant. For Kontsevich's star product, we assign to each graph Γ a multidifferential operator B_Γ and a *weight* w_Γ such that the map U that sends an n-tuple of multivector fields to the corresponding weighted sum over all possible graphs in this set of multidifferential operators is an L_∞-morphism.

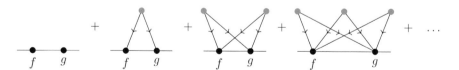

Fig. 5.1 The Moyal product $f \star g$ represented in terms of graphs. The gray vertices represent the Poisson tensor π induced by ω_0. The term with n Poisson vertices in the sum represents the n-th term in the formal power series of the Moyal product. For each gray vertex, the two outgoing arrows represent the derivatives ∂_i and ∂_j acting on f and g

5.3.1 Data for the Construction

5.3.1.1 Admissible Graphs

Definition 5.3.2 (Admissible Graphs) The set $\mathcal{G}_{n,\bar{n}}$ of *admissible graphs* consists of all connected graphs Γ such that

- The set of vertices $V(\Gamma)$ is decomposed in two ordered subsets $V_1(\Gamma)$ and $V_2(\Gamma)$ isomorphic to $\{1, \ldots, n\}$ respectively $\{\bar{1}, \ldots, \bar{n}\}$ whose elements are called vertices of *first type* respectively *second type*;
- The following inequalities involving the number of vertices of the two types are satisfied: $n \geq 0$, $\bar{n} \geq 0$ and $2n + \bar{n} - 2 \geq 0$;
- The set of edges $E(\Gamma)$ is finite and does not contain *short loops*, i.e. edges starting and ending at the same vertex;
- All edges $E(\Gamma)$ are oriented and start from a vertex of first type;
- The set of edges starting at a given vertex $v \in V_1(\Gamma)$ will be ordered. We will denote this set by Star(v).

See Figs. 5.2 and 5.3 for examples of admissible and non-admissible graphs respectively.

5.3.1.2 The Multidifferential Operators B_Γ

Let us consider pairs $(\Gamma, \xi_1 \otimes \cdots \otimes \xi_n)$ consisting of a graph $\Gamma \in \mathcal{G}_{n,\bar{n}}$ with $2n + \bar{n} - 2$ edges and of a tensor product of n multivector fields on $\mathbb{R}^d$. We want to understand

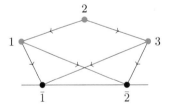
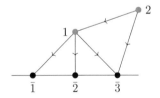

Fig. 5.2 Examples of admissible graphs

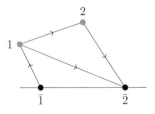
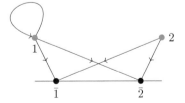

Fig. 5.3 Examples of non-admissible graphs

how we can associate to such a pair a multidifferential operator $B_\Gamma \in \mathcal{D}^{\bar{n}-1}$. This is done in the following way:

- Associate to each vertex v of first type with k outgoing arrows the skew-symmetric tensor $\xi_i^{j_1 \cdots j_k}$ corresponding to a given ξ_i via the natural identification.
- Place a function at each vertex of second type.
- Associate to the ℓ-th arrow in Star(v) a partial derivative with respect to the coordinate labeled by the ℓ-th index of ξ_i acting on the function or the tensor appearing at its endpoint.
- Multiply such elements in the order prescribed by the labeling of the graph.

Example 5.3.3 Denote by Γ_1 the first graph in Fig. 5.2. Then, to a triple of bivector fields (ξ_1, ξ_2, ξ_3) with $\xi_1 = \sum_{i_1 < i_2} \xi_1^{i_1 i_2} \partial_{i_1} \wedge \partial_{i_2}$, $\xi_2 = \sum_{j_1 < j_2} \xi_2^{j_1 j_2} \partial_{j_1} \wedge \partial_{j_2}$ and $\xi_3 = \sum_{\ell_1 < \ell_2} \xi_3^{\ell_1 \ell_2} \partial_{\ell_1} \wedge \partial_{\ell_2}$, we associate the bidifferential operator

$$U_{\Gamma_1}(\xi_1 \wedge \xi_2 \wedge \xi_3)(f \otimes g) := \xi_2^{j_1 j_2} \partial_{j_2} \xi_1^{i_1 i_2} \partial_{j_2} \xi_3^{\ell_1 \ell_2} \partial_{i_1} \partial_{\ell_1} f \partial_{i_2} \partial_{\ell_2} g.$$

Example 5.3.4 Denote by Γ_2 the second graph in Fig. 5.2. Then, to a tuple of multivector fields (χ_1, χ_2) with $\chi_1 = \sum_{i_1 < i_2} \chi_1^{i_1 i_2} \partial_{i_1} \wedge \partial_{i_2}$ and $\chi_2 = \sum_{j_1 < j_2 < j_3} \chi_2^{j_1 j_2 j_3} \partial_{j_1} \wedge \partial_{j_2} \wedge \partial_{j_3}$, we associate the tridifferential operator

$$U_{\Gamma_2}(\chi_1 \wedge \chi_2)(f \otimes g \otimes h) := \chi_1^{i_1 i_2} \partial_{i_1} \chi_2^{j_1 j_2 j_3} \partial_{j_1} f \partial_{j_2} g \partial_{j_3} \partial_{i_2} h.$$

Remark 5.3.5 In particular, for each admissible graph $\Gamma \in \mathcal{G}_{n,\bar{n}}$, we get a linear map $U_\Gamma \colon \bigwedge^n V \to \mathcal{D}$ which is equivariant with respect to the action of the symmetric group. Kontsevich's main construction was to choose *weights* $w_\Gamma \in \mathbb{R}$ such that the linear combination

$$U := \sum_{n,\bar{n} \geq 0} \sum_{\Gamma \in \mathcal{G}_{n,\bar{n}}} w_\Gamma B_\Gamma$$

becomes an L_∞-morphism.

5.3.1.3 Weights of Graphs

The weight w_Γ associated to an admissible graph $\Gamma \in \mathcal{G}_{n,\bar{n}}$ is defined by an integral of a differential form ω_Γ over (a suitable compactification of) some *configuration space* $C_{n,\bar{n}}^+$. Namely, it is given by

$$w_\Gamma := \prod_{1 \leq k \leq n} \frac{1}{|\text{Star}(k)|!} \frac{1}{(2\pi)^{2n+\bar{n}-2}} \int_{\bar{C}_{n,\bar{n}}^+} \omega_\Gamma. \tag{5.61}$$

This holds for the case when Γ has exactly $2n + \bar{n} - 2$ edges, otherwise the weight is just zero.

Let us look at the integral (5.61) more carefully.

5.3.1.4 Configuration Spaces

Consider an embedding of the graph Γ into the 2-dimensional upper half-space $\mathbb{H}^2$ (also called *upper half-plane*) as defined in (2.26) where we want to identify $\mathbb{R}^2 \cong \mathbb{C}$. Moreover, denote by $\mathbb{H}^2_+ := \{z \in \mathbb{C} \mid \Im(z) > 0\}$ where $\Im(z)$ denotes the *imaginary part* of z.

Definition 5.3.6 (Open Configuration Space) Define the *open configuration space* of the distinct $n + \bar{n}$ vertices of an admissible graph $\Gamma \in \mathcal{G}_{n,\bar{n}}$ as the smooth manifold

$$\mathrm{Conf}_{n,\bar{n}} := \Big\{(z_1, \ldots, z_n, z_{\bar{1}}, \ldots, z_{\bar{n}}) \in \mathbb{C}^{n+\bar{n}} \mid z_i \in \mathbb{H}^2_+,$$

$$z_{\bar{i}} \in \mathbb{R}, \ z_i \neq z_j \text{ for } i \neq j, \ z_{\bar{i}} \neq z_{\bar{j}} \text{ for } \bar{i} \neq \bar{j}\Big\} \qquad (5.62)$$

Remark 5.3.7 More precisely, we need to consider everything up to *scaling* and *translation*. Hence, we consider the Lie group G consisting of translations in the horizontal direction and rescaling such that the action on $z \in \mathbb{H}^2$ is given by

$$z \mapsto az + b, \qquad a \in \mathbb{R}_{\geq 0}, b \in \mathbb{R}.$$

The action of G is free whenever the number of vertices is $2n + \bar{n} - 2 \geq 0$. Thus the quotient space of $\mathrm{Conf}_{n,\bar{n}}$ with respect to the G-action, which we will denote by $C_{n,\bar{n}}$, is again a smooth manifold of (real) dimension $2n + \bar{n} - 2$. If there are no vertices of second type (i.e. $\bar{n} = 0$), one can define the open configuration space as a subset of $\mathbb{C}^n$ instead of $\mathbb{H}^n$ and one can consider the Lie group which consists of rescaling and translation in any direction. The corresponding quotient space C_n for $n \geq 2$ still a smooth manifold but now of dimension $2n - 3$.

Definition 5.3.8 (Connected Configuration Space) Define the *connected configuration space* to be the subset of $C_{n,\bar{n}}$ given by

$$C^+_{n,\bar{n}} := \Big\{(z_1, \ldots, z_n, z_{\bar{1}}, \ldots, z_{\bar{n}}) \in C_{n,\bar{n}} \mid z_{\bar{i}} < z_{\bar{j}} \text{ for } \bar{i} < \bar{j}\Big\} \qquad (5.63)$$

Remark 5.3.9 One can show that $C^+_{n,\bar{n}}$ is again a smooth manifold which is now connected.

Let us define an *angle map*

$$\phi \colon C_{2,0} \to S^1,$$

Fig. 5.4 Illustration of the
angle map ϕ

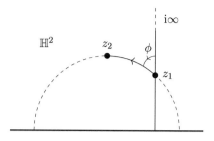

which assigns to each pair of distinct points $z_1, z_2 \in \mathbb{H}^2$ the angle between the
geodesics with respect to the *Poincaré metric* connecting z_1 to $i\infty$ and to z_2, in the
counterclockwise direction (see Fig. 5.4). Formally, we have

$$\phi(z_1, z_2) := \arg\left(\frac{z_2 - z_1}{z_2 - \bar{z}_1}\right) = \frac{1}{2i} \log\left(\frac{(z_2 - z_1)(\bar{z}_2 - z_1)}{(z_2 - \bar{z}_1)(\bar{z}_2 - \bar{z}_1)}\right). \tag{5.64}$$

The differential of ϕ is then a 1-form on $C_{2,0}$ which we can pullback to the
configuration space including points in $\mathbb{R}$ by the projection p_e, associated to each
edge $e = (z_i, z_j)$ of Γ, defined by

$$p_e \colon C_{n,\bar{n}} \to C_{2,0},$$

$$(z_1, \ldots, z_n, z_{\bar{1}}, \ldots, z_{\bar{n}}) \mapsto (z_i, z_j).$$

Hence we get a 1-form $\mathrm{d}\phi_e := p_e^* \mathrm{d}\phi \in \Omega^1(C_{n,\bar{n}})$. The map ϕ is usually called a
propagator, for consistency with the physics literature as we will see in the next
chapter. The differential form ω_Γ in (5.61) is then defined as

$$\omega_\Gamma := \bigwedge_{e \in E(\Gamma)} \mathrm{d}\phi_e.$$

Remark 5.3.10 Note that the ordering one these 1-forms is induced by the ordering
on the set of edges by the first-type vertices and the ordering on $\mathrm{Star}(v)$. Moreover,
note that we indeed obtain a top-form on $C_{n,\bar{n}}$ (resp. $C_{n,\bar{n}}^+$) as long as we consider
graphs with exactly $2n + \bar{n} - 2$ edges since $\dim C_{n,\bar{n}} = \dim C_{n,\bar{n}}^+ = 2n + \bar{n} - 2$.

Remark 5.3.11 In order to make sense of the integral (5.61), we need to make sure
that it converges. Note that, by construction of ϕ, we get that $\mathrm{d}\phi$ is not defined as
soon two points collide with each other. This is the reason why we in fact need a
suitable *compactification* $\bar{C}_{n,\bar{n}}^+$ of the connected configuration space $C_{n,\bar{n}}^+$.

Theorem 5.3.12 ([FM94, AS94]) *For any configuration space $C_{n,\bar{n}}$ (resp. C_n)
there exists a compactification $\bar{C}_{n,\bar{n}}$ (resp. $\bar{C}_n$) whose interior is the open configuration space, such that the projections p_e and the angle map ϕ (thus the differential
form ω_Γ) extend smoothly to the corresponding compactifications.*

Remark 5.3.13 This compactification is usually called *FMAS compactification* for *Fulton–MacPherson* who have proved a first result of this in the algebro-geometrical setting for configuration spaces of points in non-singular algebraic varieties [FM94] and *Axelrod–Singer* who have proved this in the smooth-geometrical setting for configuration spaces of points in smooth manifolds [AS94]. Let us give some ideas of the (smooth) Fulton–MacPherson construction. Let S be a finite set and consider the space $\mathrm{Map}(S, \Sigma)$ of maps from S to some manifold Σ. Moreover, consider the smooth blow up $B\ell(\mathrm{Map}(S, \Sigma), \Delta_S)$, where Δ_S denotes the diagonal $\Delta_S \subset \mathrm{Map}(S, \Sigma)$, consisting of constant maps $S \to \Sigma$. Denote by $\mathrm{Conf}_S(\Sigma)$ the space of embeddings of S into Σ. One can then observe that for every inclusion $K \subset S$ there are natural projections $\mathrm{Map}(S, \Sigma) \to \mathrm{Map}(K, \Sigma)$ and corresponding arrows $\mathrm{Conf}_S(\Sigma) \to \mathrm{Conf}_K(\Sigma)$ by restriction of maps from S to K as a functorial approach. Further, one can show that the inclusions $\mathrm{Conf}_S(\Sigma) \hookrightarrow \mathrm{Map}(S, \Sigma)$ can be lifted to inclusions $\mathrm{Conf}_S(\Sigma) \hookrightarrow B\ell(\mathrm{Map}(S, \Sigma), \Delta_S)$ since these sets avoid all diagonals. Thus, for a finite set X, we have a canonical inclusion

$$\mathrm{Conf}_X(\Sigma) \hookrightarrow \bigotimes_{\substack{S \subset X \\ |S| \geq 2}} B\ell(\mathrm{Map}(S, \Sigma), \Delta_S) \times \mathrm{Map}(S, \Sigma).$$

The Fulton–MacPherson compactification, denoted by $\overline{\mathrm{Conf}_X(\Sigma)}$, is then defined as the closure of $\mathrm{Conf}_X(\Sigma)$ in this embedding.

Remark 5.3.14 The compactified configuration space is a compact smooth manifold with *corners*. Moreover, it turns out that the compactified configuration space comes with equivariant functorial properties under embeddings and that the propagators do indeed extend smoothly to this compactification in certain important cases, e.g. when $\Sigma = \mathbb{R}^3$, and $\varphi_{ij} \colon C_n(\mathbb{R}^3) \to S^2$, $\varphi_{ij}(x_1, \ldots, x_n) := \frac{x_j - x_i}{\|x_j - x_i\|}$ for $1 \leq i, j \leq n$, such an extensions holds.

Definition 5.3.15 (Manifold with Corners) A manifold with corners of dimension m is a topological Hausdorff space M which is locally homeomorphic to $\mathbb{R}^{m-n} \times \mathbb{R}^n_{\geq 0}$ with $n = 0, \ldots, m$.

Remark 5.3.16 The points $x \in M$ of a manifold with corners of dimension m whose local expression in some (and hence in any) chart has the form

$$(x_1, \ldots, x_{m-n}, 0, \ldots, 0)$$

are said to be of *type n* and they form submanifolds of M which are called *strata of codimension n*.

5.3.2 Proof of Kontsevich's Formula

To show that U is indeed an L_∞-morphism, we need to show the following points:

1. The first component of the restriction of U to $\mathcal{V}$ is up to a shift in degrees given by the natural map $U_1^{(0)}$ as defined in (5.47).
2. U is a graded linear map of degree zero.
3. U satisfies the equations (5.31) for an L_∞-morphism.

5.3.2.1 U_1 Coincides with $U_1^{(0)}$

Lemma 5.3.17 *The map*

$$U_1 : \mathcal{V} \to \mathcal{D}$$

is the natural HKR map $U_1^{(0)}$ that identifies each multivector field with the corresponding multidifferential operator.

Proof Consider the set $\mathcal{G}_{1,\bar{n}}$ of admissible graphs with one vertex of first type and $\bar{n}$ vertices of second type. It is easy to see that this set only contains one element $\Gamma_{\bar{n}}$ which has $2 \cdot 1 + \bar{n} - 2 = \bar{n}$ arrows outgoing from the single vertex of first type and each arrow is incoming to a vertex of second type (see Fig. 5.5).

Hence, to a multivector field ξ of degree k we assign the multidifferential operator

$$U_{\Gamma_{\bar{n}}}(\xi) : f_{\bar{1}} \otimes \cdots \otimes f_{\bar{n}} \mapsto w_{\Gamma_{\bar{n}}} \xi^{i_{\bar{1}} \cdots i_{\bar{n}}} \partial_{i_{\bar{1}}} f_{\bar{1}} \cdots \partial_{i_{\bar{n}}} f_{\bar{n}}.$$

One can easily check that

$$\int_{\bar{C}_{1,\bar{n}}} \omega_{\Gamma_{\bar{n}}} = \int_{\bar{C}_{1,\bar{n}}} \bigwedge_{e \in E(\Gamma_{\bar{n}})} d\phi_e = (2\pi)^{\bar{n}}.$$

Fig. 5.5 The single element $\Gamma_{\bar{n}} \in \mathcal{G}_{1,\bar{n}}$

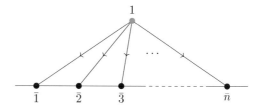

Using (5.61), this will give us

$$w_{\Gamma_{\bar{n}}} = \frac{1}{\bar{n}!}.$$

This shows that U_1 is indeed the analogue for $U_1^{(0)}$ in (5.47) which induces the HKR isomorphism (5.46).

5.3.2.2   Checking the Degrees

Lemma 5.3.18 *The n-th component*

$$U_n := \sum_{\bar{n} \geq 1} \sum_{\Gamma \in \mathcal{G}_{n,\bar{n}}} w_\Gamma B_\Gamma$$

has the correct degree for U to be an L_∞-morphism.

Proof Note that to a vertex v_i with $|\mathrm{Star}(v_i)|$ outgoing arrows, we associate an element in $\mathcal{V}^{|\mathrm{Star}(v_i)|} = \tilde{\mathcal{V}}^{|\mathrm{Star}(v_i)|+1}$. On the other hand, each graph Γ with $\bar{n}$ vertices of second type and n multivector fields $\xi_1, \dots, \xi_n$ induce a multidifferential operator $U_n(\xi_1 \wedge \cdots \wedge \xi_n)$ of degree $s = \bar{n} - 1$. Since we are only considering graphs with exactly $2n + \bar{n} - 2$ edges and since

$$|E(\Gamma)| = \sum_{1 \leq i \leq n} |\mathrm{Star}(v_i)|,$$

we can write the degree of $U_n(\xi_1 \wedge \cdots \wedge \xi_n)$ as

$$s = (2n + \bar{n} - 2) + 1 - n = |E(\Gamma)| + 1 - n = \sum_{1 \leq i \leq n} |\mathrm{Star}(v_i)| + 1 - n.$$

This is exactly the degree for the n-th component of an L_∞-morphism.

5.3.2.3   Reformulation of the L_∞-Condition in Terms of Graphs

Next we discuss the geometric proof of the formality statement. We have to extend the morphism U with a degree zero component which represents the usual multiplication between smooth maps. Therefore, we can consider the special case of the L_∞-condition (5.56) where m_0, $\tilde{m}_0$ are given by the Taylor coefficients U_n:

$$
\sum_{0\leq\ell\leq n} \sum_{-1\leq k\leq m} \sum_{0\leq i\leq m-k} \varepsilon_{kim} \sum_{\sigma\in S_{\ell,n-\ell}} \varepsilon_\xi(\sigma) U_\ell\left(\xi_{\sigma(1)}\wedge\cdots\wedge\xi_{\sigma(\ell)}\right)
$$

$$
\left(f_0\otimes\cdots\otimes f_{i-1}\otimes U_{n-\ell}\left(\xi_{\sigma(\ell+1)}\wedge\cdots\wedge\xi_{\sigma(n)}\right)\left(f_i\otimes\cdots\otimes f_{i+k}\right)\otimes f_{i+k+1}\otimes\cdots\otimes f_m\right)
$$

$$
= \sum_{1\leq i\neq j\leq n} \varepsilon_\xi^{ij} U_{n-1}\left(\xi_i\circ\xi_j\wedge\xi_1\wedge\cdots\wedge\widehat{\xi_i}\wedge\cdots\wedge\widehat{\xi_j}\wedge\cdots\wedge\xi_n\right)\left(f_0\otimes\cdots\otimes f_n\right),
$$

$$(5.65)$$

where $(\xi_j)_{j=1,\ldots,n}$ are multivector fields, $f_0,\ldots,f_m$ are smooth maps on which the multidifferential operator is acting, $S_{\ell,n-\ell}$ is the subset of S_n consisting of $(\ell, n-\ell)$-shuffle permutations, the product $\xi_i\circ\xi_j$ is defined in a way such that the Schouten–Nijenhuis bracket can be expressed in terms of this composition by a formula similar to the one relating the Gerstenhaber bracket to the analogous composition $\circ$ on $\mathcal{D}$ in (5.39) and the signs are defined as follows: $\varepsilon_{kim} := (-1)^{k(m+i)}$, $\varepsilon_\xi(\sigma)$ is the Koszul sign for the permutation σ and ε_ξ^{ij} is defined as in (5.56).

Remark 5.3.19 Note that (5.65) carries the formality condition since the left-hand-side corresponds to the Gerstenhaber bracket of multidifferential operators and the right-hand-side to (a part of) the Schouten–Nijenhuis bracket (since the differential on $\mathcal{V}$ is identical to zero). Recall that the differential on $\mathcal{D}$ is given by the Hochschild differential $[m, \ ]_G$ which in (5.65) is replaced by U_0.

We can now rewrite (5.65) in terms of admissible graphs and weights to prove that it holds. Note that the difference between the left-hand-side and right-hand-side of (5.65) can be formulated as a linear combination

$$
\sum_{\Gamma\in\mathcal{G}_{n,\bar{n}}} c_\Gamma U_\Gamma(\xi_1\wedge\cdots\wedge\xi_n)(f_0\otimes\cdots\otimes f_n). \tag{5.66}
$$

Recall that we want to consider admissible graphs Γ with exactly $2n+\bar{n}-2$ edges. In fact, (5.65) is satisfied if $c_\Gamma = 0$ for all such Γ. We will show that this is indeed the case.

5.3.2.4 The Key Is Stokes' Theorem

In order to prove that all these coefficients c_Γ vanish, we will use Stokes' theorem for manifolds with corners. Similarly as for the usual Stokes' theorem, we can replace an integral of an exact form over some manifold M as the integral of its primitive form over the boundary ∂M. In particular, we have

$$\int_{\partial \bar{C}^+_{n,\bar{n}}} \omega_\Gamma = \int_{\bar{C}^+_{n,\bar{n}}} d\omega_\Gamma = 0, \quad \forall \Gamma \in \mathcal{G}_{n,\bar{n}} \tag{5.67}$$

since $d\phi_e$ is closed and $\bar{C}^+_{n,\bar{n}}$ is compact. Let us expand the left-hand-side of (5.67) in order to show that it is exactly given by the coefficients c_Γ. Therefore, we want to take a closer look on the manifold $\partial \bar{C}^+_{n,\bar{n}}$. Recall that we have restricted the weights in (5.65) to be equal to zero if the form degree is not the same as the dimension of the underlying manifold over which we integrate. Hence, we only have to consider the codimension 1 strata of $\partial \bar{C}^+_{n,\bar{n}}$ which have dimension $2n + \bar{n} - 3$. Note that the dimension of $\partial \bar{C}^+_{n,\bar{n}}$ is equal to the number of edges and thus of the amount of 1-forms $d\phi_e$.

Remark 5.3.20 Intuitively, one can think of the boundary of $\bar{C}^+_{n,\bar{n}}$ to be represented by the degenerate configuration in which some of the $n + \bar{n}$ points *collide with each other*.

5.3.2.5 Classification of Boundary Strata

Using Remark 5.3.20, we can classify the codimension 1 strata of $\partial \bar{C}^+_{n,\bar{n}}$ as follows:

- (Strata of type S1) These are strata in which $i \geq 2$ points in $\mathbb{H}^2_+$ collide together to a point which still lies in $\mathbb{H}^2_+$. The points in such a stratum can be locally described by

$$C_i \times C_{n-i+1,\bar{n}}. \tag{5.68}$$

  The first term represents the relative position of the colliding points when we look at them under a *magnifying glass*. The second term is the space of all remaining points plus the point which appears as the point on which the first i points have collapsed to.

- (Strata of type S2) These are strata in which $i > 0$ points in $\mathbb{H}^2_+$ and $j > 0$ points in $\mathbb{R}$ with $2i + j - 2 \geq 0$ collide to a single point on $\mathbb{R}$. The points in such a stratum can be locally represented by

$$C_{i,j} \times C_{n-i,\bar{n}-j+1}. \tag{5.69}$$

  The two terms are similarly given as in (5.68).

See Figs. 5.6 and 5.7 for an illustration of a stratum of type S1 and type S2 respectively.

Now we can split the integral on the left-hand-side of (5.67) into the sum of strata of type S1 and type S2. For the strata of type S1, we will distinguish two

Fig. 5.6 Example of a stratum of type S1. Note that here $i = 3$ points have collapsed together to a point $\ell \in \mathbb{H}^2_+$

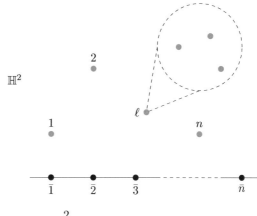

Fig. 5.7 Example of a stratum of type S2. Note that here $i = 3$ and $j = 2$ points have collapsed together to a point $\bar{\ell} \in \mathbb{R}$

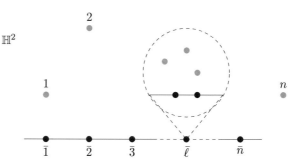

subcases for the i collapsing vertices. Since the integral vanishes if the form degree is not equal to the dimension of the underlying manifold, one can show that the only contributions come from graphs Γ whose subgraphs Γ_1, spanned by the collapsing vertices, contain exactly $2i - 3$ edges.

When $i = 2$, there is only one edge e, an hence in the first integral of the decomposition of (5.68) the differential $\mathrm{d}\phi_e$ is integrated over $C_2 \cong S^1$ and thus we get a factor of 2π which cancels the coefficient in (5.61). The remaining integral represents the weight of the corresponding quotient graph Γ_2 which is obtained from Γ after the contraction of the edge e. In particular, to the vertex j of first type, which results from this contraction, we associate the j-composition of the two multivector fields that were associated to the endpoints of e. Hence, summing over all graphs and all strata of this type, we get the right-hand-side of (5.65).

5.3.2.6 A Trick Using Logarithms

When $i \geq 3$, the integral corresponding to this stratum considers the product of $2i - 3$ angle forms over C_i. According to a *Lemma of Kontsevich*, this integral vanishes.

Fig. 5.8 Example of a
non-vanishing term

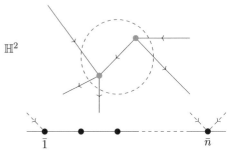

Fig. 5.9 Example of a
vanishing term

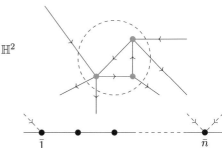

Lemma 5.3.21 (Kontsevich[Kon03]) *The integral over the configuration space C_n of $n \geq 3$ points in the upper half-plane of any* $\dim C_n := 2n - 3$ *angle forms* $d\phi_{e_i}$ *with* $i = 1, \ldots, n$ *vanishes for* $n \geq 3$ *(see e.g. Figs. 5.8 and 5.9).*

Proof of Lemma 5.3.21 First we need to restrict the integral to an even number of angle forms. This can be done by identifying C_n with the subset of $\mathbb{H}^n$, where one of the endpoints of e_1 is chosen to be the origin and the second is forced to lie on the unit circle (note that this can always be done by considering the action of the Lie group in the definition of C_n). Then the integral can be written as a product of integrals of $d\phi_e$ over S^1 and the remaining $2n - 4 =: 2N$ forms integrated over the resulting complex manifold U given by the isomorphism

$$C_n \cong S^1 \times U.$$

Then the claim follows by the following computation:

$$\int_U \bigwedge_{j=1}^{2N} d\arg(f_j) = \int_U \bigwedge_{j=1}^{2N} d\log(|f_j|) = \int_{\bar{U}} \mathcal{I}\left(d\log(|f_1|) \bigwedge_{j=2}^{2N} d\log(|f_j|)\right)$$

$$= \int_{\bar{U}} d\left(\mathcal{I}\left(\log(|f_1|) \bigwedge_{j=2}^{2N} d\log(|f_j|)\right)\right) = 0.$$

$$(5.70)$$

Let us see what is happening here. First, note that the angle function ϕ_{e_j} was expressed in terms of the argument of the (holomorphic) function f_j. In particular, f_j is just the difference of the coordinates of the endpoints of e_j. The first equality follows by the decompositions

$$d\arg(f_j) = \frac{1}{2i}\big(d\log(f_j) - d\log(\bar{f}_j)\big),$$

$$d\log(|f_j|) = \frac{1}{2}\big(d\log(f_j) + d\log(\bar{f}_j)\big).$$

Thus, the product of $2N$ of these expressions is a linear combination of products of k holomorphic and $2N - k$ anti-holomorphic forms. By a result of complex analysis, one can show that in the integration over the complex manifold U, the only terms which do not vanish are the ones where $k = N$. Then one can observe that the terms coming from the first decomposition are equal to those coming from the second decomposition.

In the second equality the integral of the differential form is replaced by an integral over a suitable 1-form with values in the space of *distributions* over the compactification $\bar{U}$ of U. One can show (this is another Lemma) that such a map $\mathcal{I}$, which sends usual 1-forms to distributional ones, commutes with the differential and thus by Stokes' theorem we get the claim.

5.3.2.7 Last Step: Vanishing Terms for Type S2 Strata

Finally, let us consider the strata of type S2. There we will have a similar dimensional argument, i.e. it analogously restricts the possible non-vanishing terms to the condition that the subgraph Γ_1, spanned by $i + j$ colliding vertices of first and second type respectively, contains exactly $2i + j - 2$ edges. Similarly as before, for the quotient graph Γ_2, obtained by contracting Γ_1, the claim is that the only non-vanishing contributions appear from the graphs for which both graphs constructed from a given Γ are admissible (see e.g. Fig. 5.10). In this case we get a decomposition of the weight w_Γ into a product $w_{\Gamma_1} \cdot w_{\Gamma_2}$ which in general, by the conditions on the amount of edges of Γ and Γ_1, will not vanish.

The only remaining thing to check is that we do not have *bad edges* by contraction (see e.g. Figs. 5.10 and 5.11). Such a situation does only appear when Γ_2 contains an edge which starts from a vertex of second type. Hence, in this situation the corresponding integral vanishes because it contains the differential of an angle map evaluated on the pair (z_1, z_2) where z_1 is constrained to lie in $\mathbb{R}$ and such maps vanish for every z_2 because the angle is measured with respect to the Poincaré metric (recall Fig. 5.4 for an intuitive picture).

This means that the only non-vanishing terms correspond to the case when we insert the differential operator associated to Γ_1 as the k-th argument of the one associated to Γ_2, where k is the vertex of the second type appearing after

Fig. 5.10 Example of an
admissible quotient

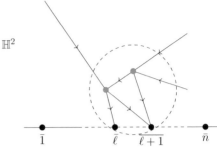

Fig. 5.11 Example of a
non-admissible quotient.
Such a term vanishes

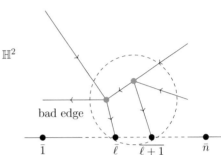

the collision. Summing over all these possibilities, the strata of type S2 exactly corresponds to the left-hand-side of (5.65).

We have thus proven that U is indeed an L_∞-morphism and since the first coefficients U_1 are given by $U_1^{(0)}$ it is also a quasi-isomorphism and hence determines uniquely a star product for any bivector field π on $\mathbb{R}^d$ by the formula (5.60).

5.3.3   Logarithmic Formality

In [Kon99], Kontsevich has proposed to replace the differential form ω_Γ in the formula of the formality map U by the differential form

$$\omega^{\log}(z_1, z_2) = \frac{1}{2\pi\mathrm{i}} \mathrm{d}\log\left(\frac{z_1 - z_2}{\bar{z}_1 - z_2}\right),$$

and claimed that a different L_∞-quasi-isomorphism $U^{\log}\colon \mathcal{V}(M) \to \mathcal{D}(M)$ is obtained in this way. This is actually not easy to see since by the presence of a logarithm, the form $\omega^{\log}$ does not extend to $\bar{C}_{2,0}^+$ because it has a pole of order one along the boundary stratum of codimension one which corresponds to the collision of its arguments to a single point on the upper half-plane. This means that the integral of the actual form we want to integrate may diverge. In [Ale+16], Alekseev,

Rossi, Torossian and Willwacher have shown that this problem can be resolved by using a version of Stokes' theorem for differential forms with singularities at the boundary and local torus actions on configuration spaces of points in the upper half-plane. They have actually proven the existence of an L_∞-quasi-isomorphism $U^{\log}$.

5.4   Globalization of Kontsevich's Star Product

Kontsevich's formula for the star product only gives a quantization for the case when $M = \mathbb{R}^d$ for a general Poisson structure π, and thus only describes a local description for the general case. Already in [Kon03], Kontsevich gave a globalization method which was very briefly mentioned, but it was explicitly constructed by Cattaneo, Felder and Tomassini in [CFT02b, CFT02a] in a similar way as Fedosov did for the symplectic case of the Moyal product [Fed94] (see also Sect. 5.1.6). Their approach uses a flat connection $\bar{D}$ on a vector bundle over M such that the algebra of the horizontal sections with respect to $\bar{D}$ is a quantization of $C^\infty(M)$. Consider the vector bundle $E_0 \to M$ of infinite jets of functions endowed with a flat connection D_0. The fiber $(E_0)_x$ over $x \in M$ is naturally a commutative algebra and carries the Poisson structure induced fiberwise by the Poisson structure on $C^\infty(M)$. The map which associates to any globally defined function its infinite jet at each point $x \in M$ is a Poisson isomorphism to the Poisson algebra of horizontal sections of E_0 with respect to D_0. Since the star product gives a deformation of the point-wise product on $C^\infty(M)$, we want to have a *quantum version* of the vector bundle and the flat connection in order to get a similar isomorphism. The vector bundle $E \to M$ is defined through a section ϕ^{aff} of the fiber bundle $M^{\text{aff}} \to M$, where M^{aff} denotes the quotient of the manifold M^{coor}, consisting of jets of coordinate systems on M, by the action of the group $\mathrm{GL}_d(\mathbb{R})$ of linear diffeomorphims given by $E := (\phi^{\text{aff}})^* \widetilde{E}$, where $\widetilde{E}$ is the bundle of $\mathbb{R}[[\hbar]]$-modules $M^{\text{coor}} \times_{\mathrm{GL}_d(\mathbb{R})} \mathbb{R}[[y^1, \ldots, y^d]][[\hbar]] \to M^{\text{aff}}$. Note that the section ϕ^{aff} can be realized explicitly by a collection of jets at zero of maps $\phi_x : \mathbb{R}^d \to M$ such that $\phi_x(0) = x$ for all $x \in M$ (modulo the action of $\mathrm{GL}_d(\mathbb{R})$). Thus, we can assume for simplicity that we have fixed a representative ϕ_x of the equivalence class for each open set of a given covering, hence assume a trivialization of E. So from now on we will identify E with the trivial bundle with fibers given by $\mathbb{R}[[y^1, \ldots, y^d]][[\hbar]]$. Therefore, E gives the desired quantization, since it is isomorphic (as a bundle of $\mathbb{R}[[\hbar]]$-modules) to the bundle $E_0[[\hbar]]$ whose elements are formal power series with coefficients given by infinite jets of functions.

5.4.1   The Product, Connection and Curvature Maps

To define the star product and the connection on E, one has to introduce new objects whose existence and properties are implicitly carried by the formality theorem. For a Poisson structure π and two vector fields ξ and η on $\mathbb{R}^d$, we define

$$P(\pi) := \sum_{j \geq 0} \frac{\hbar^j}{j!} U_j(\pi \wedge \cdots \wedge \pi), \tag{5.71}$$

$$A(\xi, \pi) := \sum_{j \geq 0} \frac{\hbar^j}{j!} U_{j+1}(\xi \wedge \pi \wedge \cdots \wedge \pi), \tag{5.72}$$

$$F(\xi, \eta, \pi) := \sum_{j \geq 0} \frac{\hbar^j}{j!} U_{j+2}(\xi \wedge \eta \wedge \pi \wedge \cdots \wedge \pi). \tag{5.73}$$

It is easy to show that the degree of the multidifferential operators on the right-hand-sides of (5.71), (5.72) and (5.73) induce that $P(\pi)$ is a (formal) bidifferential operator, $A(\xi, \pi)$ is a differential operator and $F(\xi, \eta, \pi)$ is a function. Indeed, $P(\pi)$ is actually just the star product associated the Poisson structure π. In particular, the maps P, A and F are elements of degree 0, 1 and 2 respectively of the Lie algebra cohomology complex of (formal) vector fields with values in the space of local polynomial maps, i.e. multidifferential operators which depend polynomially on π. An element of degree j of this complex is a map that assigns to $\xi_1 \wedge \cdots \wedge \xi_j$ a multidifferential operator $S(\xi_1 \wedge \cdots \wedge \xi_j \wedge \pi)$. The differential δ of this complex can then be defined as

$$\delta S(\xi_1 \wedge \cdots \wedge \xi_{j+1} \wedge \pi)$$

$$:= \sum_{1 \leq i \leq j+1} (-1)^i \frac{\partial}{\partial t}\Big|_{t=0} S\Big(\xi_1 \wedge \cdots \wedge \widehat{\xi_i} \wedge \cdots \wedge \xi_{j+1} \wedge (\Phi_t^{\xi_i})_* \pi\Big)$$

$$+ \sum_{i < \ell} (-1)^{i+\ell} S\Big([\xi_i, \xi_\ell] \wedge \xi_1 \wedge \cdots \wedge \widehat{\xi_i} \wedge \cdots \wedge \widehat{\xi_\ell} \wedge \cdots \wedge \xi_{j+1} \wedge \pi\Big),$$

where Φ_t^ξ denotes the flow of the vector field ξ. The associativity of the star product can now be expressed by

$$P \circ (P \otimes \mathrm{id} - \mathrm{id} \otimes P) = 0.$$

This follows from the formality theorem and hence the following equations do hold in a similar way:

$$P(\pi) \circ (A(\xi, \pi) \otimes \mathrm{id} + \mathrm{id} \otimes A(\xi, \pi)) = A(\xi, \pi) \circ P(\pi) + \delta P(\xi, \pi), \tag{5.74}$$

$$P(\pi) \circ (F(\xi, \eta, \pi) \otimes \mathrm{id} - \mathrm{id} \otimes F(\xi, \eta, \pi))$$

$$= -A(\xi, \pi) \circ A(\eta, \pi) + A(\eta, \pi) \circ A(\xi, \pi) + \delta A(\xi, \eta, \pi), \tag{5.75}$$

$$- A(\xi, \pi) \circ F(\eta, \zeta, \pi) - A(\eta, \pi) \circ F(\zeta, \xi, \pi)$$
$$- A(\zeta, \pi) \circ F(\xi, \eta, \pi) = \delta F(\xi, \eta, \zeta, \pi). \tag{5.76}$$

Equation (5.74) describes that under coordinate transformation, induced by the vector field ξ, the star product $P(\pi)$ is changed to an equivalent one up to terms of higher order. Equations (5.75) and (5.76) will be used for the construction of the relations between a connection 1-form A and its curvature F_A. For the explicit computation of the configuration space integrals, which appear for the weight computation in the Taylor coefficients of U_j, we can also describe the lowest order terms in the expansion of P, A and F and their action on functions:

1. $P(\pi)(f \otimes g) = fg + \hbar\pi(df, dg) + O(\hbar^2)$,
2. $A(\xi, \pi) = \xi + O(\hbar)$, where we identify ξ with a differential operator of first order on the right-hand-side,
3. $A(\xi, \pi) = \xi$, if ξ is a linear vector field,
4. $F(\xi, \eta, \pi) = O(\hbar)$,
5. $P(\pi)(1 \otimes f) = P(\pi)(f \otimes 1) = f$,
6. $A(\xi, \pi)1 = 0$.

Equations (1) and (5) have already been introduced before as the defining conditions for a star product. The equations using the connection A are needed to construct a connection D on sections on E. A section $f \in \Gamma(E)$ is locally given by a map $x \mapsto f_x$, where for every y, $f_x(y)$ is a formal power series with coefficients given by infinite jets. On this space we can introduce a deformed product $\star$ which will be the desired star product on $C^\infty(M)$ after having identified horizontal section with ordinary functions. Similarly, let us denote by π_x the push-forward by ϕ_x^{-1} of the Poisson structure π on $\mathbb{R}^d$. Then we can define the deformed product by the formal bidifferential operator $P(\pi_x)$ similarly as we did for the usual product by $P(\pi)$:

$$(f \star g)_x(y) := f_x(y)g_x(y) + \hbar\pi_x^{ij}(y)\frac{\partial f_x(y)}{\partial y^i}\frac{\partial g_x(y)}{\partial y^j} + O(\hbar^2).$$

Hence, the connection D on the space of sections $Gamma(E)$ can be defined by

$$(Df)_x := d_x f + A_x^M f,$$

where $d_x f$ denotes the de Rham differential of f regarded as a function with values in $\mathbb{R}[[y^1, \ldots, y^d]][[\hbar]]$ and the formal connection 1-form is defined through its action on some tangent vector ξ by

$$A_x^M(\xi) := A(\widehat{\xi_x}, \pi_x),$$

where A is the operator constructed as in (5.72) evaluated on the multivector fields ξ and π given through the local coordinate system defined by ϕ_x. Note that the coefficients U_j of the formality map which appear in the definition of P and A are polynomials in the derivatives of the coordinate of the arguments ξ and π, hence all results which hold for $P(\pi)$ and $A(\xi, \pi)$ are transferred to their formal extensions. In fact, Equations (1) and (5) make sure that $\star$ is an associative deformation of the point-wise product on sections and Equations (2) and (3) can be used to prove that D is independent of the choice of the formal map ϕ and hence induces a global connection on E.

5.4.2 Construction of Solutions for a Fedosov-Type Equation

We can extend D and $\star$ by the (graded) Leibniz rule to the whole complex of formal differential forms $\Omega^\bullet(E) := \Omega^\bullet(M) \otimes_{C^\infty(M)} \Gamma(E)$ and use (5.75) to get the following lemma.

Lemma 5.4.1 *Let F^M be the E-valued 2-form given by $x \mapsto F_x^M$ where $F_x^M(\xi, \eta) := F(\widehat{\xi}_x, \widehat{\eta}_x, \pi_x)$ for any pair of vector fields ξ, η. Then F^M represents the curvature of D and the two are related to each other and to the star product by the usual identities:*

(1) $D(f \star g) = D(f) \star g + f \star D(g)$,
(2) $D^2 = [F^M, \ ]_\star$,
(3) $DF^M = 0$.

Proof We can deduce these identities from the equations (5.74), (5.75), and (5.76), in which the star commutator $[f, g]_\star := f \star g - g \star f$ is already implicitly defined, after identifying the complex of formal multivector fields endowed with the differential δ with the complex of formal multidifferential operators with the de Rham differential. Such an isomorphism was explicitly given in [CFT02b].

Remark 5.4.2 A connection D which satisfies the above conditions on a bundle E of associative algebras is called *Fedosov connection* with *Weyl curvature F*. It is the kind of connection Fedosov introduced in order to give a global construction for the symplectic case (Sect. 5.1.6) [Fed94]. Note that a connection which satisfies (2) was called *abelian* in Fedosov's construction.

The final step towards a globalization is to *deform* the connection D into a new connection $\bar{D}$ which has the same properties as D and moreover has vanishing Weyl curvature, i.e. $\bar{D}^2 = 0$. Hence, we can define the complex $H_{\bar{D}}^j(E)$ and thus the (sub)algebra of horizontal sections $H_{\bar{D}}^0(E)$.

Lemma 5.4.3 *Let D be a Fedosov connection on E with Weyl curvature F and γ an E-valued 1-form. Then*

$$\bar{D} := D + [\gamma, \ ]_\star$$

is also a Fedosov connection with Weyl curvature $\bar{F} = F + D\gamma + \gamma \star \gamma$.

Proof Let f be a section in $\Gamma(E)$. Then

$$\bar{D}^2 f = [\bar{F}, f]_\star + D[\gamma, f]_\star + [\gamma, Df]_\star + [\gamma, [\gamma, f]_\star]_\star$$
$$= [F, f]_\star + [D\gamma, f]_\star + [\gamma, [\gamma, f]_\star]_\star$$
$$= \left[F + D\gamma + \frac{1}{2}[\gamma, \gamma]_\star, f\right]_\star,$$

where the last equality follows from the Jacobi identity of the star commutator $[\ ,\ ]_\star$, since each associative product induces a Lie bracket given by the commutator. If we apply $\bar{D}$ to the obtained curvature $\bar{F}$, we get

$$\bar{D}\left(F + D\gamma + \frac{1}{2}[\gamma, \gamma]_\star\right) = D^2\gamma + \frac{1}{2}[D\gamma, \gamma]_\star - \frac{1}{2}[\gamma, D\gamma]_\star + [\gamma, F + D\gamma]_\star$$

$$= [F, \gamma]_\star + [\gamma, F]_\star$$

$$= 0,$$

where we have again used the (graded) Jacobi identity and the (graded) skew-symmetry of $[\ ,\ ]_\star$.

Lemma 5.4.4 *Let D be a Fedosov connection on a bundle $E = E_0[[\hbar]]$ and F its Weyl curvature and let*

$$D = D_0 + \hbar D_1 + \cdots, \qquad F = F_0 + \hbar F_1 + \cdots$$

be their expansion as formal power series. If $F_0 = 0$ and the second cohomology of E_0 with respect to D_0 is trivial, then there exists a 1-form γ such that $\bar{D}$ has vanishing Weyl curvature.

Proof By Lemma 5.4.3, we can equivalently say that there exists a solution to the equation

$$\bar{F} = F + D\gamma + \frac{1}{2}[\gamma, \gamma]_\star = 0.$$

We can explicitly construct a solution by induction on the order of $\hbar$. We start with $\gamma_0 = 0$ and assume that $\gamma^{(j)}$ is a solution modulo $\hbar^{(j+1)}$. We can add to $\bar{F}^{(j)} = F + D\gamma^{(j)} + \frac{1}{2}[\gamma^{(j)}, \gamma^{(j)}]_\star$ the next term $\hbar^{j+1}D_0\gamma_{j+1}$ to get $\bar{F}^{j+1}$ modulo higher terms. From the equation $D\bar{F}^{(j)} + [\gamma^{(j)}, \bar{F}^{(j)}]_\star = 0$ and the induction hypothesis $\bar{F}^j = 0$ modulo $\hbar^{(j+1)}$, we get $D_0\bar{F}^{(j)} = 0$. Now since $H^2_{D_0}(E_0) = 0$, we can invert D_0 in order to define γ_{j+1} in terms of the lower order terms $\bar{F}^{(j)}$ in a way such that $\bar{F}^{(j+1)} = 0$ modulo $\hbar^{(j+2)}$. This completes the induction step.

Remark 5.4.5 Note that in our case D is a deformation of the natural flat connection D_0 on sections of the bundle of infinite jets. Thus, the hypothesis of Lemma 5.4.4

is indeed satisfied and we can find a flat connection $\bar{D}$ which is still a satisfying deformation of D_0. In the setting of formal geometry, a flat connection as D_0 is called a *Grothendieck connection* [Gro68, Kat70] which is a certain generalization of the *Gauss–Manin connection* [Man58]. Another more technical lemma, which uses the notion of *homological perturbation theory*, actually gives an isomorphism between the algebra of horizontal sections $H_{\bar{D}}^0(E)$ and its non-deformed counterpart $H_{D_0}^0(E_0)$ which is isomorphic to $C^\infty(M)$. This implies the globalization procedure.

Remark 5.4.6 In [Dol05], Dolgushev gave another proof of Kontsevich's formality theorem for general manifolds by using covariant tensors instead of ∞-jets of multidifferential operators and multivector fields which is intrinsically local. In particular, he also formulated the deformation quantization construction on *Poisson orbifolds*.

5.5 Operadic Approach to Formality and Deligne's Conjecture

5.5.1 Operads and Algebras

Definition 5.5.1 (Algebraic Operad) An (algebraic) *operad* (of vector spaces) consists of the following:

(1) A collection of vector spaces $P(n)$, $n \geq 0$,
(2) An action of the symmetric group S_n on $P(n)$ for all n,
(3) An identity element $\mathrm{id}_P \in P(1)$,
(4) Compositions $m_{n_1,\ldots,n_k}$:

$$P(k) \otimes (P(n_1) \otimes \cdots \otimes P(n_k)) \to P(n_1 + \cdots + n_k),$$

for all $k \geq 0$ and $n_1, \ldots, n_k \geq 0$ satisfying a list of axioms.

Remark 5.5.2 We want to obtain the list of axioms for an operad by looking at some examples.

Example 5.5.3 (Endomorphism Operad) Consider the very simple operad $P(n) := \mathrm{Hom}(V^{\otimes n}, V)$ for some vector space V. The action of the symmetric group and the identity element are obvious. The compositions are defined by

$$(m_{n_1,\ldots,n_k}(\phi \otimes (\psi_1 \otimes \cdots \otimes \psi_k)))(v_1 \otimes \cdots \otimes v_{n_1+\cdots+n_k})$$
$$:= \phi(\psi_1(v_1 \otimes \cdots \otimes v_{n_1}) \otimes \cdots \otimes \psi_k(v_{n_1} + \cdots + v_{n_{k-1}+1} \otimes \cdots \otimes v_{n_1+\cdots+n_k})),$$

where $\phi \in P(k) = \mathrm{Hom}(V^{\otimes k}, V)$, $\psi_i \in P(n_i) = \mathrm{Hom}(V^{\otimes n_i}, V)$ for $i = 1, \ldots, k$. This operad is called the *endomorphism operad* of a vector space V.

Remark 5.5.4 We can see from Example 5.5.3 that there should be an assoicativity axiom for multiple compositions, various compatibilities for actions of symmetric groups, and evident relations for compositions including the identity element.

Example 5.5.5 (Associative Operad) Consider the operad $P = \mathsf{Assoc}_1$. The n-th component $P(n) = \mathsf{Assoc}_1(n)$ for $n \geq 0$ is defined as the collection of all universal (functorial) n-linear operations $\mathcal{A}^{\otimes n} \to \mathcal{A}$ of associative algebras $\mathcal{A}$ with unit. The space $\mathsf{Assoc}_1(n)$ has dimension $n!$, and is spanned by the operations

$$a_1 \otimes \cdots \otimes a_n \mapsto a_{\sigma(1)} \cdots a_{\sigma(n)}, \quad \sigma \in S_n.$$

We can identify $\mathsf{Assoc}_1(n)$ with the subspace of free associative unital algebras in n generators consisting of expressions which are multilinear in each generator.

Definition 5.5.6 (Algebra Over an Operad) An *algebra over an operad P* consists of a vector space $\mathcal{A}$ and a collection of multilinear maps $f_n \colon P(n) \otimes \mathcal{A}^{\otimes n} \to \mathcal{A}$ for all $n \geq 0$ satisfying the following axioms:

(1) The map f_n is S_n-equivariant for any $n \geq 0$,
(2) We have $f_1(\mathrm{id}_P \otimes a) = a$ for all $a \in \mathcal{A}$,
(3) All compositions in P map to compositions of multilinear operations on $\mathcal{A}$.

Remark 5.5.7 We often also call an algebra $\mathcal{A}$ over an operad P a *P-algebra*.

Example 5.5.8 The algebra over the operad Assoc_1 is an associative unital algebra. If we replace the 1-dimensional space $\mathsf{Assoc}_1(0)$ by the zero space, we obtain an operad Assoc describing associative algebras possibly without unit.

Example 5.5.9 (Lie Operad) There is an operad Lie such that Lie-algebras are Lie algebras. The dimension of the n-th component $\mathsf{Lie}(n)$ is $(n-1)!$ for $n \geq 0$ and zero for $n = 0$.

Theorem 5.5.10 *Let P be an operad and V a vector space. The free P-algebra* $\mathsf{Free}_P(V)$ *generated by V is naturally isomorphic as a vector space to*

$$\bigoplus_{n \geq 0} (P(n) \otimes V^{\otimes n})_{S_n},$$

where the subscript S_n denotes the quotient space of coinvariants for the diagonal action of the symmetric group.

Remark 5.5.11 The *free* algebra $\mathsf{Free}_P(V)$ is defined by the usual categorical *adjunction property*, i.e. the set $\mathrm{Hom}_{P\text{-algebras}}(\mathsf{Free}_P(V), \mathcal{A})$ (i.e. homomorphisms in the category of P-algebras) is naturally isomorphic to the set $\mathrm{Hom}_{\text{vector spaces}}(V, \mathcal{A})$ for any P-algebra $\mathcal{A}$.

5.5.2 Topological Operads

We can replace vector spaces by *topological spaces* in the definition of an operad.
The tensor product is then replaced by the Cartesian product.

Definition 5.5.12 (Topological Operad) A *topological operad* consists of the
following:

(1) A collection of topological spaces $P(n)$ for $n \geq 0$,
(2) A continuous action of the symmetric group S_n on $P(n)$ for all n,
(3) An identity element $\mathrm{id}_P \in P(1)$,
(4) Compositions $m_{n_1,\ldots,n_k}$:

$$P(k) \times (P(n_1) \times \cdots \times P(n_k)) \rightarrow P(n_1 + \cdots + n_k),$$

which are continuous maps for all $k \geq 0$ and $n_1, \ldots, n_k \geq 0$ satisfying a list of
axioms similarly to the ones in Definition 5.5.1.

Example 5.5.13 (Topological Version of Endomorphism Operad) The analog version to the endopmorphism operad is the operad P such that for any $n \geq 0$, we get
a topological space $P(n)$ which is the space of continuous maps from $X^n \rightarrow X$,
where X is some compact topological space.

Remark 5.5.14 In general, one can define an operad and an algebra over an operad
in any arbitrary symmetric monoidal category $(\mathscr{C}, \otimes, \mathbf{1})$ (see Sect. 2.11).
   In particular, we want to consider the symmetric monoidal category **Complexes**
of $\mathbb{Z}$-*graded cochain complexes* of abelian groups (or vector spaces over some field).
Such operads are called *differential graded operads* (or also *dg-operads*). Note that
each component $P(n)$ is a cochain complex, i.e. a vector space decomposed into
a direct sum $P(n) = \bigoplus_{i \in \mathbb{Z}} P(n)^i$ and endowed with a differential $\mathrm{d} \colon P(n)^i \rightarrow$
$P(n)^{i+1}$ which is of degree $+1$ satisfying $\mathrm{d}^2 = 0$.

Remark 5.5.15 We can construct an operad of cochain complexes out of a topological operad by considering a version of the *singular chain complex*. For a
topological space X we denote by $\mathsf{Chains}(X)$ the complex concentrated in negative
degrees, whose $(-k)$-th component for $k \geq 0$ consists of the formal finite additive
combinations

$$\sum_{1 \leq i \leq N} n_i f_i, \quad n_i \in \mathbb{Z},\ N \in \mathbb{Z}_{\geq 0}.$$

of continuous maps $f_i \colon [0, 1]^k \rightarrow X$ (*singular cubes in X*) modulo the following
relations:

(1) $f \circ \sigma = \mathrm{sign}(\sigma) f$ for any $\sigma \in S_k$ acting on the standard cube $[0, 1]^k$ by
    permutations of coordinates,

(2) $f' \circ \mathrm{pr}_{k \to k-1} = 0$, where $\mathrm{pr}_{k \to k-1} \colon [0, 1]^k \to [0, 1]^{k-1}$ is the projection onto the first $(k - 1)$ coordinates and $f' \colon [0, 1]^{k-1} \to X$ is a continuous map.

The boundary operator is defined similarly as for singular chains. Note that, in contrast to singular chains, for cubical chains we have an external product map

$$\bigotimes_{i \in I} \mathsf{Chains}(X_i) \to \mathsf{Chains}\left(\prod_{i \in I} X_i\right).$$

If P is a topological operad, then $\mathsf{Chains}(P(n))$ for $n \geq 0$ has a natural operad structure in the category of complexes of abelian groups. The compositions are then given in terms of the external tensor product of cubical chains. On the level of cohomology, we obtain an operad $H(P)$ of $\mathbb{Z}$-graded abelian groups which are complexes endowed with the zero differential. This is called the *homology* of the operad P.

5.5.3   The Little Disks Operad

Let $d \geq 1$ and denote by G_d the $(d + 1)$-dimensional Lie group acting on $\mathbb{R}^d$ by affine transformations $u \mapsto \lambda u + v$ where $\lambda > 0$ is a real number and $v \in \mathbb{R}^d$. This group acts simply and transitively on the space of closed disks in $\mathbb{R}^d$ endowed with the usual Euclidean metric. The disk with center v and radius λ is given by a transformation of G_d with parameters (λ, v) of the unit disk

$$D_0 := \{(x_1, \dots, x_d) \in \mathbb{R}^d \mid x_1^2 + \cdots + x_d^2 \leq 1\}.$$

Definition 5.5.16 (Little Disks Operad) The *little disks operad* C_d is a topological operad with the following structure:

(1) $C_d(0) = \varnothing$,
(2) $C_d(1) = \{\mathrm{id}_{C_d}\}$,
(3) The space $C_d(n)$ is the space of configurations of n disjoint disks $(D_i)_{1 \leq i \leq n}$ inside the standard disk D_0 for $n \geq 2$.

The composition

$$C_d(k) \times (C_d(n_1) \times \cdots \times C_d(n_k)) \to C_d(n_1 + \cdots + n_k)$$

is obtained by applying elements from G_d associated with disks $(D_i)_{1 \leq i \leq n}$ in the configuration in $C_d(k)$ to configurations in all $C_d(n_i)$, $i = 1, \dots, k$ and putting the resulting configurations together. The action of the symmetric group S_n on $C_d(n)$ is given by renumerations of indices of disks $(D_i)_{1 \leq i \leq n}$ (see Fig. 5.12).

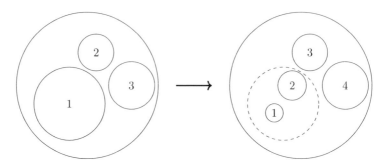

Fig. 5.12 Example for the compositions of little disks

Remark 5.5.17 Note that $C_d(n)$ is homotopy equivalent to the configuration space $\mathrm{Conf}_n(\mathbb{R}^d)$ of n pairwise distinct points in $\mathbb{R}^d$. Obviously, we can consider the map $C_d(n) \to \mathrm{Conf}_n(\mathbb{R}^d)$ which takes a collection of disjoint disks to the collection of the centers. Note that in particular, we have that $\mathrm{Conf}_2(\mathbb{R}^d)$ (and hence $C_2(n)$) is homotopy equivalent to the $(d-1)$-dimensional sphere S^{d-1}. The homotopy equivalence is given by the map

$$(v_1, v_2) \mapsto \frac{v_1 - v_2}{\|v_1 - v_2\|} \in S^{d-1} \subset \mathbb{R}^d.$$

Remark 5.5.18 Since we are abusing notation, one should not confuse the operad C_d with the configuration space C_n as in Remark 5.3.7.

5.5.4 Deligne's Conjecture

Recall that the Hochschild complex of an associative algebra $\mathcal{A}$ concentrated in positive degree is given by

$$C^\bullet(\mathcal{A}, \mathcal{A}) := \bigoplus_{n \geq 0} C^n(\mathcal{A}, \mathcal{A}), \quad C^n(\mathcal{A}, \mathcal{A}) := \mathrm{Hom}(\mathcal{A}^{\otimes n}, \mathcal{A})$$

and the Hochschild differential $\mathsf{d_m}$ is given by the formula

$$(\mathsf{d_m}\phi)(a_1 \otimes \cdots \otimes a_{n+1}) := a_1 \phi(a_2 \otimes \cdots \otimes a_n)$$

$$+ \sum_{1 \leq i \leq n} (-1)^i \phi(a_1 \otimes \cdots \otimes \mathsf{m}(a_i \otimes a_{i+1}) \otimes \cdots \otimes a_{n+1}) + \cdots$$

$$\cdots + (-1)^{n+1} \mathsf{m}(\phi(a_1 \otimes \cdots \otimes a_n) \otimes a_{n+1}), \quad \forall \phi \in C^n(\mathcal{A}, \mathcal{A})$$

for some multiplication map $\mathsf{m} \in C^1(\mathcal{A}, \mathcal{A})$.

Conjecture 5.5.19 (Deligne) *There exists a natural action of the operad* $\mathsf{Chains}(C_2)$ *on the Hochschild complex* $C^\bullet(A, A)$ *for any associative algebra* A.

Remark 5.5.20 The proof of this conjecture was given by a combination of results of [Tam98, Tam03] and (a higher version) also in [Kon99]. See also [MS99, KS00].

5.5.5 Formality of Chain Operads and Relation to Deformation Quantization

Theorem 5.5.21 (Kontsevich–Tamarkin[Tam03, Kon99]) *The operad*

$$\mathsf{Chains}(C_d) \otimes \mathbb{R}$$

of complexes of real vector spaces is quasi-isomorphic to its cohomology operad endowed with the zero differential.

Sketch of the Proof Fix a dimension $d \geq 1$. According to the Fulton–MacPherson compactification as discussed in Remark 5.3.13, we can consider a modification of the topological operad C_d which we will denote by FM_d (for Fulton–MacPherson) and call it the *Fulton–MacPherson operad*. The components of the operad FM_d are given as follows:

(1) $FM_d(0) := \varnothing$,
(2) $FM_d(1) = \mathrm{pt}$,
(3) $FM_d(2) = \tilde{C}_d(2) = S^{d-1}$, where $\tilde{C}_d(n)$ denotes the quotie nt space of the configuration space $\mathrm{Conf}_n(\mathbb{R}^d)$ of n points in $\mathbb{R}^d$ modulo $G_d = \{x \mapsto ax + b \mid a \in \mathbb{R}_{\geq 0}, v \in \mathbb{R}^d\}$. Note that the space $\tilde{C}_d(n)$ is a smooth manifold of dimension $nd - d - 1$.
(4) for $n \geq 3$, the space $FM_d(n)$ is a manifold with corners with interior given by $C_d(n')$ and all boundary strata are products of copies of $\tilde{C}_d(n')$ for $n' < n$.

More rigorously, for $n \geq 2$, we can define the manifold with corners $FM_d(n)$ as the closure of the image of $\tilde{C}_d(n)$ in the compact manifold $(S^{d-1})^{\frac{n(n-1)}{2}}$ of the map

$$G_d \cdot (x_1, \ldots, x_n) \mapsto \left(\frac{x_j - x_i}{\|x_j - x_i\|} \right)_{1 \leq i < j \leq n}.$$

One can then define another topological operad FM'_d with two mophisms

$$f_1 \colon C_d \to FM'_d,$$
$$f_2 \colon FM_d \to FM'_d$$

such that for all $n \geq 0$, the maps

$$f_1(n) \colon C_d(n) \to FM'_d(n),$$
$$f_2(n) \colon FM_d(n) \to FM'_d(n)$$

are homotopy equivalences. If we use the functor Chains, we get two quasi-isomorphims of operads of complexes

$$\mathsf{Chains}(C_d) \to \mathsf{Chains}(FM'_d),$$

$$\mathsf{Chains}(FM_d) \to \mathsf{Chains}(FM'_d).$$

If we have $d \geq 2$, we can define operads of complexes and construct a chain of quasi-isomorphisms between them. In particular, we get the following total diagram:

$$\mathsf{Chains}(FM_d) \leftarrow \mathsf{SemiAlgChains}(FM_d),$$

$$\mathsf{SemiAlgChains}(FM_d) \otimes \mathbb{R} \to \mathsf{Graphs}_d \widehat{\otimes} \mathbb{R},$$

$$\mathsf{Graphs}_d \to \mathsf{Forests}_d \leftarrow H_\bullet(\mathsf{Forests}) = H_\bullet(C_d).$$

$$(5.77)$$

Let us briefly look at the ingredients of the total diagram (5.77). The operad $\mathsf{SemiAlgChains}(FM_d)$ is the suboperad of $\mathsf{Chains}(FM_d)$ consisting of combinations of maps $[0, 1]^d \to FM_d(n)$ with graphs being real semi-algebraic sets. This is indeed well-defined since $FM_d(n)$ can be obtained by algebraic equations and inequalities. Moreover, it is a closed semi-algebraic subset in $(S^d)^{\frac{n(n-1)}{2}}$ and the inclusion of semi-algebraic chains into all continuous chains is a quasi-isomorphism of operads.

Let us now define the admissible graphs (slightly different than the ones in Definition 5.3.2 though) to be graphs with parameters (n, m, k), for $n \geq 1$ and $m \geq 0$, to be finite graphs Γ such that it has no multiple edges, it contains no short loops (tadpoles), it contains $n + m$ vertices numbered from 1 to $n + m$, it contains k edges numbered from 1 to k, any vertex can be connected by a path with a vertex whose label lies in $\{1, \ldots, n\}$, any vertex with label in $\{n + 1, \ldots, n + m\}$ has valency ≥ 3, for every edge e of Γ we choose an orientation, i.e. the two-element set on which the edges are based on is ordered and finally, if $n = 1$, then the graph Γ consists of just one vertex and has no edges and the parameters are $(1, 0, 0)$.

We define now, for an admissible graph Γ, the differential form ω_Γ on $FM_d(n)$ as

$$\omega_\Gamma := (\pi_1)_* \circ \pi_2^* \left(\bigwedge_{e \in E(\Gamma)} d\mathrm{Vol}_{S^{d-1}} \right),$$

where $\pi_1 \colon FM_d(n + m) \to FM_d(n)$ is the projection map which forgets about the last m points in the configuration of $n + m$ points in $\mathbb{R}^d$ and $\pi_2 \colon FM_d(n + m) \to (FM_d(2))^k$ is the product of forgetful maps $FM_d(n + m) \to FM_d(2) = S^{d-1}$ associated with the edges of Γ and $d\mathrm{Vol}_{S^{d-1}}$ is the volume form on S^{d-1} which is invariant under the action of $SO_d(\mathbb{R})$ such that $\int_{S^{d-1}} d\mathrm{Vol}_{S^{d-1}} = 1$. Note that the form-degree of ω_Γ is given by

$$(d-1)k - dm = \dim F M_d(2)^k - (\dim F M_d(n+m) - \dim F M_d(n)).$$

For each $n \geq 1$, we define $\mathsf{Graphs}_d(n)$ to be the $\mathbb{Z}$-graded vector space of rational-valued functions on the set of equivalence classes of admissible graphs with parameters (n, m, k) with m and k such that if we change the labeling of edges, or the labeling of the vertices with labels from $n + 1$ to $n + m$, the value of the function will be multiplied by an appropriate sign. The $\mathbb{Z}$-grading of a function, concentrated on one given equivalence class $[\Gamma]$ with parameters (n, m, k) is defined to be $dm - (d-1)k$.

Remark 5.5.22 One can check that the forms ω_Γ are not smooth on the boundary of $F M_d(n)$ but still the integral $\int_C \omega_\Gamma$ is absolutely convergent for any semi-algebraic chain C. This is true since the computation of the integral reduces to computing the volume form over a compact semi-algebraic chain of top degree in $(S^d)^{\frac{n(n-1)}{2}}$. Since the multiplicity of a semi-algebraic map is bounded, the total volume will be finite. Hence, the integral of the volume form is convergent.

By Remark 5.5.22 we get a well-defined functional on $\mathsf{SemiAlgChains}(F M_d)$ induced by the form ω_Γ. This means that any semi-algebraic chain gives a functional on the set of equivalence classes of admissible graphs which is $\mathbb{R} \widehat{\otimes} \mathsf{Graphs}_d(n)$.

Using a more general version of Lemma 5.3.21 for $d \geq 3$, also proved by Kontsevich, we can see that the graded space $\mathsf{Graphs}_d(n)$, consisting of functions on graphs, is endowed with a naturally defined differential and thus defines a complex of $\mathbb{Q}$-vector spaces. Also, the restrictions of the forms ω_Γ to irreducible components of the boundary of the manifold $F M_d(n)$ are finite linear combinations of products of similar forms for simpler graphs. In particular, $\mathsf{Graphs}_d(n)$ is an operad of complexes and integration defines a morphism of operads

$$\mathbb{R} \otimes \mathsf{SemiAlgChains}(F M_d) \to \mathbb{R} \widehat{\otimes} \mathsf{Graphs}_d. \tag{5.78}$$

Theorem 5.5.23 (Kontsevich[Kon99]) *The morphism of operads* (5.78) *is a quasi-isomorphism.*

Definition 5.5.24 (Forest) An admissible graph Γ is called a *forest* if and only if it contains no non-trivial closed paths.

Definition 5.5.25 (Forest Operad) The operad $\mathsf{Forests}_d$ is a suboperad of Graphs_d consisting of functions vanishing on all non-forest graphs.

Remark 5.5.26 Note that if Γ is not a forest, we can write $d\omega_\Gamma$ in terms of a linear combination of forms $\omega_{\Gamma'}$ with Γ' not being a forest. Moreover, when restricting ω_Γ to irreducible components of the boundary of $F M_d(n)$ is a linear combination of products $\omega_{\Gamma_1} \times \omega_{\Gamma_2}$ where at least one of the graphs Γ_1, Γ_2 is not a forest. This makes sure that Definition 5.5.25 indeed defines an operad.

Remark 5.5.27 There is a *spectral sequence* (see [BT82] for a detailed discussion of spectral sequences) which shows that $\mathsf{Forests}_d \to \mathsf{Graphs}_d$ is actually a quasi-isomorphism. This can be seen as follows: Note that any non-forest graph contains a non-empty maximal subgraph Γ_{core} such that the valency of each vertex of Γ_{core} is at least two. We can obtain the graph Γ out of Γ_{core} by attaching trees to vertices of Γ_{core} and also a forest which is not connected with Γ_{core}. The result is then a consequence of the vanishing of the first term of the spectral sequence associated with the filtration by the number of vertices in Γ_{core} on the quotient complex $\mathsf{Graphs}_d(n)/\mathsf{Forests}_d(n)$.

Note that any admissible graph Γ with parameters (n, m, k) gives a partition of $\{1, \ldots, n\}$ into parts which correspond to the connected components of Γ. All graphs Γ' which appear in the decomposition of $d\omega_\Gamma$ will give the same partition of $\{1, \ldots, n\}$ as Γ. Thus, the complex $\mathsf{Forests}_d(n)$ splits naturally for any n into a direct sum of subcomplexes corresponding to partitions of the set $\{1, \ldots, n\}$. The subcomplex $\mathsf{Forests}_d(n)$ associated to a partition is the tensor product of tree-complexes over pieces of this partition. Trees here are non-empty connected forests. The tree-complex is actually the dual complex to the complex given by differential forms ω_Γ associated to trees Γ. However, the tree-complex has non-zero cohomology only in the lowest degree. This means that there is a natural map from the cohomology of the tree-complex to the tree-complex itself. Hence, there is a natural morphism from the cohomology of the forest-complex to the forest-complex itself and this map is a map of operads defining a quasi-isomorphism

$$H^\bullet(\mathsf{Forests}_d) \to \mathsf{Forests}_d.$$

It is then not hard to see that

$$H^\bullet(\mathsf{Forests}_d) = H_\bullet(C_d) = H_\bullet(FM_d).$$

Finally, since we know that $H^\bullet(\mathsf{Graphs}_d) = H^\bullet(\mathsf{Forests}_d)$, we get a chain of quasi-isomorphisms as claimed.

The following theorem, due to Tamarkin, gives the relation of the operadic constructions to deformation quantization.

Theorem 5.5.28 (Tamarkin[Tam03]) *Let $\mathcal{A} := \mathbb{R}[x_1, \ldots, x_n]$ be the algebra of polynomials considered as an associative algebra. Then the Hochschild complex $C^\bullet(\mathcal{A}, \mathcal{A})$ is quasi-isomorphic as a 2-algebra to its cohomology*

$$H^\bullet(C(\mathcal{A}, \mathcal{A})) = \mathrm{Sym}(\mathcal{V}^\bullet(\mathbb{R}^n))$$

considered as a Gerstenhaber algebra, hence a 2-algebra.

Remark 5.5.29 Theorem 5.5.28 implies directly Kontsevich's formality theorem.

Chapter 6
Quantum Field Theoretic Approach to Deformation Quantization

Kontsevich's formula is in its nature a pure algebraic construction. However, the concept of deformation quantization should give a physical quantization procedure which opens the question whether it is possible to naturally extract the star product out of a quantum field theory, i.e. a perturbative *Feynman path integral quantization* [Fey42, Fey49, Fey50, FH65, Pol05]. The appearance of graphs in Kontsevich's construction might already inspire to regard them as Feynman diagrams of a certain field theory. The important field theory for this is given by the *Poisson sigma model* [Ike94, SS94, CF01c], which is a 2-dimensional topological field theory that can be actually regarded as a *bosonic string theory*. In particular, its perturbative functional integral quantization produces the desired outcome. Since the Poisson sigma model is a theory with *symmetries* (also called a *gauge theory*), i.e. there is a Lie group whose action leaves the theory invariant, the usual methods for perturbative evaluation of functional integrals (stationary phase expansion) over critical points of the action functional of the given theory fails and we need to understand how this can be resolved. The first common formalism to treat gauge theories is the method of *Faddeev–Popov ghosts* [FP67]. This procedure can be extended to a graded configuration of fields and cohomological setting, called the *BRST* formalism [BRS74, BRS75, Tyu76]. However, both methods do only work as long as the theory has a natural linear behaviour, which is unfortunately not the case for the general Poisson sigma model, since the perturbation part is highly non-linear in general. Thus, it turns out that the BRST method fails and hence the more sophisticated symplectic cohomological *Batalin–Vilkovisky (BV) formalism* [BV77, BV81, BV83, Sch93, Mne19] is actually needed in order to deal with this particular sigma model. In this chapter we will start by giving a functorial definition of a quantum field theory introduced by Atiyah [Ati88] (for the topological case) and Segal [Seg88] (for the case with geometric structure, usually in the setting of *conformal structure*). We will then describe the mathematical framework for functional integrals and their perturbative evaluation (finite- and infinite-dimensional), construct the Moyal product out of a functional integral

© The Author(s), under exclusive license to Springer Nature Switzerland AG 2022
N. Moshayedi, *Kontsevich's Deformation Quantization and Quantum Field Theory*,
Lecture Notes in Mathematics 2311, https://doi.org/10.1007/978-3-031-05122-7_6

quantization, describe the Faddeev–Popov ghost method and the BRST formalism in order to deal with the Poisson sigma model. At first, we will explicitly show how the perturbative quantization of the Poisson sigma model for the easier case of *linear* Poisson structures produces Kontsevich's star product which only requires the BRST formalism. We then introduce the more complicated machinery of odd symplectic supermanifolds and the Batalin–Vilkovisky gauge formalism in order to discuss the situation of a general Poisson structure on $\mathbb{R}^d$. Finally, we state the Cattaneo–Felder theorem [CF00, CF01c] for general Poisson structures on $\mathbb{R}^d$, which tells us how the perturbative quantization of the Poisson sigma model yields Kontsevich's star product on $\mathbb{R}^d$ and sketch the proof by using the previously developed Batalin–Vilkovisky formalism. This chapter is based on [Zin94, Pol05, Cat+05, CF00, CF01c, CF01b, CF01a, CF01d, Kon94, Mne19, Fed96, BV77, BV81, BV83, Ale+97, FH65, Ati88, Seg88].

6.1 The Atiyah–Segal Definition of a Functorial Quantum Field Theory

The definition of a general quantum field theory turns out to be a rather difficult issue since the mathematical formulation of such does indeed very heavily depend on the different structures appearing in the given type of theory. Moreover, there are different (but equivalent) ways of how one can approach this problem. One way is from the point of view of category theory (see Sect. 2.11), usually called *functorial field theory*. Another way is from the point of view of functional (Feynman path) integrals, usually called *perturbative field theory*. The first approach is the subject of this section, whereas the second approach will be the relevant approach for the relation to deformation quantization.

In [Ati88], Atiyah formulated a set of axioms in order to give a mathematically rigorous definition for a topological quantum field theory.

We need to define a category which will model the space-time manifold for a field theory. This category will be called the category of *n-cobordisms*, denoted by $\mathbf{Cob}_n$. Its objects are given by closed $(n-1)$-manifolds equipped with an additional orientation structure which we call "in" and "out". The morphisms are given by the bounding n-manifolds between them (see Fig. 6.1). The composition of morphisms is given by gluing of $(n-1)$-manifols (see Fig. 6.2). It is easy to see that the category $\mathbf{Cob}_n$ is a symmetric monoidal category together with the symmetric monoidal product $\sqcup$ given by disjoint union of topological spaces.

Definition 6.1.1 (Topological Field Theory) Let $(\mathscr{C}, \otimes_\mathscr{C}, \mathbf{1}_\mathscr{C})$ be any symmetric monoidal category. An *n-dimensional oriented closed topological field theory (TFT)* is a symmetric monoidal functor

$$Z : (\mathbf{Cob}_n, \sqcup) \to (\mathscr{C}, \otimes_\mathscr{C}). \tag{6.1.1}$$

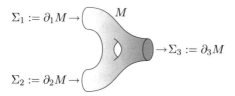

Fig. 6.1 A cobordism M represented by a pair of pants of genus 1 with boundary components $\Sigma_1, \Sigma_2, \Sigma_3$. The "in" and "out" structure of the objects is indicated with an incoming arrow (for "in"-boundary components) and an outgoing arrow (for "out"-boundary components)

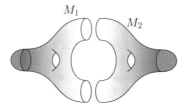

Fig. 6.2 Gluing of two cobordisms M_1 and M_2 along a common boundary Σ

Definition 6.1.2 (Topological Quantum Field Theory) An n-dimensional oriented topological quantum field theory (TQFT) is a symmetric monoidal functor

$$Z : (\mathbf{Cob}_n, \sqcup) \to (\mathbf{Vect}_{\mathbb{C}}, \otimes_{\mathbb{C}}). \tag{6.1.2}$$

Remark 6.1.3 Note that the target category contains also infinite-dimensional vector spaces. However, one can actually show that the state spaces have to be finite-dimensional.

Example 6.1.4 Let us consider the cobordism represented by some pair of pants with genus 1 as in Fig. 6.1. The TQFT Z assigns to each boundary component a vector space, i.e. $Z(\Sigma_k) = \mathcal{H}_k$ for $k = 1, 2, 3$. Since Z is a symmetric monoidal functor, we have $Z(\Sigma_1 \sqcup \Sigma_2 \sqcup \Sigma_3) = \mathcal{H}_1^* \otimes_{\mathbb{C}} \mathcal{H}_2^* \otimes_{\mathbb{C}} \mathcal{H}_3$. As said before, each cobordism comes with a certain orientation: Σ_1 as well as Σ_2 are incoming boundaries, while Σ_3 is an outgoing boundary. Associated to Σ_1 and Σ_2, we have an incoming vector space $\mathcal{H}_{\mathrm{in}} := \mathcal{H}_1^* \otimes_{\mathbb{C}} \mathcal{H}_2^* \cong \mathcal{H}_1 \otimes_{\mathbb{C}} \mathcal{H}_2$ and an outgoing vector space $\mathcal{H}_{\mathrm{out}} := \mathcal{H}_3$ associated to Σ_3. The state ψ corresponding to this cobordism and the given TQFT is given as the value of the morphism represented by the genus 1 pair of pants above (i.e. the bounding manifold M) under Z. In particular, it is given by a morphism $Z(M) \in \mathrm{Hom}(\mathcal{H}_1^* \otimes_{\mathbb{C}} \mathcal{H}_2^* \otimes_{\mathbb{C}} \mathcal{H}_3) \cong \mathcal{H}_1 \otimes_{\mathbb{C}} \mathcal{H}_2 \otimes_{\mathbb{C}} \mathcal{H}_3$.

Remark 6.1.5 The category of smooth oriented cobordisms is usually used to describe a TQFT. However, cobordisms may carry other geometric structure such as *conformal structure, spin structure, framing, boundaries*, etc. Consequently, the associated field theory will be a conformal QFT, spin or framed TQFT, etc. For example, for Yang-Mills theories and sigma models, the source category is the category of smooth Riemannian manifolds with a collar at the boundary. For the case where the cobordism is equipped with a geometric structure, Segal gave a way of defining a functorial field theory in the conformal setting in [Seg88]. We will not consider this here since the field theory we are interested in will be purely topological, but we refer to the original reference for the interested reader.

Example 6.1.6 A closed manifold M can be regarded as a cobordism $\varnothing \to \varnothing$. We can cut it in two disjoint manifolds M_1 and M_2 along a common boundary Σ, i.e. $M = M_1 \sqcup_\Sigma M_2$. Then we can assign an opposite orientation to $\partial_1 M_1$ with respect to the orientation of $\partial_1 M_2$. The same can be done to $\partial_2 M_1$ with respect to the orientation of $\partial_2 M_2$. The two manifolds with boundary M_1 and M_2 can be glued back together to recover the partition function of the closed manifold M.

Remark 6.1.7 It is important to highlight that the functorial approach to TQFT is not based on any perturbative framework, therefore its nature is intrinsically non-perturbative. Beyond dimension two, the degree of complexity increases tremendously: for $n \geq 3$ an explicit classification is not known yet.

Remark 6.1.8 (Higher Categories) Successively, in [BD95], Baez and Dolan suggested to extend Atiyah's notion of TQFT to a functor from the (∞, n)-extension of the cobordism category (see [Lur17] for the definition and concepts of higher category theory). The idea here is to allow gluing as well as cutting with higher codimension data. Moreover, they conjectured that these TQFTs can be completely classified. This conjecture is known today as the *cobordism hypothesis*. In [Lur09], Lurie provided a complete classification result for *fully extended* TQFTs formulated in the language of (∞, n)-categories, a higher generalization of the notion of a category.

Remark 6.1.9 The categorical (Functorial) approach to TQFT gives insights and relations to many aspects of low-dimensional topology and the concept of *quantum groups* which we will not touch here. However, we want to mention the important work of Reshetikhin and Turaev providing, given a certain type of category, a 3-dimensional TQFT which induces a topological invariant of the underlying 3-manifold [RT91]. This is a famous construction using methods of quantum groups and polynomial invariants of knots and links.

6.2 Feynman Path Integrals and Perturbative Quantum Field Theory

6.2.1 *Functional Integrals and Expectation Values*

For a perturbative field theory construction we want to consider a *space of fields* $\mathcal{M}$ which in most cases is given by the space of sections for some vector bundle and an *action function* $S\colon \mathcal{M} \to \mathbb{R}$. The action is usually given *locally* (see also

Definition 6.3.24), i.e., roughly, as an integral over a density $\mathscr{L}$ on some manifold M. Namely, it is of the form $S(\phi) = \int_M \mathscr{L}(\phi, \partial\phi, \ldots, \partial^N\phi)$ for $N \in \mathbb{Z}_{>0}$.

Definition 6.2.1 (Expectation Value) For a function $\mathcal{O}$ on M we define the *expectation value*

$$\langle \mathcal{O} \rangle := \frac{\displaystyle\int_M \exp\left(iS/\hbar\right) \mathcal{O}}{\displaystyle\int_M \exp\left(iS/\hbar\right)}.$$

Definition 6.2.2 (Observable) An *observable* is a function $\mathcal{O}$ on M whose expectation value is well-defined.

Remark 6.2.3 Integrals as in Definition 6.2.1 are called *functional integrals* or *path integrals*. One can consider a perturbative evaluation of such integrals by expanding S around a non-degenerate critical point and defining the integral as a formal power series in $\hbar$ with coefficients given by Gaussian expectation values. If S carries certain *symmetries*, the critical points will always be degenerate. In this case we speak of a *gauge theory*. There are different methods to deal with such theories such a the *Faddeev–Popov ghost method* [FP67], the *BRST method* [BRS74, BRS75, Tyu76] and the *Batalin–Vilkovisky method* [BV77, BV81, BV83]. Interestingly, it was shown that the field theoretic construction of Kontsevich's star product for general Poisson structures requires the Batalin–Vilkovisky formalism to deal with the gauge theory given by the Poisson sigma model [CF00].

6.2.2 Gaussian Integrals

Let A be a positive-definite symmetric matrix on $\mathbb{R}^n$ and assume that it is endowed with the Lebesgue measure $d^n x$ and the Euclidean inner product $\langle \, , \, \rangle$ (note that n has to be even). Then

$$I(\lambda) := \int_{\mathbb{R}^n} \exp\left(-\frac{\lambda}{2}\langle x, Ax \rangle\right) d^n x = \frac{(2\pi)^{\frac{n}{2}}}{\lambda^{\frac{n}{2}}} \frac{1}{\sqrt{\det A}}, \quad \lambda > 0.$$

If we continue I to the whole complex plane without the negative real axis, we get

$$I(-i) = (2\pi)^{\frac{n}{2}} \exp\left(\frac{i\pi n}{4}\right) \frac{1}{\sqrt{\det A}}, \qquad I(i) = (2\pi)^{\frac{n}{2}} \exp\left(\frac{-i\pi n}{4}\right) \frac{1}{\sqrt{\det A}}.$$

Thus, when A is negative-definite, we can define the integral by

$$\int_{\mathbb{R}^n} \exp\left(\frac{i}{2}\langle x, Ax \rangle\right) d^n x = \int_{\mathbb{R}^n} \exp\left(-\frac{i}{2}(-\langle x, Ax \rangle)\right) d^n x = (2\pi)^{\frac{n}{2}} \exp\left(\frac{-i\pi n}{4}\right) \frac{1}{\sqrt{|\det A|}}.$$

For the case when A is non-degenerate (not necessarily positive- or negative-definite), we get

$$\int_{\mathbb{R}^n} \exp\left(\frac{i}{2}\langle x, Ax\rangle\right) d^n x = (2\pi)^{\frac{n}{2}} \exp\left(\frac{i\pi \operatorname{sign} A}{4}\right) \frac{1}{\sqrt{|\det A|}},$$

where sign A denotes the *signature* of A. From now on we will use the notation that when $A = (A_{ij})$ then $A^{-1} = (A^{ij})$.

We want to compute expectation values with respect to a Gaussian distribution. Let us denote such an expectation value by $\langle\ ,\ \rangle_0$. Define first the generating function

$$Z(J) := \int_{\mathbb{R}^n} \exp\left(\frac{i}{2}\langle x, Ax\rangle + \frac{i}{2}\langle J, x\rangle\right) d^n x$$

$$= (2\pi)^{\frac{n}{2}} \exp\left(\frac{i\pi \operatorname{sign} A}{4}\right) \frac{1}{\sqrt{|\det A|}} \exp\left(\frac{i}{2}\langle J, A^{-1} J\rangle\right).$$

Then we get

$$\langle x^{i_1} \cdots x^{i_k}\rangle_0 = \frac{\displaystyle\int_{\mathbb{R}^n} \exp\left(\frac{i}{2}\langle x, Ax\rangle\right) x^{i_1} \cdots x^{i_k} d^n x}{\displaystyle\int_{\mathbb{R}^n} \exp\left(\frac{i}{2}\langle x, Ax\rangle\right) d^n x}$$

$$= \frac{\frac{\partial}{\partial J^{i_1}} \cdots \frac{\partial}{\partial J^{i_k}} Z(J)|_{J=0}}{Z(0)} = \frac{\partial}{\partial J^{i_1}} \cdots \frac{\partial}{\partial J^{i_k}} \exp\left(\frac{i}{2}\langle J, A^{-1} J\rangle\right)\Bigg|_{J=0}.$$

Remark 6.2.4 Note that $\langle x^{i_1} \cdots x^{i_k}\rangle_0$ vanishes if k is odd and is a sum of products of matrix elements of the inverse of A if k is even. For example, if $k = 2$ we have $\langle x^i x^j\rangle_0 = iA^{ij}$ and if $k = 2s$, we get

$$\langle x^{i_1} \cdots x^{i_{2s}}\rangle_0 = i^s \sum_{\sigma \in S_{2s}} \frac{1}{2^s s!} A^{i_{\sigma(1)} i_{\sigma(2)}} \cdots A^{i_{\sigma(2s-1)} i_{\sigma(2s)}}.$$

Theorem 6.2.5 (Wick) *Denote by* $P(s)$ *the set of* pairings, *i.e. permutations* $\sigma \in S_{2s}$ *with the property that* $\sigma(2i - 1) < \sigma(2i)$ *for* $i = 1, \ldots, s$ *and* $\sigma(1) < \sigma(3) < \cdots < \sigma(2s - 3) < \sigma(2s - 1)$. *Then we have*

$$\langle x^{i_1} \cdots x^{i_{2s}}\rangle_0 = i^s \sum_{\sigma \in P(s)} A^{i_{\sigma(1)} i_{\sigma(2)}} \cdots A^{i_{\sigma(2s-1)} i_{\sigma(2s)}}. \tag{6.2.1}$$

6.2.2.1 Green's Functions

A powerful method of solving certain linear differential equations with unique solutions defined through appropriate boundary conditions is given by the method of *Green's functions*. In particular, we want to solve inhomogeneous linear differential equations of the form

$$Dg = f,$$

for smooth functions f, g, where D is some linear differential operator. This can be done by studying the *integral kernel* representing the inverse operator D^{-1}. We can express D^{-1} by

$$(D^{-1})_{x,y} = G(x, y),$$

such that

$$D_x G(x, y) = \delta(x - y),$$

where D_x means that it acts only on the first argument of G. Then we have

$$g(x) = \int G(x, y) f(y) \mathrm{d}y$$

and hence

$$D_x g = \int D_x G(x, y) f(y) \mathrm{d}y = \int \delta(x - y) f(y) \mathrm{d}y = f(x).$$

There are several ways of how to construct the Green's function $G(x, y)$. The following points have to be satisfied:

(1) The function $G^y(x) := G(x, y)$ is discontinuous on the diagonal, i.e. for $x = y$, since it has to generate the delta function.
(2) Away from the diagonal, we need to have $DG^y = 0$,
(3) The function $G^y(x)$ must obey the homogeneous boundary conditions which hold for g on the boundary of the domain.

Example 6.2.6 (Laplacian) The Green's function for the Laplacian $\Delta = \frac{\partial^2}{\partial x^2} + \frac{\partial^2}{\partial y^2} + \frac{\partial^2}{\partial z^2}$ in $\mathbb{R}^3$ without boundary conditions is given by

$$G(\mathbf{x}, \mathbf{x}') = -\frac{1}{4\pi} \frac{1}{\|\mathbf{x} - \mathbf{x}'\|}, \quad \mathbf{x}, \mathbf{x}' \in \mathbb{R}^3.$$

6.2.2.2 Infinite-Dimensional Case

By (6.2.1), the infinite-dimensional case only makes sense if A is invertible. However, usually A will be a differential operator and thus $G := A^{-1}$ will denote the distributional kernel of its inverse, i.e. its *Green's function*. So if e.g. A is a differential operator on functions on some manifold Σ, we get

$$
\langle \phi(x_1) \cdots \phi(x_{2s}) \rangle_0 = \frac{\displaystyle\int \exp\left(\frac{i}{2\hbar}\int_\Sigma \phi A\phi\right) \phi(x_1)\cdots\phi(x_{2s}) \mathscr{D}[\phi]}{\displaystyle\int \exp\left(\frac{i}{2\hbar}\int_\Sigma \phi A\phi\right) \mathscr{D}[\phi]}
$$

$$
:= (i\hbar)^s \sum_{\sigma \in P(s)} G\big(x_{\sigma(1)}, x_{\sigma(2)}\big) \cdots G\big(x_{\sigma(2s-1)}, x_{\sigma(2s)}\big),
$$

$$(6.2.2)$$

where ϕ denotes a function on Σ, $\mathscr{D}[\phi]$ denotes the *formal Lebesgue measure* on the space of functions and $x_1, \ldots, x_{2s}$ are distinct points in Σ.

Remark 6.2.7 (Normal Ordering) Equation (6.2.2) is usually extended to the singular case when points coincide by restricting the sum to pairings with the property that $x_{\sigma(2i-1)} \neq x_{\sigma(2i)}$ for all i. This is called *normal ordering* since it corresponds to the usual normal ordering in the operator formulation of Gaussian field theories.

Remark 6.2.8 (Propagator) For Gaussian integrals all expectation values are expressed through the expectation value of the quadratic monomials. These are usually called *2-point functions* or *propagators*. For the case of a Gaussian field theory defined by a differential operator, the word propagator is just another name for the Green's function. We will usually denote propagators either by G or by $\mathscr{P}$.

Remark 6.2.9 (Derivatives) One can extend the definition of expectation values to derivatives of fields by linearity. In particular, for multi-indices $I_1, \ldots, I_{2s}$ we set

$$
\langle \partial_{I_1}\phi(x_1) \cdots \partial_{I_{2s}}\phi(x_{2s}) \rangle_0 = (i\hbar)^s \frac{\partial^{|I_1|}}{\partial x_1^{I_1}} \cdots \frac{\partial^{|I_{2s}|}}{\partial x_{2s}^{I_{2s}}} \sum_{\sigma \in P(s)} G\big(x_{\sigma(1)}, x_{\sigma(2)}\big) \cdots G\big(x_{\sigma(2s-1)}, x_{\sigma(2s)}\big),
$$

$$(6.2.3)$$

where the derivatives on the right-hand-side are meant in the sense of distributions.

6.2.3 Integration of Grassmann Variables

Let V be a vector space. Recall that the exterior algebra $\bigwedge V^*$ is regarded as the algebra of functions on the odd vector space ΠV (recall the definition of ΠV from Sect. 5.2.7). Choosing a basis and orientation, we can identify $\bigwedge^{\text{top}} V^*$ with $\mathbb{R}$. The composition of this isomorphism with the projection $\bigwedge V^* \to \bigwedge^{\text{top}} V^*$ gives a map

$\bigwedge V^* \to \mathbb{R}$ which will be denoted by $\int_{\Pi V}$ and we call it the *integral on* ΠV. Let $B \in \mathrm{End}(V)$ and regard it as an element of $V^* \otimes V$ and thus as a function on

$$\Pi V^* \times \Pi V := \Pi(V^* \oplus V).$$

There is a natural identification of $\bigwedge^{\mathrm{top}}(V^* \oplus V)$ with $\mathbb{R}$. Thus, we have

$$\int_{\Pi V^* \times \Pi V} \exp(B) = \det B.$$

If B is non-degenerate, we define the expectation value of a function f on $\Pi V^* \times \Pi V$ by

$$\langle f \rangle_0 := \frac{\displaystyle\int_{\Pi V^* \times \Pi V} \exp(B) f}{\displaystyle\int_{\Pi V^* \times \Pi V} \exp(B)}.$$

Let $\{e_i\}$ be a basis of V and denote by $\{\bar{e}^i\}$ the corresponding dual basis. Then $\bigwedge(V^* \oplus V)$ can be identified with the Grassmann algebra generated by the anticommuting *coordinate functions* $\bar{e}^i$ and e_j. Functions on $\Pi V^* \times \Pi V$ are given by linear combinations of monomials $e_{j_1} \cdots e_{j_r} \bar{e}^{i_1} \cdots \bar{e}^{i_s}$. To B we associate the function

$$\langle \bar{e}, Be \rangle = \bar{e}^j B^i_j e_i,$$

where $\langle \; , \; \rangle$ denotes the canonical pairing between V^* and V. Hence, we can write

$$\int \exp(\langle \bar{e}, Be \rangle) = \det B,$$

and

$$\langle e_{j_1} \cdots e_{j_r} \bar{e}^{i_1} \cdots \bar{e}^{i_s} \rangle_0 = \frac{\displaystyle\int \exp(\langle \bar{e}, Be \rangle) e_{j_1} \cdots e_{j_r} \bar{e}^{i_1} \cdots \bar{e}^{i_s}}{\displaystyle\int \exp\left(\bar{e}^j B^i_j e_i\right)}.$$

It is easy to see that $\langle e_{j_1} \cdots e_{j_r} \bar{e}^{i_1} \cdots \bar{e}^{i_s} \rangle_0$ vanishes if $r \neq s$ and that

$$\langle e_{j_1} \cdots e_{j_s} \bar{e}^{i_1} \cdots \bar{e}^{i_s} \rangle_0 = \sum_{\sigma \in S_s} \mathrm{sign}(\sigma)(B^{-1})^{i_{\sigma(1)}}_{j_1} \cdots (B^{-1})^{i_{\sigma(s)}}_{j_s}.$$

Remark 6.2.10 (Odd Vector Fields) A vector field on ΠV is by definition a graded derivation of $\bigwedge V^*$. In particular, an endomorphism X of $\bigwedge V^*$ is a vector field of degree $|X|$ whenever

$$X(fg) = X(f)g + (-1)^{|X|r} fX(g), \quad \forall f \in \bigwedge^r V^*, \forall g \in \bigwedge V^*, \forall r.$$

A *right* vector field X of degree $|X|$ is an endomorphism $f \mapsto (f)X$ of $\bigwedge V^*$ which satisfies

$$(fg)X = f(g)X + (-1)^{|X|s}(f)Xg, \quad \forall f \in \bigwedge V^*, \forall g \in \bigwedge^s V^*, \forall s.$$

We can identify the vector space of all vector fields on ΠV with $\bigwedge V^* \otimes V$. Note that the elements of V can be regarded as constant vector fields. Integration gives us

$$\int_{\Pi V} X(f) = 0, \quad \forall f,$$

if X is a constant vector field. In general, one can define the *divergence* div X of X by

$$\int_{\Pi V} X(f) = \int_{\Pi V} \text{div } Xf, \quad \forall f.$$

If $X = g \otimes v$ for $g \in \bigwedge^s V^*$ and $v \in V$, we actually get

$$\text{div } X = (-1)^{s+1} \iota_v g.$$

6.3 The Moyal Product as a Path Integral Quantization

Consider the symplectic manifold $T^*\mathbb{R}^n$ endowed with the canonical symplectic form $\omega_0 = \sum_{1 \le i \le n} dp_i \wedge dq^i$, where $(q^1, \ldots, q^n, p_1, \ldots, p_n) \in T^*\mathbb{R}^n \cong \mathbb{R}^{2n}$. Denote by $\alpha = \sum_{1 \le i \le n} p_i dq^i$ the Liouville 1-form such that $\omega_0 = d\alpha$. Consider a path $\gamma: I \to T^*\mathbb{R}^n$, where I is a 1-dimensional manifold, and define an action $S(\gamma) = \int_I \gamma^*\alpha$. Let us write $\gamma(t) = (Q(t), P(t))$ for $t \in I$. Then we can rewrite the action as

$$S(Q, P) = \int_{t \in I} P_i \frac{d}{dt} Q^i \, dt. \tag{6.3.1}$$

Given a Hamiltonian function H, one can deform the action to

$$S_H(Q, P) = \int_{t \in I} \left(P_i \frac{d}{dt} Q^i + H(Q(t), P(t), t) \right) dt. \tag{6.3.2}$$

Let now $I = S^1$ and, in order to make the quadratic form non-degenerate, we choose a basepoint $\infty \in S^1$. Let

$$\mathcal{M} := \{(Q, P) \in C^\infty(S^1, T^*\mathbb{R}^n)\}$$

and

$$\mathcal{M}(q, p) := \{(Q, P) \in C^\infty(S^1, T^*\mathbb{R}^n) \mid Q(\infty) = q, \; P(\infty) = p\}.$$

The path integral can then be defined by using Fubini's theorem:

$$\int_{\mathcal{M}} (\cdots) := \int_{(q,p) \in T^*\mathbb{R}^n} \mu(q, p) \int_{\mathcal{M}(q,p)} (\cdots) \tag{6.3.3}$$

where μ denotes a measure on $T^*\mathbb{R}^n$. One can check that the quadratic form in S is non-degenerate when restricted to $\mathcal{M}(q, p)$. Hence, we can compute

$$\langle \mathcal{O} \rangle_0(q, p) := \frac{\displaystyle\int_{\mathcal{M}(q,p)} \exp(iS/\hbar)\, \mathcal{O}}{\displaystyle\int_{\mathcal{M}(q,p)} \exp(iS/\hbar)},$$

where $\mathcal{O}$ is some function on $\mathcal{M}$ that is either polynomial or a formal power series in Q and P. Using the fact that the denominator is constant (infinite though), we can use (6.3.3) to write

$$\langle \mathcal{O} \rangle_0 := \frac{\displaystyle\int_{(q,p) \in T^*\mathbb{R}^n} \mu(q, p)\langle \mathcal{O} \rangle_0(q, p)}{\displaystyle\int_{(q,p) \in T^*\mathbb{R}^n} \mu(q, p)}$$

whenever the functions $\langle \mathcal{O} \rangle_0(q, p)$ and 1 are integrable. The condition that the function 1 is integrable does not allow us to choose μ to be the Liouville volume form $\frac{\omega^n}{n!}$. However, at this point one can also forget about the denominator and define

$$\langle \mathcal{O} \rangle_0' := \int_{(q,p) \in T^*\mathbb{R}^n} \mu(q, p)\langle \mathcal{O} \rangle_0(q, p)$$

such that the Liouville volume is allowed. In fact, we could also choose μ to be the delta measure peaked at a point $(q, p) \in T^*\mathbb{R}^n$. In this case we get

$$\langle \mathcal{O} \rangle_0 = \langle \mathcal{O} \rangle_0' = \langle \mathcal{O} \rangle_0(q, p).$$

If we fix $(q, p) \in T^*\mathbb{R}^n$, we can can use the *change of variables* $Q = q + \tilde{Q}$ and $P = p + \tilde{P}$, where $(\tilde{Q}, \tilde{P})$ is a map from S^1 to $T^*\mathbb{R}^n$ which vanishes at ∞. This is the same as a map $\mathbb{R} \to T^*\mathbb{R}^n$ which vanishes at ∞. The action is then given by

$$S(Q, P) = S(q + \tilde{Q}, p + \tilde{P}) = \int_{\mathbb{R}} \tilde{P}_i \frac{\mathrm{d}}{\mathrm{d}t} \tilde{Q}^i \, \mathrm{d}t.$$

Expectation values are given by

$$\langle \mathcal{O}(Q, P) \rangle_0 (q, p) = \left\langle \mathcal{O}(q + \tilde{Q}, p + \tilde{P}) \right\rangle_0^{\sim} := \frac{\int \exp(\mathrm{i}S/\hbar)\, \mathcal{O}(q + \tilde{Q}, p + \tilde{P}) \mathscr{D}[\tilde{P}]\mathscr{D}[\tilde{Q}]}{\int \exp(\mathrm{i}S/\hbar)\, \mathscr{D}[\tilde{P}]\mathscr{D}[\tilde{Q}]}.$$

6.3.1 The Propagator

We need to find the Green's function for the differential operator $\frac{\mathrm{d}}{\mathrm{d}t}$. It is given in terms of the *sign function*:

$$\left(\frac{\mathrm{d}}{\mathrm{d}t}\right)^{-1}(u, v) = \theta(u - v) = \frac{1}{2}\mathrm{sig}(u - v) := \begin{cases} \frac{1}{2}, & u > v \\ -\frac{1}{2}, & u < v \end{cases}$$

As a consequence, we get

$$\left\langle \tilde{P}_i(u)\tilde{Q}^j(v) \right\rangle_0^{\sim} = \mathrm{i}\hbar\,\theta(v - u)\delta_i^j$$

and more generally

$$\left\langle \tilde{P}_{i_1}(u_1)\cdots\tilde{P}_{i_s}(u_s)\tilde{Q}^{j_1}(v_1)\cdots\tilde{Q}^{j_s}(v_s) \right\rangle_0^{\sim} = (\mathrm{i}\hbar)^s \sum_{\sigma \in S_s} \theta(v_{\sigma(1)} - u_1)\cdots\theta(v_{\sigma(s)} - u_s)\delta_{i_1}^{j_{\sigma(1)}}\cdots\delta_{i_s}^{j_{\sigma(s)}}.$$

The normal ordering can be implemented by setting $\theta(0) = 0$ which is compatible with the skew-symmetry of $\frac{\mathrm{d}}{\mathrm{d}t}$.

6.3.2 Expectation Values

Let us consider the observable on $\mathcal{M}$ which is given by evaluation of a smooth function f on $T^*\mathbb{R}^n$ at some point u in the path. Define

$$\mathcal{O}_{f;u}(Q, P) := f(Q(u), P(u)), \quad f \in C^\infty(T^*\mathbb{R}^n), \ u \in S^1 \setminus \{\infty\}.$$

in order to compute the expectation value of this observable, we need to introduce some notation. Let $I = (i_1, \ldots, i_r)$ be a multi-index and set $|I| = r$, $p_I := p_{i_1} \cdots p_{i_r}$, $q^I := q^{i_1} \cdots q^{i_r}$ (similarly for $\tilde{P}_I$ and $\tilde{Q}^I$). Moreover, define

$$\partial_I := \frac{\partial}{\partial q^{i_1}} \cdots \frac{\partial}{\partial q^{i_r}}, \quad \partial^I := \frac{\partial}{\partial p_{i_1}} \cdots \frac{\partial}{\partial p_{i_r}}.$$

We can then extend everything to the case of $|I| = 0$ by setting $p_I = q^I = 1$ and $\partial_I = \partial^I = \mathrm{id}$. If we use the change of variables as before and use the Taylor expansion of f, we get

$$f(Q(u), P(u)) = f(q, \tilde{Q}(u), p + \tilde{P}(u)) = \sum_{r,s \geq 0} \frac{1}{r!s!} \sum_{\substack{|I|=r \\ |J|=s}} \tilde{P}_I(u)\tilde{Q}^J(u)\partial^I \partial_J f(q, p).$$

Hence

$$\langle \mathcal{O}_{f;u} \rangle_0 (q, p) = \left\langle \mathcal{O}_{f;u}(q + \tilde{Q}, p + \tilde{P}) \right\rangle_0^{\sim} = f(q, p).$$

If f is integrable we can also define

$$\langle \mathcal{O}_{f;u} \rangle_0' = \int_{T^*\mathbb{R}^n} \mu(q, p) f(q, p).$$

Remark 6.3.1 Note that these expectation values do not depend on the point u.

Consider now the observable given by

$$\mathcal{O}_{f,g;u,v} = f(Q(u), P(u))g(Q(v), P(v)), \quad f, g \in C^\infty(T^*\mathbb{R}^n), \ u, v \in S^1 \setminus \{\infty\} \cong \mathbb{R}, \ u < v. \tag{6.3.4}$$

Its expectation value can then be computed as

$$\langle \mathcal{O}_{f,g;u,v} \rangle_0 (q, p) = \left\langle f(q + \tilde{Q}(u), p + \tilde{P}(u))g(q + \tilde{Q}(v), p + \tilde{P}(v)) \right\rangle_0^{\sim}$$

$$= \sum_{r_1,s_1,r_2,s_2 \geq 0} \frac{1}{r_1!s_1!r_2!s_2!} \sum_{\substack{|I_1|=r_1 \\ |J_1|=s_1}} \sum_{\substack{|I_2|=r_2 \\ |J_2|=s_2}} \left\langle \tilde{P}_{I_1}(u)\tilde{Q}^{J_1}(u)\tilde{P}_{I_2}(v)\tilde{Q}^{J_2}(v) \right\rangle_0^{\sim} \partial^{I_1} \partial_{J_1} f(q, p)\partial^{I_2} \partial_{J_2} g(q, p)$$

$$= \sum_{r,s \geq 0} \frac{1}{r!s!} \left(\frac{i\hbar}{2}\right)^{r+s} (-1)^s \sum_{\substack{|I|=r \\ |J|=s}} \partial^I \partial_J f(q, p)\partial^J \partial_I g(q, p) = f \star g(q, p).$$

Here we have denoted by $\star$ the Moyal product induced by the Poisson bivector field $\frac{\partial}{\partial p_i} \wedge \frac{\partial}{\partial q^i}$.

Remark 6.3.2 Note also here that the expectation is independent of u and v.

Fig. 6.3 The points
$u_1, \ldots, u_k$ on S^1

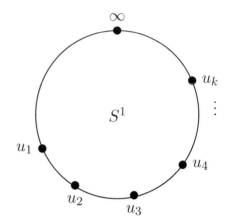

If $f \star g$ is integrable, we can consider

$$\langle \mathcal{O}_{f,g;u,v} \rangle_0' = \int_{T^*\mathbb{R}^n} \mu f \star g.$$

If μ is given by the Liouville volume $\frac{\omega_0^n}{n!}$, we can define a trace map (see also [Fed94, NT95, Fed96] and Exercise 5.1.21)

$$\mathrm{Tr}(f \star g) := \frac{1}{n!} \int_{T^*\mathbb{R}^n} f \star g \omega_0^n = \frac{1}{n!} \int_{T^*\mathbb{R}^n} f g \omega_0^n,$$

where the second equality follows from using integration by parts in the correction terms to the commutative product (see also [CF10] for a similar trace formula for the Poisson manifold $\mathbb{R}^d$ endowed with any Poisson structure and [Mos19] for the formulation on general Poisson manifolds). More generally, we can define an observable (see Fig. 6.3)

$$\mathcal{O}_{f_1,\ldots,f_k;u_1,\ldots,u_k} = f_1(Q(u_1), P(u_1)) \cdots f_k(Q(u_k), P(u_k)),$$

$$f_1, \ldots, f_k \in C^\infty(T^*\mathbb{R}^n), \ u_1, \ldots, u_k \in S^1 \setminus \{\infty\} \cong \mathbb{R}, \ u_1 < \cdots < u_k.$$

Exercise 6.3.3 Show that

$$\langle \mathcal{O}_{f_1,\ldots,f_k;u_1,\ldots,u_k} \rangle_0(q, p) = f_1 \star \cdots \star f_k(q, p).$$

6.3.3   Digression on the Divergence of Vector Fields

The normal ordering condition implies that local vector fields are in fact divergence-free. Consider a vector at $(\tilde{Q}, \tilde{P}) \in \mathcal{M}(q, p)$ which can be identified with a smooth map $\mathbb{R} \to T^*\mathbb{R}^n$ that vanishes at ∞. Note that a vector field X on $\mathcal{M}(q, p)$ is an association of a map $X(\tilde{Q}, \tilde{P})$ to each path $(\tilde{Q}, \tilde{P})$. The vector field is said to be *local* if $X(\tilde{Q}, \tilde{P})(t)$ is a function of $\tilde{Q}(t)$ and $\tilde{P}(t)$ for all $t \in \mathbb{R}$.

Formally, we want to express the divergence of a vector field in terms of path integrals similarly to the description using ordinary integrals. In particular, we would like to have

$$\int_{\mathcal{M}(q,p)} X\,(\exp{(\mathrm{i}S/\hbar)}\,\mathcal{O}) = \int_{\mathcal{M}(q,p)} \operatorname{div} X \exp{(\mathrm{i}S/\hbar)}\,\mathcal{O},$$

for all observables $\mathcal{O}$.

Definition 6.3.4 (Divergence) If there is an observable div X such that for any observable $\mathcal{O}$ we have

$$\langle X(S)\mathcal{O}\rangle_0(q, p) = \mathrm{i}\hbar\langle X(\mathcal{O})\rangle_0(q, p) - \mathrm{i}\hbar\langle \operatorname{div} X\mathcal{O}\rangle_0(q, p),$$

we say that div X is the *divergence* of X.

Lemma 6.3.5 *If a vector field X is local, then it has vanishing divergence, i.e.* div X.

Proof Consider a local vector field X and the induced map $X(\tilde{Q}, \tilde{P})(t) = (X_q(\tilde{Q}(t), \tilde{P}(t)), X_p(\tilde{Q}(t), \tilde{P}(t)))$. Then

$$X(S) = \int \left(X_p \frac{\mathrm{d}}{\mathrm{d}t}\tilde{Q} - X_q \frac{\mathrm{d}}{\mathrm{d}t}\tilde{P} \right) \mathrm{d}t.$$

By normal ordering and the locality of X, when computing the expectation value $\langle X(S)\mathcal{O}\rangle_0(q, p)$ we only have to contract the $\tilde{P}$'s ($\tilde{Q}$'s) in $\mathcal{O}$ with the $\tilde{Q}$'s ($\tilde{P}$'s) in $X(S)$ and replace each pair by a propagator. Whenever a $\tilde{P}$ ($\tilde{Q}$) in $\mathcal{O}$ is contracted with $\frac{\mathrm{d}}{\mathrm{d}t}\tilde{Q}$ ($\frac{\mathrm{d}}{\mathrm{d}t}\tilde{P}$) in $X(S)$, we get the identity operator times $\mathrm{i}\hbar$ ($-\mathrm{i}\hbar$). Summing everything up, we get that this is the same as taking the expectation value of $X(\mathcal{O})$ multiplied with $\mathrm{i}\hbar$. This completes the proof. □

We can also consider the divergence of vector fields on $\mathcal{M}$. In fact, a local vector field X on $\mathcal{M}$ can be uniquely written as the sum of a vector field X_∞ on $T^*\mathbb{R}^n$ and a section $\tilde{X}$ of local vector fields (i.e. an association of a local vector field $\tilde{X}(q, p)$ on $\mathcal{M}(q, p)$ to all $(q, p) \in T^*\mathbb{R}^n$). By Lemma 6.3.5 we get

$$\langle X(S)\mathcal{O}\rangle_0' = \mathrm{i}\hbar\langle X(\mathcal{O})\rangle_0' - \mathrm{i}\hbar\langle \operatorname{div}_\mu X_\infty\mathcal{O}\rangle_0',$$

where $\operatorname{div}_\mu X_\infty$ denotes the ordinary divergence of the vector field X_∞ with respect to the measure μ. One can then define $\operatorname{div}_\mu X_\infty$ to be the *divergence* of the local vector field X on $\mathcal{M}$.

Exercise 6.3.6 Using the local vector field $X(\tilde{Q}, \tilde{P})(t) := (\tilde{P}(t), 0)$, show that

$$\hbar \frac{\mathrm{d}}{\mathrm{d}\hbar}(f \star g) = (p_i \partial^i f) \star g + f \star (p_i \partial^i g) - p_i \partial^i (f \star g).$$

6.3.4 Independence of Evaluation Points

Note that in the previous computations we have considered observables whose definition depends on the choice of points in $S^1 \setminus \{\infty\}$. However, computing the expectation values we have seen that they are in fact independent of these points. In particular, we have the following Proposition:

Proposition 6.3.7 *The expectation values are invariant under (pointed) diffeomorphism of the source manifold S^1 on which the field theory is defined.*

Remark 6.3.8 A quantum field theory with such a property is usually called *topological quantum field theory (TQFT)*.

For a function $f \in C^\infty(T^*\mathbb{R}^n)$ and two points $a, b \in \mathbb{R}$, we actually have

$$f(Q(b), P(b)) - f(Q(a), P(a))$$

$$= \int_a^b \left(\partial_i f(Q(t), P(t)) \frac{\mathrm{d}}{\mathrm{d}t} Q^i(t) + \partial^i f(Q(t), P(t)) \frac{\mathrm{d}}{\mathrm{d}t} P_i(t) \right) \mathrm{d}t$$

$$= \lim_{r \to \infty} \int_{-\infty}^{+\infty} \left(\partial_i f(Q(t), P(t)) \frac{\mathrm{d}}{\mathrm{d}t} Q^i(t) + \partial^i f(Q(t), P(t)) \frac{\mathrm{d}}{\mathrm{d}t} P_i(t) \right) \lambda_r(t) \mathrm{d}t$$

$$= \lim_{r \to \infty} \tilde{X}_{f,r}(S),$$

where (λ_r) is a sequence of smooth, compactly supported functions that converges almost everywhere to the characteristic function of the interval $[a, b]$ and $\tilde{X}_{f,r}$ is the local vector field given by

$$\tilde{X}_{f,r}(t) = (-\partial^i f(Q(t), P(t)), \partial_i f(Q(t), P(t))) \lambda_r(t).$$

Remark 6.3.9 Recall that we have chosen $\mathcal{M}$ to be a space of maps and thus a vector field on the target $T^*\mathbb{R}^n$ will generate a local vector field on $\mathcal{M}$. As a special case, if we consider $\tilde{X}_f$, i.e. the local vector field induced by the Hamiltonian vector field X_f of f, then we can construct the local vector field $\tilde{X}_{f,r}$ as the product of $\tilde{X}_f$ and λ_r.

If we restrict $\tilde{X}_{f,r}$ to $\mathcal{M}(q, p)$, we can observe by Lemma 6.3.5 that it is divergence-free. Hence, we get

$$\langle (f(Q(b), P(b)) - f(Q(a), P(a)))\mathcal{O}\rangle_0(q, p) = \lim_{r \to \infty} \langle \tilde{X}_{f,r}(S)\mathcal{O}\rangle_0 (q, p)$$

$$= i\hbar \lim_{r \to \infty} \langle \tilde{X}_{f,r}(\mathcal{O})\rangle_0 (q, p),$$

which vanishes if $\mathcal{O}$ does not depend on $(\tilde{Q}(t), \tilde{P}(t))$ for $t \in [a, b]$. Therefore we can move the evaluation point at least as long as we do not approach any other evaluation point.

6.3.5 Associativity

Let us prove the associativity of the Moyal product by using the path integral description considering the expectation value of the observable $\mathcal{O}_{f,g,h;u,v,w}$. There will be three propagators appearing which correspond to the three different ways of pairing u, v, w. Note that the function θ will not see the difference, since

$$\theta(w - u) = \theta(w - v) = \theta(v - u) = \frac{1}{2}.$$

We group the propagators by considering first only those between u and v and subsequently the other ones. This will imply that

$$\langle \mathcal{O}_{f,g,h;u,v,w}\rangle_0 = (f \star g) \star h.$$

If we group the propagators such that we first consider those between v and w and then all the other pairs, we get

$$\langle \mathcal{O}_{f,g,h;u,v,w}\rangle_0 = f \star (g \star h).$$

This proves associativity of the Moyal product.

Remark 6.3.10 We can also prove this result by using the independence of evaluation points. Hence, we get

$$\lim_{v \to u^+} \langle \mathcal{O}_{f,g,h;u,v,w}\rangle_0(q, p) = \lim_{v \to w^-} \langle \mathcal{O}_{f,g,h;u,v,w}\rangle_0(q, p).$$

The left-hand-side corresponds to first consider the evaluation of the expectation value of $\mathcal{O}_{f,g;u,v}$ with putting the result afterwards at u and computing the expectation value of $\mathcal{O}_{\langle \mathcal{O}_{f,g;u,v}\rangle_0,h;w}$. This will produce the result $(f \star g) \star h$. Repeating this computation on the right-hand-side, we get associativity.

6.3.6   The Evolution Operator as an Application

Let us consider the deformed action S_H as in (6.3.2). We restrict ourselves to a time-independent Hamiltonian h defined on $T^*\mathbb{R}^n$ and let it act from the time a to the time b with $a < b$. Hence, we consider the action S_H with

$$H(q, p, t) = h(q, p)\chi_{[a,b]}(t), \tag{6.3.5}$$

where $\chi_{[a,b]}$ denotes the characteristic function on the interval $[a, b]$. The aim is to compute the *evolution operator*

$$U(q, p, T) := \frac{\displaystyle\int_{\mathcal{M}(q,p)} \exp{(\mathrm{i}S_H/\hbar)}}{\displaystyle\int_{\mathcal{M}(q,p)} \exp{(\mathrm{i}S/\hbar)}}, \qquad T := b - a.$$

Observe that

$$U(q, p, T) = \left\langle \exp\left(\frac{\mathrm{i}}{\hbar}\int H(Q(t), P(t), t)\mathrm{d}t\right) \right\rangle_0 (q, p)$$

$$= \left\langle \exp\left(\frac{\mathrm{i}}{\hbar}\int_a^b h(Q(t), P(t), t)\mathrm{d}t\right) \right\rangle_0 (q, p).$$

Now we can express the integral in terms of Riemann sums as

$$\int_a^b h(Q(t), P(t), t)\mathrm{d}t = \lim_{N\to\infty} \frac{T}{N} \sum_{1\leq r\leq N} h(Q(a + rT/N), P(a + rT/N)).$$

Hence, we get

$$U(q, p, T) = \lim_{N\to\infty} \left\langle \prod_{1\leq r\leq N} \exp\left(\frac{\mathrm{i}}{\hbar}\frac{T}{N}h(Q(a + rT/N), P(a + rT/N))\right) \right\rangle_0 (q, p)$$

$$= \lim_{N\to\infty} \left\langle \mathcal{O}_{\exp\left(\frac{\mathrm{i}}{\hbar}\frac{T}{N}h\right),\ldots,\exp\left(\frac{\mathrm{i}}{\hbar}\frac{T}{N}h\right);a+\frac{T}{N},a+2\frac{T}{N},\ldots,a+T} \right\rangle_0 (q, p)$$

$$= \lim_{N\to\infty} \left(\exp\left(\frac{\mathrm{i}}{\hbar}\frac{T}{N}h\right)\right)^{\star N} (q, p) = \exp_\star\left(\frac{\mathrm{i}}{\hbar}h\right)(q, p).$$

Remark 6.3.11 Note that the previous result also includes negative powers of $\hbar$. However, everything is well-defined since each term in the power series expansion of $\exp_\star$ is actually a *Laurent series* in $\hbar$.

6.3.7 Perturbative Evaluation of Integrals

Let $\mathcal{M}$ be an n-dimensional manifold and $S \in C^\infty(\mathcal{M})$. We want to understand how we can compute an integral of the form $Z := \int_\mathcal{M} \exp(iS/\hbar)d^n x$. Consider first the case where S has a unique critical point $x_0 \in \mathcal{M}$ which is non-degenerate. Then we can write

$$S(x_0 + \sqrt{\hbar}x) = S(x_0) + \frac{\hbar}{2}d_{x_0}^2 S(x) + R_{x_0}(\sqrt{\hbar}x),$$

where R_{x_0} is a formal powers series given by the Taylor expansion of S starting with the cubic term in $\sqrt{\hbar}x$ with $x \in T_{x_0}\mathcal{M}$. Let A be the *Hessian* of S at x_0 with respect to the Euclidean metric $\langle \ , \ \rangle$. In particular, we write

$$d_{x_0}^2 S(x) = \langle x, Ax \rangle.$$

Theorem 6.3.12 (Stationary Phase Expansion) *The* stationary phase expansion *(or also* saddle-point approximation*) of the integral $Z := \int_\mathcal{M} \exp(iS/\hbar)d^n x$ is given by*

$$Z = \hbar^{\frac{n}{2}} \exp(iS(x_0)/\hbar) \int_\mathcal{M} \exp\left(\frac{i}{2}\langle x, Ax \rangle\right) \sum_{r \geq 0} \frac{1}{r!} R_{x_0}^r(\sqrt{\hbar}x)d^n x$$

$$= (2\pi\hbar)^{\frac{n}{2}} \exp(iS(x_0)/\hbar) \exp\left(\frac{i\pi \ \mathrm{sign}\ A}{4}\right) \frac{1}{\sqrt{|\det A|}} \sum_{r \geq 0} \frac{1}{r!} \left\langle R_{x_0}^r(\sqrt{\hbar}x) \right\rangle_0 ,$$

where $\langle \ \rangle_0$ denotes the Gaussian expectation value with respect to the non-degenerate symmetric matrix A.

Expectation values as in Definition 6.2.1 are then given by

$$\langle \mathcal{O} \rangle = \frac{\sum_{r \geq 0} \frac{1}{r!} \left\langle \mathcal{O}(x_0 + \sqrt{\hbar}x) R_{x_0}^r(\sqrt{\hbar}x) \right\rangle_0}{\sum_{r \geq 0} \frac{1}{r!} \left\langle R_{x_0}^r(\sqrt{\hbar}x) \right\rangle_0}. \tag{6.3.6}$$

Remark 6.3.13 Note that the denominator is of the form $1 + O(\hbar)$ and thus the expectation value can be computed as a formal power series in $\hbar$. One can consider the computation in terms of graphs by associating a vertex of valence k to the term of degree k in R_{x_0} and a vertex of valence ℓ to the term of degree ℓ in $\mathcal{O}$. By Wick's theorem (Theorem 6.2.5), one has to connect pairs of vertices in all possible ways the half-edges emerging from each vertex. The result of this graphical illustration is a collection of *Feynman diagrams* with a *weight* associated to each of them. The expectation value is then obtained by summing all the weights of the appearing Feynman diagrams. A Feynman diagram with vertices corresponding only to R_{x_0},

i.e. a Feynman diagram appearing only in the denominator of (6.3.6), is called a
vacuum diagram. A combinatorial fact of (6.3.6) is that an expectation value is
given by the sum over graphs which do not have a connected component which is
a vacuum diagram (see [Fey49, FH65, Zin94, Pol05] for more details on Feynman
diagrams).

6.3.8 Feynman Diagrams and Perturbative Expansion

We want to describe the notion of *Feynman diagrams* more in detail. The way of
expressing a diagrammatic interpretation for the matrix elements as they appear
in Theorem 6.2.5 for the Gaussian integral plays a crucial role in the general
development. Let us first give the definition of a graph in this setting:

Definition 6.3.14 (Graph) A *graph* consists of a set V of *vertices*, a set HE of
half-edges, an *incidence* map $i : HE \to V$ and a *perfect matching* E on HE, i.e.
some partition of the set E of *edges* into subsets consisting of two-elements.

Remark 6.3.15 All graphs considered for our purposes will be finite, i.e. all the sets,
V and HE are finite.

If we have two graphs Γ and Γ', we can look at a *graph isomophism* $\Gamma \xrightarrow{\sim} \Gamma'$
which is a pair of bijections $\sigma_V : V \xrightarrow{\sim} V'$ and $\sigma_{HE} : HE \xrightarrow{\sim} HE'$ commuting
with the incidence maps, i.e. we have $i' \circ \sigma_{HE} = \sigma_V \circ i$ and moreover, preserving
the partition into edges. E.g. consider the vertices $V = \{a, b, c\}$ and half-edges
$HE = \{1, 1', 2, 2', 3, 3'\}$ with the incidence i given by

$$1 \mapsto a, \quad 1' \mapsto a, \quad 2 \mapsto b, \quad 2' \mapsto b, \quad 3 \mapsto c, \quad 3' \mapsto c.$$

and edges $E = \{1, 2'\} \cup \{2, 3'\} \cup \{3, 1'\}$. Note that there is an involution such that
$1 \leftrightarrow 2', 2 \leftrightarrow 3'$ and $3 \leftrightarrow 1'$. An automorphism of the graph in Fig. 6.4 is given by

$$\sigma_V : (a, b, c) \mapsto (b, a, c), \quad \sigma_{HE} : (1, 1', 2, 2', 3, 3') \mapsto (2', 2, 1', 1, 3', 3).$$

Fig. 6.4 Triangle graph

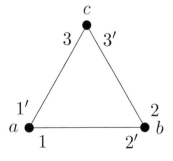

Fig. 6.5 Polygonal graph
with $n = 6$

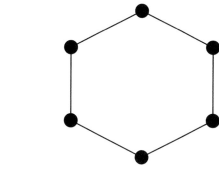

Fig. 6.6 Theta graph and
figure-eight graph

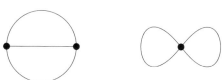

It is then easy to see that these permutations do indeed commute with the incidence maps and the involutions. This is an important example when considering the automorphism group $\mathrm{Aut}(\Gamma)$ of a graph Γ. Let us consider some more examples:

(1) Consider a *polygonal graph* Γ with $n \geq 3$ vertices and n edges as in Fig. 6.5. Then $\mathrm{Aut}(\Gamma) = \mathbb{Z}_2 \ltimes \mathbb{Z}_n$.
(2) Consider the *theta graph* Γ as in Fig. 6.6. Then $\mathrm{Aut}(\Gamma) = \mathbb{Z}_2 \times S_3$.
(3) Consider the *figure-eight* graph Γ as in Fig. 6.6. Then we have $\mathrm{Aut}(\Gamma) = \mathbb{Z}_2 \ltimes (\mathbb{Z}_2 \times \mathbb{Z}_2)$.

Now we want to see how we can associate such graphs to Gaussian expectation values for the special case of homogeneous polynomials. Let $V = \mathbb{R}^n$ and consider $A \in \mathrm{Sym}^2(V^*)$. Moreover, consider a collection $\{p_a\}$ of homogeneous polynomials on V with $p_a \in \mathrm{Sym}^{d_a}(V^*)$ for $a = 1, \ldots, r$ of degree $d_1, \ldots, d_r$. In coordinates, we have

$$p_a = \sum_{1 \leq i_1, \ldots, i_{d_a} \leq n} (p_a)_{i_1 \cdots i_{d_a}} x_{i_1} \cdots x_{i_{d_a}}.$$

Then, using Theorem 6.2.5, we get that the Gaussian expectation value is given by

$$\left\langle \frac{1}{d_1!} p_1 \cdots \frac{1}{d_r!} p_r \right\rangle_0 = \frac{1}{d_1! \cdots d_r!} \sum_{\sigma \in P(2m)} \left\langle \sigma \circ (A^{-1})^{\otimes m}, p_1 \otimes \cdots \otimes p_r \right\rangle,$$

$$(6.3.7)$$

where $2m = \sum_{1 \leq a \leq r} d_a$ and $P(2m)$ is defined as in Theorem 6.2.5. Let us look at different examples which explain the output as in (6.3.7) in a graphical manner.

Example 6.3.16 Let $p = \sum_{1 \le i,j,k,l \le n} p_{ijkl} x_i x_j x_k x_l \in \mathrm{Sym}^4(V^*)$. Then we get

$$\left\langle \frac{1}{4!} p \right\rangle_0 = \frac{1}{4!} \sum_{\sigma \in P(4)} \left\langle \sigma \circ (A^{-1})^{\otimes 2}, p \right\rangle$$

$$= \frac{3}{4!} \left\langle A^{-1} \underset{i \quad\quad l}{\overset{j \quad\quad k}{\bigotimes_{p}}} A^{-1} \right\rangle = \frac{1}{8} \sum_{1 \le i,j,k,l \le n} p_{ijkl} A^{ij} A^{kl}.$$

$$(6.3.8)$$

Example 6.3.17 Let $p_1 = \sum_{1 \le i,j,k \le n} (p_1)_{ijk} x_i x_j x_k$ and $p_2 = \sum_{1 \le i',j',k' \le n} (p_2)_{i'j'k'} x_{i'} x_{j'} x_{k'} \in \mathrm{Sym}^3(V^*)$. Then we get

$$\left\langle \frac{1}{3!} p_1 \frac{1}{3!} p_2 \right\rangle_0 = \frac{1}{3!3!} \sum_{\sigma \in P(6)} \left\langle \sigma \circ (A^{-1})^{\otimes 3}, p_1 \otimes p_2 \right\rangle$$

$$= \frac{6}{3!3!} \left\langle p_1 \underset{k \quad\quad k'}{\overset{i \quad\quad i'}{\bigcirc}} p_2 \right\rangle + \frac{9}{3!3!} \left\langle A^{-1} \, p_1 \underset{k \quad i' k'}{\overset{j \quad\quad j'}{\bigcirc}} p_2 \, A^{-1} \right\rangle$$

$$= \frac{1}{6} \sum_{1 \le i,j,k,i',j',k' \le n} (p_1)_{ijk} (p_2)_{i'j'k'} A^{ii'} A^{jj'} A^{kk'} + \frac{1}{4} \sum_{1 \le i,j,k,i',j',k' \le n} (p_1)_{ijk} (p_2)_{i'j'k'} A^{ii'} A^{jk} A^{j'k'}.$$

$$(6.3.9)$$

Consider now a propagator $\mathscr{P} = \sum_{1 \leq i,j \leq n} \mathscr{P}_{ij} e_i \otimes e_j \in \mathrm{Sym}^2(V)$ with $\{e_i\}$ the standard basis in $\mathbb{R}^n$ and *vertex functions* of valency d:

$$p_d = \sum_{1 \leq i_1,\dots,l_d \leq n} (p_d)_{i_1 \cdots i_d} x_{i_1} \cdots x_{i_d} \in \mathrm{Sym}^d(V^*), \quad d = 0, \dots, D.$$

Here we have denoted by $\{x_i\}$ the dual basis to $\{e_i\}$. We define the *state sum* $\Phi_{\mathscr{P};p_0,\dots,p_D}(\Gamma)$ of a graph Γ as follows:

(1) A *state* s on Γ is defined as a decoration of all half-edges of Γ by numbers in $\{1, \dots, n\}$.
(2) To each state $s : HE \to \{1, \dots, n\}$, one assigns a *weight*

$$w_s := \prod_{\text{edges } e=(h,h')} \mathscr{P}_{s(h)s(h')} \times \prod_{\text{vertices } v} (p_d)_{s(h_1) \cdots s(h_d)}.$$

We have denoted by h and h' the two half-edges of the edge e. Moreover, d is the valency of the vertex v and $h_1, \dots, h_d$ are the half-edges adjacent to v.

(3) Finally, we define Φ to be the sum over all states on Γ:

$$\Phi_{\mathscr{P};p_0,\dots,p_D}(\Gamma) := \sum_{\text{states } s : HE \to \{1,\dots,n\}} w_s.$$

Definition 6.3.18 (Feynman Weight) The *Feynman weight* of a graph Γ is defined to be

$$\frac{1}{|\mathrm{Aut}(\Gamma)|} \Phi_{\mathscr{P};p_0,\dots,p_D}(\Gamma).$$

Theorem 6.3.19 (Feynman) *Let A be a positive-definite quadratic form on $\mathbb{R}^n$ and let $p(x) = \sum_{1 \leq d \leq D} \frac{g_d}{d!} P_d(x)$ be a polynomial with homogeneous terms $P_d \in \mathrm{Sym}^d((\mathbb{R}^n)^*)$ and some formal parameters $g_0, \dots, g_D$ (coupling constants). Then, perturbatively, we get*

$$\int_{\mathbb{R}^n} \exp\left(-\frac{1}{2}\langle x, Ax \rangle + p(x)\right) d^n x$$

$$= (2\pi)^{\frac{n}{2}} (\det A)^{-\frac{1}{2}} \sum_{\text{Graphs } \Gamma} \frac{1}{|\mathrm{Aut}(\Gamma)|} \Phi_{A^{-1};g_0 P_0,\dots,g_D P_D}(\Gamma), \quad (6.3.10)$$

Example 6.3.20 (Theta Graph) Let us consider the theta graph where each half-edges are labeled by $\{A, B, C, D, E, F\}$.

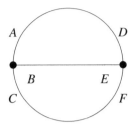

We can now map the labels of the half-edges to numbers from 1 to n by using a state $s: \{A, B, C, D, E, F\} \to \{i, j, k, i', j', k'\}$ on Γ.

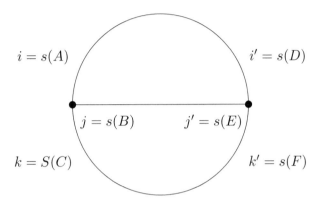

Hence, the weight is given by

$$w_s = \mathscr{P}_{ii'} \mathscr{P}_{jj'} \mathscr{P}_{kk'} (p_3)_{ijk} (p_3)_{i'j'k'}.$$

The Feynman weight of the theta graph is then given by

$$\frac{1}{12} \sum_{s:\, \{A,B,C,D,E,F\} \to \{i,j,k,i',j',k'\}} w_s = \frac{1}{12} \sum_{1 \le i,j,k,i',j',k' \le n} \mathscr{P}_{ii'} \mathscr{P}_{jj'} \mathscr{P}_{kk'} (p_3)_{ijk} (p_3)_{i'j'k'}$$

Example 6.3.21 Let us compute the Feynman weight of the graph below.

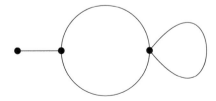

For this graph, we get the expression

$$
\frac{g_1 g_3 g_4}{|\mathrm{Aut}(\Gamma)|}\,\Phi\left(\begin{array}{c}\text{graph}\end{array}\right)
$$

$$
= \frac{g_1 g_3 g_4}{4}\sum_{1\le i,j,k,l,m,n,o,p\le n} A^{kl}A^{im}A^{jn}A^{op}(P_3)_{ijk}(P_1)_l(P_4)_{mnop}.
$$

$$(6.3.11)$$

Let us also consider an example which goes the other way around.

Example 6.3.22 We want to determine the perturbative expansion of the integral

$$
I(\lambda) = \int_{\mathbb{R}} \exp\left(\frac{x^2}{2} + \frac{\lambda}{4!}x^4\right) dx
$$

Using the formula of Theorem 6.3.19, we get

$$
\int_{\mathbb{R}} \exp\left(\frac{x^2}{2} + \frac{\lambda}{4!}x^4\right) dx = \sqrt{2\pi}\sum_{\text{4-valent graphs } \Gamma}\frac{\lambda^{|V(\Gamma)|}}{|\mathrm{Aut}(\Gamma)|}
$$

$$
= \sqrt{2\pi}\left(1 + \frac{1}{8}\lambda + \left(\frac{1}{2\cdot 8^2}\lambda^2 + \frac{1}{2\cdot 4!}\lambda^2 + \frac{1}{16}\lambda^2\right) + \cdots\right).\qquad(6.3.12)
$$

The first contributing graphs are then given by the empty graph and the following graphs:

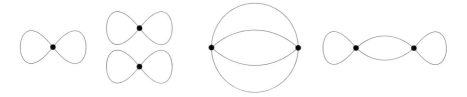

Remark 6.3.23 It is important to say that for any perturbation $p(x)$ as in Theorem 6.3.19, the perturbative integral is well-defined, whereas the measure-theoretic integral might not exist. Consider e.g. the integral

$$\int_{\mathbb{R}} \exp\left(-\frac{x^2}{2} + \frac{\alpha}{3!}x^3\right) dx.$$

Then this integral diverges for any non-zero coefficient α, except when α is purely imaginary, i.e. $\alpha \in i\mathbb{R}$. Similarly for the integral

$$\int_{\mathbb{R}} \exp\left(-\frac{x^2}{2} + \frac{\lambda}{4!}x^4\right) dx.$$

It will only converge for $\lambda \leq 0$, or more generally, whenever the real part of λ is ≤ 0.

6.3.9 Infinite-Dimensional Case

Let us look at the situation of computing (6.3.6) in the case when $\mathcal{M}$ is infinite-dimensional. We want to recall the following definition.

Definition 6.3.24 (Local Function) A function on a space of fields on some manifold M is *local* if it is the integral on M of a function that depends at each point on finite jets of fields at that point.

Remark 6.3.25 If the action is a local function, the matrix A will be a differential operator. For our purpose, we will need its Green's function (the propagator) and in the Gaussian expectation values of the observables $\mathcal{O}R^r$ and R^r we will use (6.2.3). The difference is that we will have an integral over Cartesian products of the source manifold instead of a sum over indices. Due to the normal ordering condition, we can exclude all graphs containing edges that start and end at the same vertex (these edges are called *tadpoles* or *short loops*) and thus integration is restricted to the configuration space of the source manifold. However, this is usually not enough to make the integrals converge as the Green's functions are usually pretty singular on the diagonal and thus we can get strange infinities. The general procedure to get rid of divergencies is called *renormalization*. Anyway, for our case (i.e. the case of a TQFT), the configuration-space integrals associated to Feynman diagrams without tadpoles do indeed converge.

Remark 6.3.26 If the action S has several critical points and all of them are non-degenerate, the asymptotic expansion is obtained by computing the stationary phase expansion (Theorem 6.3.12) around each critical point and then summing all these terms together. The expression for the expectation value is then no longer given by (6.3.6). It can happen that one of the critical points dominates the others so that one can forget them. In concrete theories this is usually done by considering the function

$\exp(iS/\hbar)$ as the analytic continuation of the exponential of minus a positive-definite function, called *Euclidean action*, such that the dominating critical point is the absolute minimum. Unfortunately, there are a lot of physical theories, where such an approach is not possible since all critical points do actually appear as saddle points (degenerate points). In such a case, one can consider another option which consists of selecting one particular critical point (called the *sector*) and expanding only around this point. In this case, (6.3.6) is again the correct expression for the infinite-dimensional extension.

As already mentioned, it can often happen that the critical points are degenerate. A simple case is given when critical points are parametrized by a finite-dimensional manifold $\mathcal{M}_{\text{crit}}$. Then, using Fubini's theorem for the finite-dimensional case, we can rewrite the integral similarly as in (6.3.3) as

$$\int_{\mathcal{M}} (\cdots) = \int_{x_0 \in \mathcal{M}_{\text{crit}}} \mu(x_0) \int_{\mathcal{M}(x_0)} (\cdots)$$

where $\int_{\mathcal{M}(x_0)}$ denotes the asymptotic expansion of the integral in the complement to $T_{x_0}\mathcal{M}_{\text{crit}}$ of a formal neighborhood of x_0, while μ denotes a measure on $\mathcal{M}_{\text{crit}}$ (which is determined in the finite-dimensional case and has to be chosen in the infinite-dimensional case as part of the definition). If the Hessian is constant on $\mathcal{M}_{\text{crit}}$, we can express the expectation value as

$$\langle \mathcal{O} \rangle = \frac{\int_{x_0 \in \mathcal{M}_{\text{crit}}} \mu(x_0) \sum_{r \geq 0} \frac{1}{r!} \left\langle \mathcal{O}(x_0 + \sqrt{\hbar}x) R_{x_0}^r(\sqrt{\hbar}x) \right\rangle_0}{\int_{x_0 \in \mathcal{M}_{\text{crit}}} \mu(x_0) \sum_{r \geq 0} \frac{1}{r!} \left\langle R_{x_0}^r(\sqrt{\hbar}x) \right\rangle_0},$$

where $\langle \ \rangle_0(x_0)$ denotes the Gaussian expectation value which is computed by expanding around x_0 orthogonally to $T_{x_0}\mathcal{M}_{\text{crit}}$. A possible choice for the measure μ is the delta measure peaked at some point x_0. In this case, as we already did before, the expectation value will be denoted by $\langle \ \rangle(x_0)$.

6.3.10 Generalizing the Expansion

Note that we have only considered the perturbative expansion with formal expansion parameter $\hbar$. It might happen that there is another small expansion parameter, or some coefficient in S which is much smaller than $\hbar$ or in our setting is an element of $\hbar^2 \mathbb{R}[[\hbar]]$. In such a case, we can define the Gaussian part by using the quadratic $\hbar$-independent term of $S/\hbar$ and view all the other terms as the perturbation R. It can also happen that the perturbation term R contains quadratic and linear terms as well. In particular, we can have an action function of the form

$$S(y, z) = \langle y, Bz \rangle + f(y, z), \tag{6.3.13}$$

where B is a non-degenerate matrix and f is a function quadratic in z. If we expand around the critical point $y = z = 0$, we can rescale z by $\hbar$ to get

$$S(y, \hbar z) = \hbar \langle y, Bz \rangle + \hbar^2 f(y, z)$$

and consider f as the perturbation to the Gaussian theory defined by B.

6.3.10.1　Quantum Mechanics

Recall that the topological action (6.3.1) defines the Moyal product in terms of the expectation value of the observable (6.3.4), which is independent of the evaluation points. Consider a Hamiltonian H as in (6.3.5) and define the expectation

$$f \hat{\star}_{a,b;u,v} g(q, p) := \frac{\int_{\mathcal{M}(q,p)} \exp(iS_H)/\hbar) \mathcal{O}_{f,g;u,v}}{\int_{\mathcal{M}(q,p)} \exp(iS_H/\hbar)} = \frac{\left\langle \exp\left(\frac{i}{\hbar} \int_a^b h dt\right) \mathcal{O}_{f,g;u,v} \right\rangle_0}{\left\langle \exp\left(\frac{i}{\hbar} \int_a^b h dt\right) \right\rangle_0} (q, p)$$

which will no longer be independent from u and v ($a < u < v < b$) and thus will not define an associative product. By the computations before, we get

$$f \hat{\star}_{a,b;u,v} g = \frac{\exp_\star\left(\frac{i}{\hbar}(u-a)h\right) \star f \star \exp_\star\left(\frac{i}{\hbar}(v-u)h\right) \star g \star \exp_\star\left(\frac{i}{\hbar}(b-v)h\right)}{\exp_\star\left(\frac{i}{\hbar}(b-a)h\right)}$$

Let us consider the perturbative computation for the Hamiltonian

$$h(q, p) = \frac{1}{2} G^{ij}(q) p_i p_j,$$

at $p = 0$ and for f and g only depending on q. By defining $Q(t) = q + \tilde{Q}(t)$ and $P(t) = p + \tilde{P}(t)$, we get

$$S_H = \hbar \int_{-\infty}^{+\infty} \tilde{P}(t) \frac{d}{dt} \tilde{Q}(t) dt + \frac{\hbar^2}{2} \int_a^b G^{ij}(q + \tilde{Q}(t)) \tilde{P}_i(t) \tilde{P}_j(t) dt.$$

Hence, $f \hat{\star}_{a,b;u,v} g(q, 0)$ is the ratio of the expectation value of $\exp\left(\frac{i\hbar}{2} \int_a^b G^{ij}\right.$ $\left.(q + \tilde{Q}) \tilde{P}_i \tilde{P}_j dt\right) \mathcal{O}_{f,g;u,v}$ and the expectation value of $\exp\left(\frac{i\hbar}{2} \int_a^b G^{ij}\right.$ $\left.(q + \tilde{Q}) \tilde{P}_i \tilde{P}_j dt\right)$. The propagator pairs the $\tilde{P}$'s to $\tilde{Q}$'s, so it is better to consider oriented graphs. The Feynman diagrams will thus be oriented graphs such that

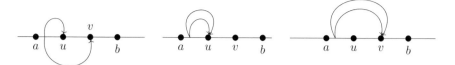

Fig. 6.7 Feynman diagrams of order one

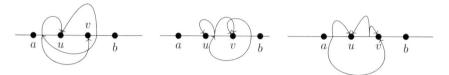

Fig. 6.8 Feynman diagrams of order two

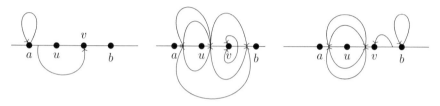

Fig. 6.9 Feynman diagrams which are not allowed: tadpoles, vacuum subdiagrams and both

(1) a vertex at u with no outgoing arrows,
(2) a vertex at v with no outgoing arrows,
(3) vertices between a and b with exactly two outgoing arrows.

Moreover, graphs containing tadpoles or vacuum subgraphs will not be allowed. See Figs. 6.7, 6.8 and 6.9 for examples of allowed and not allowed Feynman diagrams. Note that to a vertex of first type with r incoming arrows we associate an r-th derivative of f, to a vertex of second type with r incoming arrows we associate an r-th derivative of g and to a vertex of third type with r incoming arrows we associate an r-th derivative of G. The order in $\hbar$ is given by the number of vertices of third type. So, at order zero, we have f placed at u and g placed at v. This yields the point-wise product of f and g. At first order, we have the graphs of Fig. 6.7. Hence, we get

$$f \hat{\star}_{a,b;u,v} g(q,0) = f(q)g(q) - \frac{i\hbar}{4}(b - a + 2u - 2v)G^{ij}(q)\partial_i f(g)\partial_j g(q)$$

$$- \frac{i\hbar}{4}(b - a)G^{ij}(q)[\partial_i \partial_j f(q)g(q) + f(q)\partial_i \partial_j g(q)] + O(\hbar^2).$$

Exercise 6.3.27 Compute some of the further orders.

6.4   Symmetries and Gauge Formalisms

If the action S has symmetries encoded in a free action of a Lie group G (gauge theory with gauge group G), the critical points will never be non-degenerate since they always will appear in G-orbits. This is also the case for the Poisson sigma model for linear Poisson structures and hence we need to understand how to deal with these type of theories.

6.4.1   The Main Construction: Faddeev–Popov Ghost Method

Let $\mathcal{M}$ be a finite-dimensional manifold together with a measure μ and let G be a compact Lie group which is endowed with an invariant measure (Haar measure) together with a measure-preserving free action on $\mathcal{M}$. Moreover, we assume that the quotient $\mathcal{M}/G$ is a manifold. An invariant function f is then the pullback of a function $\underline{f} \in \mathcal{M}/G$ and

$$I := \frac{1}{\mathrm{Vol}(G)} \int_{\mathcal{M}} f \mu = \int_{\mathcal{M}/G} \underline{f}\, \underline{\mu}$$

where $\underline{\mu}$ is the measure induced by μ on the quotient. If we consider a section of a principal bundle $\pi \colon \mathcal{M} \to \mathcal{M}/G$, we can rewrite I as an integral on the image of this section. Assume that this integral is locally given by the zero locus of a function $F \colon \mathcal{M}' \to \mathfrak{g}$, where $\mathcal{M}' \subset \mathcal{M}$ is such that $\pi(\mathcal{M}') = \mathcal{M}/G$ and $\mathfrak{g}$ denotes the Lie algebra of G. In the physics literature the condition $F = 0$ is usually called *gauge fixing* and F is usually called *gauge-fixing function*. For $x \in \mathcal{M}'$, let $A(x)$ be the differential $\mathrm{d}F(x)$ restricted to the vertical tangent space at x. This can be identified with the Lie algebra $\mathfrak{g}$ and thus we can regard $A(x)$ as an endomorphism of $\mathfrak{g}$ and denote by J its determinant. In the physics literature J is usually called *Faddeev–Popov determinant*. In fact, we get

$$I = \int_{\mathcal{M}'} f \delta_0(F) J \mu,$$

where δ_0 denotes the delta function at $0 \in \mathfrak{g}$. Let us now rewrite I such that it is suitable for the stationary phase expansion formula. We need to write $\delta_0(F)$ and J in terms of an exponential map. Therefore, we use the Fourier transform of the delta function and the Grassmann (odd) integration as in Sect. 6.2.3. Denote by $\langle\ ,\ \rangle$ the canonical pairing between $\mathfrak{g}^*$ and $\mathfrak{g}$. Then we have

$$I = C \int f(x) \mu(x) \exp\left(\frac{\mathrm{i}}{\hbar}\langle \lambda, F(x) \rangle\right) \omega(\lambda) \exp\left(\frac{\mathrm{i}}{\hbar}\langle \bar{c}, A(x)c \rangle\right),$$

where we integrate over $x \in \mathcal{M}'$, $\lambda \in \mathfrak{g}^*$, $c \in \Pi\mathfrak{g}$, $\bar{c} \in \Pi\mathfrak{g}^*$ and where C is a constant depending on $\hbar$ and on the choice of the top form $\omega \in \bigwedge^{\text{top}} \mathfrak{g}^*$. Note that we have added the prefactors $\frac{i}{\hbar}$ for later purposes.

Remark 6.4.1 In the physics literature c is usually called *ghost* and $\bar{c}$ is usually called *anti-ghost*. λ is usually called *Lagrange multiplier*.

Note that if we choose a basis $\{c^i\}$ of $\mathfrak{g}^*$, then $\bigwedge \mathfrak{g}^*$ can be identified with the algebra generated by the c^is with the relations

$$c^i c^j = -c^j c^i, \qquad \forall i, j.$$

The generators $c^i \in \mathfrak{g}^*$ are usually called *ghost variables*. Similarly, one can introduce *anti-ghost variables* $\bar{c}_i \in \mathfrak{g}$.

Remark 6.4.2 The determinant of A can also be obtained in terms of an ordinary Gaussian integral over $\mathfrak{g} \times \mathfrak{g}^*$. However, then one needs to put A^{-1} into the exponential. In the infinite-dimensional case we can use the techniques of Feynman diagrams, but only if we consider local functions. If the action of the group is local, the matrix A will be a differential operator, so $\langle \bar{c}, Ac \rangle$ will be a local function, in contrary to the quadratic function defined in terms of the Green's function A^{-1}.

The expectation value of an invariant function g with respect to a given invariant action S can be written as

$$\langle g \rangle = \frac{\displaystyle\int_{\mathcal{M}} \exp(iS/\hbar) g\mu}{\displaystyle\int_{\mathcal{M}} \exp(iS/\hbar)\mu} = \frac{\displaystyle\int_{\widetilde{\mathcal{M}}} \exp(iS_F/\hbar) g\widetilde{\mu}}{\displaystyle\int_{\widetilde{\mathcal{M}}} \exp(iS/\hbar)\widetilde{\mu}} =: \langle g \rangle_F, \qquad (6.4.1)$$

where

$$\widetilde{\mathcal{M}} := \mathcal{M}' \times \Pi\mathfrak{g} \times \mathfrak{g}^* \times \Pi\mathfrak{g}^*, \qquad (6.4.2)$$

$$S_F := S + \langle \lambda, F \rangle - \langle \bar{c}, Ac \rangle, \qquad (6.4.3)$$

and $\widetilde{\mu} := \mu\omega$. The function S_F is usually called the *gauge-fixed action*.

6.4.2 The BRST Formalism

Note that, by construction, the right-hand-side of (6.4.1) is independent of F. Moreover, the assumptions we had are quite restrictive, namely, we want to have a measure-preserving action on $\mathcal{M}$ by a compact Lie group G and we have to assume that the principal bundle $\mathcal{M} \to \mathcal{M}/G$ is trivial. On the other hand, to define

$$\langle g \rangle_F = \frac{\displaystyle\int_{\widetilde{\mathcal{M}}} \exp(\mathrm{i}S_F/\hbar)g\widetilde{\mu}}{\displaystyle\int_{\widetilde{\mathcal{M}}} \exp(\mathrm{i}S/\hbar)\widetilde{\mu}} \qquad (6.4.4)$$

we only need the infinitesimal action

$$X : \mathfrak{g} \to \mathfrak{X}(\mathcal{M})$$

$$\gamma \mapsto X_\gamma$$

of a Lie algebra $\mathfrak{g}$ on $\mathcal{M}$. In this case A is simply given by

$$A_\gamma = L_{X_\gamma} F, \quad \gamma \in \mathfrak{g}.$$

Moreover, we want to relax the condition that $F^{-1}(0)$ defines a section and rather require that $A(x)$ should be non-degenerate for all $x \in \mathcal{M}'$. Of course, we also require that the integrals are well-defined and that the denominator of (6.4.1) does not vanish.

Let $\mathcal{F} \subset C^\infty(\mathcal{M}', \mathfrak{g})$ be the space of all the allowed gauge-fixing functions. As in this case $\langle g \rangle_F$ is well-defined, it makes sense to consider it at more general instances. Note also that if the Lie group is not compact, an invariant function is not a test function. It is then understood that it is replaced by a test function that in a neighborhood of the zeros of F coincides with the given function. To avoid cumbersome notation, we will never explicitly change the function, so we will write e.g. $\int_{\mathbb{R}} \delta_0(x)\mathrm{d}x = 1$, without mentioning that the constant function 1 is replaced by a test function which is one in a neighborhood of zero. However, since the gauge-fixing function F is arbitrarily chosen, we want the conditions on $\langle g \rangle_F$ to be locally independent of F.

Definition 6.4.3 (Gauge-Fixing Independence) A locally constant function on $\mathcal{F}$ is called *gauge-fixing independent*.

Theorem 6.4.4 *Let $X : \mathfrak{g} \to \mathcal{M}$ be an infinitesimal action of the Lie algebra $\mathfrak{g}$ on the manifold $\mathcal{M}$. If S and g are invariant functions and*

$$\mathrm{div}\, X_\gamma + \mathrm{Tr}\, \mathrm{ad}_\gamma = 0, \qquad \forall \gamma \in \mathfrak{g}, \qquad (6.4.5)$$

then $\langle g \rangle_F$ is gauge-fixing independent.

The condition in Theorem 6.4.4 basically tells us that the divergence of X is a constant function. In particular, for the case when the Lie algebra is *unimodular* (i.e. $\mathrm{Tr}\, \mathrm{ad}_\gamma = 0$ for all γ), the condition says that the infinitesimal action must be measure-preserving. The discussions before are covered by Theorem 6.4.4 since the Lie algebra of a compact Lie group is indeed unimodular.

6.4.2.1 The BRST Operator and the Proof of Theorem 6.4.4

The infinitesimal action of $\mathfrak{g}$ on $\mathcal{M}$ gives $C^\infty(\mathcal{M})$ the structure of a $\mathfrak{g}$-module. Therefore, we can consider the Lie algebra complex $\bigwedge \mathfrak{g}^* \otimes C^\infty(\mathcal{M})$. Note that the Lie algebra differential δ is in particular a derivation on $\bigwedge \mathfrak{g}^* \otimes C^\infty(\mathcal{M})$ and moreover defines a vector field on $\mathcal{M} \times \Pi\mathfrak{g}$. Recall that a vector field on the superspace $\Pi\mathfrak{g}$ is an element of $\bigwedge \mathfrak{g} \otimes \mathfrak{g}$. If $\mathfrak{g}$ is a Lie algebra, the commutator can be regarded as an element of $\bigwedge^2 \mathfrak{g}^* \otimes \mathfrak{g}$ and as such it is a vector field on $\Pi\mathfrak{g}$. Vector fields on $\mathcal{M} \times \Pi\mathfrak{g}$ can then be identified with elements of $\bigwedge \mathfrak{g}^* \otimes \mathfrak{g} \otimes C^\infty(\mathcal{M}) \oplus \bigwedge \mathfrak{g}^* \otimes \mathfrak{X}(\mathcal{M})$. The commutator tensor of the constant function 1 is an element of $\bigwedge^2 \mathfrak{g}^* \otimes \mathfrak{g} \otimes C^\infty(\mathcal{M})$ while the infinitesimal action of $\mathfrak{g}$ on $\mathcal{M}$ is an element of $\mathfrak{g}^* \otimes \mathfrak{X}(\mathcal{M})$. Thus they define vector fields on $\mathcal{M} \times \Pi\mathfrak{g}$ and δ is their sum. If we choose a basis $\{e_i\}$ of $\mathfrak{g}$ and denote by $\{f_{ij}^k\}$ the corresponding structure constants, the algebra of functions on $\Pi\mathfrak{g}$ can be identified with the graded commutative algebra with odd generators c^i (ghost variables) and we have

$$\delta c^k = -\frac{1}{2} f_{ij}^k c^i c^j.$$

On functions $f \in C^\infty(\mathcal{M})$, we get that δ acts by

$$\delta f = c^i L_{X_{e_i}} f.$$

The vector field δ can be extended to $\widetilde{\mathcal{M}}$ by adding it to the vector field on $\mathfrak{g}^* \otimes \Pi\mathfrak{g}^*$. Moreover, using the dual basis $\{e^i\}$ of $\mathfrak{g}^*$, the algebra of functions can be identified with the graded commutative algebra with odd generators $\bar{c}_i$ and even generators λ_i. There we define

$$\delta \bar{c}_i = \lambda_i, \qquad \delta \lambda_i = 0, \quad \forall i. \tag{6.4.6}$$

More abstractly, note that a polynomial vector field on $\mathfrak{g}^* \times \Pi\mathfrak{g}^*$ is an element of $\bigwedge \mathfrak{g} \otimes \mathrm{Sym}(\mathfrak{g}) \otimes (\mathfrak{g}^* \oplus \mathfrak{g}^*)$. The identity operator can then be seen as an element of $\mathfrak{g} \otimes \mathfrak{g}^*$. Using the inclusion map

$$i : \mathfrak{g} \otimes \mathfrak{g}^* \hookrightarrow \bigwedge \mathfrak{g} \otimes \mathrm{Sym}(\mathfrak{g}) \otimes (\mathfrak{g}^* \oplus \mathfrak{g}^*),$$

$$a \otimes b \mapsto 1 \otimes a \otimes (b \oplus 0),$$

we can think of it as a vector field on $\mathfrak{g}^* \times \Pi\mathfrak{g}^*$. Moreover, note that this vector field corresponds to the de Rham differential on $\Pi\mathfrak{g}^*$.

Lemma 6.4.5 *The map δ is a differential (i.e. an odd derivation that squares to zero) on $\bigwedge \mathfrak{g}^* \otimes C^\infty(\mathcal{M}) \otimes \bigwedge \mathfrak{g} \otimes \mathrm{Sym}(\mathfrak{g})$. It has degree one with respect to the grading*

$$\deg(\alpha \otimes f \otimes \beta \otimes \gamma) := \deg(\alpha) - \deg(\beta).$$

In fact, we can rewrite Lemma 6.4.5 by saying that δ is a cohomological vector field on $\widetilde{\mathcal{M}}$. In the physics literature, δ is usually called the *BRST operator*. A function on $\mathcal{M}$ can also be considered as a function on $\widetilde{\mathcal{M}}$. By definition of δ, a function f is invariant if and only if $\delta f = 0$. Note that mathematically, the cochain complex constructed by the Lie algebra $\mathfrak{g}$ corresponds to the *Chevalley–Eilenberg complex* as in Definition 2.7.48.

Exercise 6.4.6 Show that

$$\operatorname{div} \delta = \operatorname{div} X_c + \operatorname{Tr} \operatorname{ad}_c.$$

Note that some gauge-fixing function $F \colon \mathcal{M} \to \mathfrak{g}$ can be regarded as an element of $C^\infty(\mathcal{M}) \otimes \mathfrak{g}$. Using the inclusion

$$C^\infty(\mathcal{M}) \otimes \mathfrak{g} \hookrightarrow C^\infty(\mathcal{M}) \otimes \bigwedge \mathfrak{g} \hookrightarrow \bigwedge \mathfrak{g}^* \otimes C^\infty(\mathcal{M}) \otimes \bigwedge \mathfrak{g} \otimes \operatorname{Sym}(\mathfrak{g}),$$

we can associate to F a function Ψ_F on $\widetilde{\mathcal{M}}$. With the same notations as above, we have

$$\Psi_F = \bar{c}_i F^i.$$

The odd function Ψ_F is usually called *gauge-fixing fermion*.

Exercise 6.4.7 Show that

$$S_F = S + \delta\Psi_F.$$

Assume now that the action S is invariant, i.e. $\delta S = 0$.

Lemma 6.4.8 *Let g be a function on $\widetilde{\mathcal{M}}$. If $\delta g = 0$ and δ is divergence-free, then*

$$I_F := \int_{\widetilde{\mathcal{M}}} \exp(\mathrm{i}S_F/\hbar) g \widetilde{\mu},$$

is gauge-fixing independent.

Proof Let (F_t) be a smooth family of gauge-fixing functions. Then

$$\frac{\mathrm{d}}{\mathrm{d}t} I_{F_t} = \frac{\mathrm{i}}{\hbar} \int_{\widetilde{\mathcal{M}}} \delta\left(\frac{\mathrm{d}}{\mathrm{d}t}\Psi_{F_t}\right) \exp(\mathrm{i}S_F/\hbar) g \widetilde{\mu}$$

$$= \frac{\mathrm{i}}{\hbar} \int_{\widetilde{\mathcal{M}}} \delta\left(\frac{\mathrm{d}}{\mathrm{d}t}\Psi_{F_t} \exp(\mathrm{i}S_F/\hbar) g\right) \widetilde{\mu} = \frac{\mathrm{i}}{\hbar} \int_{\widetilde{\mathcal{M}}} \operatorname{div} \delta \exp(\mathrm{i}S_F/\hbar) g \widetilde{\mu} = 0.$$

$\square$

Proof of Theorem 6.4.4 Using Lemma 6.4.8 together with Exercise 6.4.6 we immediately get the proof. $\square$

An particular case of a δ-closed function is a δ-exact function. These functions are irrelevant for the computation of any expectation values. In fact, we get

$$\int_{\widetilde{\mathcal{M}}} \exp(iS_F/\hbar)\delta h\widetilde{\mu} = \int_{\widetilde{\mathcal{M}}} \delta\,(\exp(iS_F/\hbar)h)\,\widetilde{\mu} = 0$$

whenever S is invariant and δ is divergence-free. We can now extend Theorem 6.4.4 to the following theorem.

Theorem 6.4.9 *Let $X: \mathfrak{g} \to \mathcal{M}$ be an infinitesimal action of the Lie algebra $\mathfrak{g}$ on the manifold $\mathcal{M}$. If the action S is invariant and the BRST operator δ is divergence-free (i.e. (6.4.5) holds), then*

(1) $\langle g\rangle_F$ is gauge-fixing independent for all $g \in \ker \delta$,
(2) $\langle g\rangle_F$ for all $g \in \operatorname{im} \delta$.

Hence, the expectation value defines a linear function on the δ-cohomology.

Remark 6.4.10 (Ward Identities) Note that point (2) produces identities relating expectation values of different quantities. Such identities as called *Ward identities* and usually have non-trivial content.

Example 6.4.11 (Translations) Let $\mathcal{M} = \mathfrak{g} = \mathbb{R}$. Moreover, let $\mathfrak{g}$ act by infinitesimal translations. If we denote by x the coordinate on $\mathcal{M}$, we have $\delta x = c$ and $\delta c = 0$. Assume S and g are constant. Then

$$\langle g\rangle_F = \frac{\exp(iS/\hbar)\displaystyle\int \delta_0(F(x))F'(x)g}{\exp(iS/\hbar)\displaystyle\int \delta_0(F(x))F'(x)} = g$$

if the denominator is different from zero. Clearly, $\langle g\rangle_F$ is gauge-fixing independent. Similarly, one can treat rotation-invariant functions on $\mathcal{M} = S^1$. A section here is just a point. We define $\mathcal{M}'$ to be a neighborhood of this point. After identifying $\mathcal{M}'$ with $\mathbb{R}$, we can proceed as above with the gauge-fixing function F being any function with a single non-degenerate zero corresponding to the image of the section.

Example 6.4.12 (Plane Rotations) Let $\mathcal{M} = \mathbb{R}^2 \setminus \{0\}$ and $\mathfrak{g} = \mathfrak{so}_2(\mathbb{R})$ acting by infinitesimal rotations. If we denote by x and y the coordinates on $\mathbb{R}^2$, we have

$$\delta x = yc, \quad \delta y = -xc, \quad \delta c = 0.$$

Define then $\mathcal{M}' := \{x > 0\}$. A possible choice for the gauge-fixing function F is then the function $F(x, y) = y$. Thus, the gauge-fixed action is given by

$$S_F(x, y, c, \lambda, \bar{c}) = S(x, y) + \lambda y + \bar{c}xc,$$

where S is the given rotation-invariant action. Hence, we get that

$$\langle g \rangle_F = \frac{\int_0^{+\infty} \exp(iS(x,0)/\hbar)g(x,0)x\,dx}{\int_0^{+\infty} \exp(iS(x,0)/\hbar)x\,dx},$$

which is equal to the expected expression

$$\frac{\int_\mathcal{M} \exp(iS(x,y)/\hbar)g(x,y)dx\,dy}{\int_\mathcal{M} \exp(iS(x,y)/\hbar)dx\,dy}$$

for a $\mathfrak{so}_2(\mathbb{R})$-equivariant function g.

6.4.3 Infinite-Dimensional Case

In the infinite-dimensional case, we want to consider (6.4.4) whenever it makes sense as a perturbative expansion. In this case $\mathcal{M}$ is an infinite-dimensional manifold, $\mathfrak{g}$ an infinite-dimensional Lie algebra acting freely on $\mathcal{M}$ and S some local action function on $\mathcal{M}$. The space $\widetilde{\mathcal{M}}$ and the BRST operator δ will be exactly defined as above and δ will still be a cohomological vector field on $\widetilde{\mathcal{M}}$. A gauge-fixing function F will be allowed whenever the corresponding A is non-degenerate and the critical point of the action at a zero of F is also non-degenerate. Then, for suitable functions g, we will be able to define $\langle g \rangle_F$ as a perturbative expansion using Feynman diagrams. Note that an observable in this setting is a δ-cohomology class g for which $\langle g \rangle_F$ is well-defined. In fact, Theorem 6.4.9 will hold whenever $\mathrm{div}\,\delta = 0$. Note that, however, the divergence of δ is not defined a priori and has to be understood in terms of expectation values. The usual way to proceed is to assume Theorem 6.4.9 to hold and to derive from it properties of the expectation values. Once they are properly defined in terms of Feynman diagrams, one can check whether the identities hold and any deviation is called an *anomaly*.

Note also that if our aims are of mathematical nature, Theorem 6.4.9 provides a source for a lot of interesting conjectures (which, fortunately, in most cases turn out to be true).

6.4.3.1 The Trivial Poisson Sigma Model on the Plane

Consider a 2-dimensional generalization of Sect. 6.3 which is also the basis for the study of the Poisson sigma model. Let ξ and η be a 0-form and 1-form on the plane respectively. We assume that they vanish at infinity sufficiently fast (e.g., as Schwarz functions) so that we may define the action

$$S := \int_\Sigma \eta \, d\xi, \tag{6.4.7}$$

with $\Sigma = \mathbb{R}^2$. Note that we have dropped the wedge product to avoid cumbersome notation and we will stick to this convention for the rest of the discussion. The space of fields is given by $\mathcal{M} = \Omega_0^0(\mathbb{R}^2) \oplus \Omega_0^1(\mathbb{R}^2)$. On $\mathcal{M}$ we have an action of the abelian Lie algebra $\mathfrak{g} = \Omega_0^0(\mathbb{R}^2)$, given by the monomorphism $i \circ d$, where

$$\mathfrak{g} \overset{d}{\to} \Omega_0^1(\mathbb{R}^2) \overset{i}{\hookrightarrow} \mathcal{M}.$$

The action function S is clearly invariant. The BRST differential δ is given on coordinates by

$$\delta\xi = 0, \qquad \delta\eta = dc.$$

In order to define a gauge-fixing function, we want to choose a Riemannian metric on $\mathbb{R}^2$. Denote by $*$ the *Hodge star operator* induced by the Riemannian metric as in Sect. 2.6.1. Then, using the Hodge star, we can define the pairing as in Definition 2.6.7

$$\langle \alpha, \beta \rangle_* := \int_{\mathbb{R}^2} (*\alpha)\beta, \quad \alpha \in \Omega^j(\mathbb{R}^2), \, \beta \in \Omega^k(\mathbb{R}^2), \quad j, k = 0, 1, 2. \tag{6.4.8}$$

Let $d^* := *d*$ be the formal adjoint of the de Rham differential and choose the gauge-fixing function $F(\xi, \eta) = d^*\eta$. Note that different metrics will give different gauge-fixing functions. One can show that the corresponding operator A is then given by the Laplacian on 0-forms which is invertible for the given conditions at infinity. There is also a unique critical point, i.e. a solution to $d\xi = d\eta = 0$, satisfying the gauge-fixing condition $d^*\eta = 0$, in particular, $\xi = \eta = 0$. Hence, this gauge-fixing is indeed allowed.

Using integration, we can identify $\mathfrak{g}^*$ with $\Omega^2(\mathbb{R}^2)$. The corresponding gauge-fixing fermion is

$$\Psi_F = \int_{\mathbb{R}^2} \bar{c} \, d^*\eta.$$

This gives the gauge-fixed action

$$S_F = \int_{\mathbb{R}^2} \left(\eta \, d\xi + \lambda \, d^*\eta - \bar{c} \, d^*dc \right). \tag{6.4.9}$$

Using the pairing as in (6.4.8), we can rewrite the gauge-fixed action as

$$S_F = \frac{1}{2}\langle \phi, \mathbf{d}\phi \rangle_* - \langle *\bar{c}, \Delta c \rangle_*,$$

where $\Delta := d^*d + dd^*$ denotes the *Hodge Laplacian*,

$$\phi := \begin{pmatrix} \xi \\ \eta \\ \lambda \end{pmatrix} \in \Omega_0^0(\mathbb{R}^2) \oplus \Omega_0^1(\mathbb{R}^2) \otimes \Omega_0^2(\mathbb{R}^2)$$

and

$$\mathbf{d} := \begin{pmatrix} 0 & *d & 0 \\ *d & 0 & d* \\ 0 & d* & 0 \end{pmatrix}.$$

The propagators between ϕ and c or $\bar{c}$ are clearly zero. Hence, we get

$$\langle *\bar{c}(z)c(w) \rangle_0 = -i\hbar G_0(w, z),$$

where G_0 denotes the Green's function of the Hodge Laplacian acting on functions. Now in order to get the propagator between two fields ϕ, we need to invert the symmetric operator $\mathbf{d}$. For this, we first compute its square

$$\mathbf{d}^2 = \begin{pmatrix} \Delta & 0 & 0 \\ 0 & \Delta & 0 \\ 0 & 0 & \Delta \end{pmatrix}.$$

Then, we can observe that $\mathbf{d}^{-1} = \mathbf{d}\mathbf{d}^{-2}$ and hence

$$\mathbf{d}^{-1} = \begin{pmatrix} 0 & *d\Delta^{-1} & 0 \\ *d\Delta^{-1} & 0 & d*\Delta^{-1} \\ 0 & d*\Delta^{-1} & 0 \end{pmatrix}.$$

In particular, we have

$$\langle \xi(z)\eta(w) \rangle_0 = i\hbar *_z d_z G_1(z, w) = i\hbar *_w d_w G_0(w, z),$$

where G_1 denotes the Green's function of the Hodge Laplacian acting on 1-forms. Let us introduce the *superfields*

$$\boldsymbol{\xi} := \xi - d^*\bar{c},$$
$$\boldsymbol{\eta} := c + \eta. \tag{6.4.10}$$

Then we can define a *superpropagator*

$$i\hbar\theta(z, w) := \langle \boldsymbol{\xi}(z)\boldsymbol{\eta}(w)\rangle_0 = \langle \xi(z)\eta(w)\rangle_0 - \langle \mathrm{d}^*\bar{c}(z)c(w)\rangle_0$$

$$= i\hbar(*_z\mathrm{d}_z + *_w\mathrm{d}_w)G_0(w, z) \in \Omega^1(C_2(\mathbb{R}^2)), \qquad (6.4.11)$$

where $C_2(\mathbb{R}^2)$ denotes the configuration space of two points in $\mathbb{R}^2$ as in Remark 5.3.7.

Lemma 6.4.13 *If we choose the Euclidean metric, then*

$$\theta = \frac{\mathrm{d}\phi_E}{2\pi},$$

where d *denotes the differential on* $C_2(\mathbb{R}^2)$ *and* $\phi_E(z, w)$ *the Euclidean angle between a fixed reference line and the line passing through z and w.*

Proof The Green's function for the Euclidean Laplacian in two dimensions is given by

$$G_0(z, w) = \frac{1}{2\pi}\log(|z - w|),$$

where $|\ |$ denotes the Euclidean norm. In complex coordinates we have

$$G_0(z, w) = \frac{1}{4\pi}\log((z - w)(\bar{z} - \bar{w})).$$

Thus, we get

$$\mathrm{d}_z G_0(z, w) = \frac{1}{4\pi}\left(\frac{\mathrm{d}z}{z - w} + \frac{\mathrm{d}\bar{z}}{\bar{z} - \bar{w}}\right).$$

The Euclidean Hodge star operator in complex coordinates gives $*\mathrm{d}z = i\mathrm{d}z$ and $*\mathrm{d}\bar{z} = i\mathrm{d}\bar{z}$. Hence, we have

$$*_z\mathrm{d}_z G_0(z, w) = \frac{1}{4\pi i}\left(\frac{\mathrm{d}w}{w - z} - \frac{\mathrm{d}\bar{w}}{\bar{w} - \bar{z}}\right).$$

Summing everything up, we get

$$\theta = \frac{1}{4\pi i}\left(\frac{\mathrm{d}z - \mathrm{d}w}{z - w} - \frac{\mathrm{d}\bar{z} - \mathrm{d}\bar{w}}{\bar{z} - \bar{w}}\right) = \frac{1}{4\pi i}\mathrm{d}\log\left(\frac{z - w}{\bar{z} - \bar{w}}\right).$$

On the other hand, we have $z - w = |w - z| \exp(i\phi)$, which gives

$$\phi = \frac{1}{2i} \log \left(\frac{z - w}{\bar{z} - \bar{w}} \right),$$

and hence we get the claim.   □

Remark 6.4.14 The cohomology class of θ is in fact the generator of $H^1(C_2(\mathbb{R}^2), \mathbb{Z})$. It is not difficult to see that other choices of metric will still give the same cohomolgy class. One can easily note that $*_w \theta(z, w)$ is the Green's function of the operator $P := *d\Delta^{-1}*$ which is a *parametrix* for the de Rham differential on forms that vanish at infinity, in particular,

$$dP + Pd = \mathrm{id}.$$

The convolution relating P to θ can be written as

$$P\alpha = -\pi_2(\theta \pi_1^* \alpha), \quad \alpha \in \Omega_0^j(\mathbb{R}^2), \quad j = 0, 1, 2,$$

with π_1 and π_2 being the projections onto $\mathbb{R}^2$. Then

$$dP\alpha - Pd\alpha = -(\pi_2)_*(d\theta \pi_1^* \alpha) + \pi_*^\partial(\theta)\alpha,$$

where $\pi_*^\partial(\theta)(w)$ denotes the integral of θ along a limiting small circle around w. Since P is a parametrix and α is arbitrary, we can see that in general θ is closed and integrates to 1 along the generators of $H_1(C_2(\mathbb{R}^2), \mathbb{Z})$.

6.4.3.2   Expectation values

Note that any function of ξ is BRST-invariant, i.e. we can consider the evaluation of ξ at some point u. A function of $\int_\gamma \eta$ is also invariant for any closed curve γ. Hence, the expectation value

$$\left\langle \xi(u) \int_\gamma \eta \right\rangle_0 =: i\hbar W_\gamma(u), \quad u \notin \mathrm{im}\,\gamma,$$

is independent of the gauge-fixing. Since we also have

$$i\hbar W_\gamma(u) = \left\langle \xi(u) \int_\gamma \eta \right\rangle_0,$$

we can immediately see that W_γ is in fact the winding number of γ around u. This number is in fact invariant under deformations of γ or displacements of u, which indicates that the theory is topological. For example, let us deform γ to γ'. Denoting by σ a 2-chain whose boundary is $\gamma - \gamma'$, we get

$$W_\gamma(u) - W_{\gamma'}(u) = \left\langle \xi(u) \int_\sigma d\eta \right\rangle_0 .$$

Moreover, introduce the sequence of divergence-free vector fields $X_r(\xi, \eta) = \lambda_r \oplus 0$, where (λ_r) is a sequence of functions that converges almost everywhere to the characteristic function of the image of σ. Then we get

$$W_\gamma(u) - W_{\gamma'}(u) = \lim_{r \to \infty} \langle \xi(u) X_r(S) \rangle_0 = i\hbar \lim_{r \to \infty} \langle X_r(\xi(u)) \rangle_0 = 0,$$

under the assumption that u does not belong to σ.

6.4.3.3   The Trivial Poisson Sigma Model on the Upper Half-Plane

Consider now the action (6.4.7) where $\Sigma = \mathbb{H}^2$. As a boundary condition, we impose that the 1-form η vanishes when restricted to the boundary $\partial \mathbb{H}^2 = \mathbb{R} \times \{0\}$. The Lie algebra acting on the space of fields is given by 0-forms on $\mathbb{H}^2$ vanishing on $\partial \mathbb{H}^2$. The BRST complex can then be defined exactly as before and we can choose the same gauge-fixing function. We define the superpropagator as in (6.8.51) with the difference that we will denote it by ϑ instead of θ in order to avoid any confusion.

Lemma 6.4.15 *If we choose the Euclidean metric, then*

$$\vartheta = \frac{d\phi_h}{2\pi},$$

where d *denotes the differential on* $C_2(\mathbb{H}^2)$ *and* $\phi_h(z, w)$ *denotes the angle between the vertical line through* w *and the geodesic connecting* w *and* z *in the hyperbolic Poincaré metric (recall Fig. 5.4).*

Proof The Green's function $G_0^{\mathbb{H}^2}$ of the Laplacian on $\mathbb{H}^2$ is the restriction to $\mathbb{H}^2$ of the Green's function of the Laplacian on $\mathbb{R}^2$ plus a harmonic function such that the sum satisfies the boundary conditions. In complex coordinates, we need $G_0^{\mathbb{H}^2}(w, z) = 0$ whenever w is real. This can be done by setting

$$G_0^{\mathbb{H}^2}(w, z) = G_0(w, z) - G_0(\bar{w}, z).$$

Then we get

$$\vartheta(z, w) = \theta(z, w) - \theta(z, \bar{w}).$$

Since the hyperbolic angle is given by

$$\phi_h(z, w) = \frac{1}{2i} \log \left(\frac{(z - w)(\bar{z} - w)}{(\bar{z} - \bar{w})(z - \bar{w})} \right),$$

we conclude the claim.   □

Remark 6.4.16 Note that ϑ is the generator of $H^1(C_2(\mathbb{H}^2), \mathbb{H}^2 \times \partial\mathbb{H}^2; \mathbb{Z})$.

6.4.3.4 Some Generalizations

We can also consider a collection of n 0-forms ξ^i and n 1-forms η_i with $i = 1, \ldots, n$. Then we can look at the action

$$\int_\Sigma \eta_i \, d\xi^i.$$

We can also think of ξ and η as forms taking values in $\mathbb{R}^n$. The Lie algebra $\mathfrak{g}$ of symmetries will then consist of the direct sum of n copies of the previous one; in other words it will be the abelian Lie algebra of $\mathbb{R}^n$-valued 0-forms. Let c_i for $i = 1, \ldots, n$ denote the generators of the algebra of functions of $\Pi\mathfrak{g}$. Then we can define the BRST operator through $\delta\xi^i$ and $\delta\eta_i = c_i$. If we choose the gauge-fixing function to be $F_i(\xi, \eta) = d^*\eta_i$, everything remains the same as before. In particular, we can again introduce superfields $\boldsymbol{\xi}^i := \xi^i - d^*\bar{c}^i$ and $\boldsymbol{\eta}_i := c_i + \eta_i$ and compute the *superpropagator*

$$\left\langle \boldsymbol{\xi}^i(z)\boldsymbol{\eta}_j(w) \right\rangle_0 := \begin{cases} i\hbar\theta(z, w)\delta^i_j, & \text{on } \mathbb{R}^2, \\ i\hbar\vartheta(z, w)\delta^i_j, & \text{on } \mathbb{H}^2. \end{cases} \tag{6.4.12}$$

Another generalization can be done by dropping the assumption that the 0-form field vanishes at ∞. More precisely, we denote by X^i a collection of maps to $\mathbb{R}^n$ with no conditions on the boundary or at ∞ and consider the action

$$S = \int_\Sigma \eta_i \, dX^i,$$

where Σ is either $\mathbb{R}^2$ or $\mathbb{H}^2$ and the η_is are 1-forms vanishing on the boundary and at ∞. Critical points are then pairs of constant maps together with closed 1-forms. They will be also degenerate modulo the action of the abelian Lie algebra of 0-forms. However, the degeneracy will be of a very simple type as it is parametrized by the finite-dimensional manifold $\mathbb{R}^n$. In fact, it is enough to choose a measure on $\mathbb{R}^n$ and use Fubini's theorem. We will choose a delta-measure peaked at a point $x \in \mathbb{R}^n$ and require X to map the point ∞ to x. Note that if we write $X^i = x^i + \xi^i$, the ξ^i vanish at infinity and everything is reduced to the previous case.

A final generalization is to replace $\mathbb{R}^n$ by a general manifold M. We think of X^i as a local coordinate expression of a map $X \colon \Sigma \to M$. For the action to be covariant, we need to assume that $\eta_i(u)$ for $u \in \Sigma$ is the local coordinate expression of a 1-form on Σ taking values in the cotangent space of M at $X(u)$. In particular, we assume $\eta \in \Gamma(\Sigma, T^*\Sigma \otimes X^*T^*M)$. The space of fields $\mathcal{M}$ can then be identified with the space of vector bundle maps $T\Sigma \to T^*M$ and the action can be invariantly written as

$$S = \int_\Sigma \langle \eta, \mathrm{d}X \rangle,$$

where $\langle \ , \ \rangle$ denotes the canonical pairing between the tangent and cotangent bundle of M and $\mathrm{d}X$ denotes the differential of the map X regarded as a section of $T^*\Sigma \otimes X^*TM$. If we also require X to map the point ∞ to a given point $x \in M$, we can expand around critical points by setting $X = x + \xi$ with $\xi \colon \Sigma \to T_x M$ and by regarding η as a 1-form taking values in T_x^*M. Choosing local coordinates, we can identify $T_x M$ with $\mathbb{R}^n$, where $n = \dim M$, and reduce everything to the previous case.

We also allow Σ to be any 2-manifold. The previous discussion will change drastically if Σ is not simply connected, as the space of solutions modulo symmetries, with $X = x$, will now be parametrized by $H^1(\Sigma, T_x^*M)$ and one has to choose a measure on this vector space as well.

6.5 The Poisson Sigma Model

Next we want to look at deformations of the trivial Poisson sigma model as discussed before. Here, we will describe how the action of the Poisson sigma model is expressed in order to derive Kontsevich's star product out of it.

6.5.1 Formulation of the Model

We want to formulate a deformation of the action functional without introducing extra structure on Σ (which is either $\mathbb{R}^2$ or $\mathbb{H}^2$). Therefore, the terms we are allowed to add must be 2-forms on Σ given in terms of the fields X^i and η_i. In particular, they need to be linear combinations of terms of the form $\alpha^{ij}(X)\eta_i\eta_j$, $\beta^i_j(X)\eta_i\,\mathrm{d}x^i$ and $\gamma_{ij}(X)\mathrm{d}X^i\mathrm{d}X^j$. Note that we are not considering a term of the form $\phi^i(X)\mathrm{d}\eta_i$, since integration by parts reduces it to a term of the second type. The second and third terms can be absorbed by a redefinition of η adding to it terms which are linear in η and $\mathrm{d}X$. Hence, modulo field redefinitions, the most general deformation of the action is of the form

$$S = \int_{\Sigma} \left(\eta_i \, dX^i + \frac{1}{2} \epsilon \alpha^{ij}(X) \eta_i \eta_j \right) + O(\epsilon^2),$$

where ϵ is the deformation parameter (here typically $\epsilon = \frac{i\hbar}{2}$) and α^{ij} is assumed to be skew-symmetric.

Remark 6.5.1 We want to show that it makes sense to only consider those deformations in which the α^{ij} are the components of a Poisson bivector field and that the BRST formalism is only available if the Poisson structure is affine.

Recall that the BRST operator before acted by $\delta X^i = 0$, $\delta \eta_i = dc_i$ and $\delta c_i = 0$, with $c \in \Pi\mathfrak{g}$ and $\mathfrak{g} = \Omega_0^0(\Sigma, \mathbb{R}^n)$. We would like to deform the trivial δ such that $\delta S = O(\epsilon^2)$ for the new S. Note that we will only consider the restriction of δ to $\mathcal{M} \times \Pi\mathfrak{g}$ as its restriction to $\mathfrak{g}^* \times \Pi\mathfrak{g}$ needs no deformation.

Lemma 6.5.2 *Modulo field redefinitions, there is a unique BRST operator deforming the trivial one such that $\delta S = O(\epsilon^2)$ and $\delta^2 = O(\epsilon^2) + R$ with R vanishing at critical points. It acts by*

$$\delta X^i = -\epsilon \alpha^{ij}(X) c_j + O(\epsilon^2),$$

$$\delta \eta_i = dc_i + \epsilon \partial_i \alpha^{jk}(X) \eta_j c_k + O(\epsilon^2),$$

$$\delta c_i = -\frac{1}{2} \epsilon \partial_i \alpha^{jk}(X) c_j c_k + O(\epsilon^2).$$

Moreover, R vanishes on the whole $\mathcal{M} \times \Pi\mathfrak{g}$ if α is at most linear.

Proof Recall that δ applied to X or η must be linear in the ghost variables c. Hence, the most general deformation of δ (without adding any extra structure on Σ) is of the form

$$\delta X^i = \epsilon v^{ij}(X) c_j + O(\epsilon^2),$$

$$\delta \eta_i = dc_i + \epsilon \left(a_i^{jk}(X) \eta_j c_k + b_i^j(X) dc_j + d_{ij}^k(X) dX^j c_k \right) + O(\epsilon^2),$$

for some functions v^{ij}, a_i^{jk}, b_i^j and d_{ij}^k on $\mathbb{R}^n$. Thus, we get

$$\delta S = \epsilon \int_{\Sigma} \left((a_i^{jk}(X) \eta_j c_k + b_i^j(X) dc_j + d_{ij}^k(X) dX^j c_k) dX^i \right.$$

$$\left. + \eta_i (dX^r \partial_r v^{ij}(X) c_j + v^{ij}(X) dc_j) + \alpha^{ij}(X) dc_i \eta_j \right) + O(\epsilon^2).$$

As the identity $\delta S = O(\epsilon^2)$ must hold for any η, we get the following two equations

$$a_i^{jk}(X)c_k dX^i + dX^r \partial_r v^{jk}(X)c_k + v^{jk}(X)dc_k + \alpha^{kj}(X)dc_k = 0, \qquad (6.5.1)$$

$$\int_\Sigma \left(b_i^j(X)dc_j + d_{ij}^k(X)dX^j c_k \right) dX^i = 0. \qquad (6.5.2)$$

In particular, if we choose X to be the constant map (with value x), we deduce from (6.5.1) that

$$v^{jk}(x)dc_k + \alpha^{kj}(x)dc_k = 0,$$

and since this has to hold for any c, we have

$$\alpha^{jk} = v^{jk}.$$

Plugging into (6.5.2), we get

$$a_i^{jk}(X)c_k dX^i - dX^i \partial_i \alpha^{jk}(X)c_k = 0,$$

and since this has to hold for all c and X, we get

$$a_i^{jk} = \partial_i \alpha^{jk}.$$

Using integration by parts, (6.5.2) gives

$$\int_\Sigma \left(-dX^r \partial_r b_i^j(X)c_j + d_{ij}^k(X)dX^j c_k \right) dX^i = 0,$$

and since this has to hold for all X and c, we finally get

$$d_{ij}^k = \partial_j b_i^k.$$

Hence, we have shown that

$$\delta X^i = -\epsilon \alpha^{ij}(X)c_j + O(\epsilon^2),$$

$$\delta \eta_i = dc_i + \epsilon \left(\partial_i \alpha^{jk}(X)\eta_j c_k + d(b_i^j(X)c_j) \right) + O(\epsilon^2),$$

which, after redefinition $c_i \mapsto c_i - \epsilon b_i^j(X)c_j + O(\epsilon^2)$, gives the first two equations in Lemma 6.5.2. For the last equation, we recall that the BRST operator on c must be quadratic in c, so its general form is

$$\delta c_i = \frac{1}{2}\epsilon f_i^{jk}(X)c_j c_k + O(\epsilon^2).$$

To determine the *structure* functions f_i^{jk}, we can compute δ^2. Note that $\delta^2 X^i = \delta^2 c_i = O(\epsilon^2)$. On the other hand,

$$\delta^2 \eta_i = \epsilon \left(\frac{1}{2} \mathrm{d}\big(f_i^{jk}(X) c_j c_i \big) + \partial_i \alpha^{jk}(X) \mathrm{d}c_j c_k \right) + O(\epsilon^2)$$

$$= \epsilon \left(\big(f_i^{jk}(X) + \partial_i \alpha^{jk}(X) \big) \mathrm{d}c_j c_k + \frac{1}{2} \mathrm{d}X^r \partial_r f_i^{jk}(X) c_j c_k \right) + O(\epsilon^2).$$

At a critical point (where $\mathrm{d}X^i = O(\epsilon^2)$) the third summand of the last equation vanishes. Thus, $\delta^2 = O(\epsilon^2)$ at critical points implies that

$$f_i^{jk} = -\partial_i \alpha^{jk},$$

which proves the last equation in Lemma 6.5.2, Note also that

$$\delta^2 \eta_i = -\frac{1}{2} \epsilon \mathrm{d}x^r \partial_r \partial_i \alpha^{jk}(X) c_j c_k,$$

which is zero (not only at critical points) whenever α is at most linear.   □

Next, we want to extend deformations beyond the first order in ϵ. Even without knowing the following terms, we can already state the following Lemma.

Lemma 6.5.3 $\delta^2 = O(\epsilon^3)$ *at critical points only if α is Poisson.*

Proof Note that we have

$$\delta^2 X^i = -\epsilon^2 \left(\alpha^{rk}(X) c_k \partial_r \alpha^{ij}(X) c_j + \frac{1}{2} \alpha^{ij}(X) \partial_i \alpha^{rj}(X) c_r c_j \right) + O(\epsilon^3).$$

Since this has to hold for all c, we get the Jacobi identity for α.   □

Remark 6.5.4 It is in fact possible to prove, under the assumption that α is Poisson, that this deformation is not only infinitesimal. In particular, we have the following theorem.

Theorem 6.5.5 *Given a Poisson bivector field α, the odd vector field*

$$\delta X^i = -\epsilon \alpha^{ij}(X) c_j,$$

$$\delta \eta_i = \mathrm{d}c_i + \epsilon \partial_i \alpha^{jk}(X) \eta_j c_k,$$

$$\delta c_i = -\frac{1}{2} \epsilon \partial_i \alpha^{jk}(X) c_j c_k,$$

is *cohomological for* α *at most linear or at critical points for all* ϵ. *Moreover,*

$$S := \int_\Sigma \left(\eta_i \, \mathrm{d}X^i + \frac{1}{2}\epsilon\alpha^{ij}(X)\eta_i\eta_j \right)$$

is δ*-closed for all* ϵ.

Exercise 6.5.6 Prove Theorem 6.5.5.

Remark 6.5.7 The geometrical meaning of Theorem 6.5.5 is that there is a distribution of vector fields on $\mathcal{M}$ under which the action is invariant. In general, this distribution is involutive only on the submanifold of critical points of S. It is involutive on whole $\mathcal{M}$ whenever α is at most linear and in this case it can be regarded as the free, infinitesimal action of a Lie algebra. The action S can be generalized to the case when one wants to consider a Poisson manifold (M, α) instead of $\mathbb{R}^n$. For this, one regards X as a map $\Sigma \to M$ and, for a given map X, η is taken to be a section of $T^*\Sigma \otimes X^*T^*M$. If $\langle \ , \ \rangle$ denotes the canonical pairing between the tangent and cotangent bundle of M and by $\alpha^\sharp$ the bundle map $T^*M \to TM$ induced by the Poisson bivector field α (see also Sect. 4.6.1), we can write

$$S = \int_\Sigma \left(\langle \eta, \mathrm{d}X \rangle + \frac{1}{2}\epsilon\langle \eta, \alpha^\sharp\eta \rangle \right).$$

Remark 6.5.8 (Dirac and Courant Sigma Model) Similarly as for Poisson manifolds, there is a way how one can associate a sigma model to manifolds endowed with a *Dirac* or *Courant* structure such as in Sect. 4.6 (see [KSS05, Roy07, CQZ10]). There exists also a notion for the deformation quantization of Dirac manifolds as it was shown by Ševera in [Šev05].

6.5.2 Observables

For the case when $\Sigma = \mathbb{H}^2$, c has to vanish on the boundary. This implies that $\delta X(u) = 0$ for $u \in \partial\Sigma$. Hence, we get

$$\mathcal{O}_{f_1,\ldots,f_k;u_1,\ldots,u_k} := f_1(X(u_1))\cdots f_k(X(u_k)),$$
$$f_1,\ldots,f_k \in C^\infty(\mathbb{R}^n), \quad u_1,\ldots,u_k \in \partial\Sigma \cong \mathbb{R}, \quad u_1 < \cdots < u_k,$$

are observables, i.e. δ-closed functions. In [CF00] it was shown that with the gauge-fixing $\mathrm{d}^*\eta = 0$ for the Euclidean metric on Σ, one has

$$\langle \mathcal{O}_{f_1,\ldots,f_k;u_1,\ldots,u_k} \rangle(x) = f_1 \star \cdots \star f_k(x),$$

where $\langle\ \rangle(x)$ denotes the expectation value for $X(\infty) = x$ (and expanding only around the trivial critical solution $X = x$, $\eta = 0$) while $\star$ denotes Kontsevich's star product for the given Poisson structure. In Sect. 6.7 we will derive this result for the case when α is at most linear so that the BRST formalism is available.

6.6   Phase Space Geometry and Symplectic Groupoids

In [CF01d] Cattaneo and Felder have constructed a canonical symplectic groupoid $\mathcal{G}$ as the phase space of the Poisson sigma model for any n-dimensional Poisson manifold (M, α) where α here denotes the Poisson structure. Formulating the Poisson sigma model in the Hamiltonian formulation it defines an infinite-dimensional Hamiltonian system with constraints. One can then observe that the given constraints generate Hamiltonian vector fields which form an integrable distribution of tangent spaces of codimension $2 \dim M$ on the constraint surface. One can then take the phase space $\mathcal{G}$ to be the leaf space of the corresponding foliation.

6.6.1   Hamiltonian Formulation of the Poisson Sigma Model

We want to consider the Hamiltonian formulation of the Poisson sigma model by considering Σ to be a rectangle $[-T, T] \times I$ with coordinates (t, u). Here we will always consider the interval $I = [0, 1]$. The action is then given by

$$S = \int_\Sigma \left(-\langle \eta_u, \partial_t X \rangle + \langle \partial_u X + \alpha \eta_u, \eta_t \rangle \right) \mathrm{d}u\mathrm{d}t.$$

The boundary conditions are then $\eta_t|_{[-T,T] \times \partial I} = 0$. Thus, the first part of the action S defines a symplectic structure on the space of vector bundle morphisms $TI \rightarrow T^*M$. Moreover, the coefficient of the Lagrange multiplier η_t gives a system of constraints generating a distribution of subspaces spanned by Hamiltonian vector fields. We can then obtain the phase space of the Poisson sigma model by Hamiltonian reduction, i.e. the set of integral manifolds of the mentioned distribution contained in the set of zeros of the constraints. Another point of view is to consider it as a Marsden–Weinstein reduction (see Theorem 3.9.12) for the symplectic action of some infinite-dimensional Lie algebra on the cotangent bundle of the path space $PM := \{\text{paths } \gamma : I \rightarrow M\}$. We can consider the cotangent bundle of the phase space T^*PM to be given by continuous vector bundle morphisms $(X, \eta) : TI \rightarrow T^*M$ with base map $X : I \rightarrow M$. The fiber $T^*_X PM$ at a map X can be considered as the space of continuous 1-forms on I with values in X^*T^*M.

Note that we have a non-degenerate symplectic pairing

$$T_X^* PM \times T_X PM \to \mathbb{R},$$

$$(\eta, V) \mapsto \int_0^1 \langle \eta(u), V(u) \rangle.$$

We can then define a canonical symplectic form ω on $T^* PM$ by the differential of the 1-form

$$\theta_{(X,\eta)}(V) := - \int_0^1 \langle \eta(u), \pi_* V(u) \rangle, \quad V \in T_{(X,\eta)} T^* PM.$$

Here, we have denoted by $\pi : T^* PM \to PM$ the bundle map. In local coordinates we have

$$\omega(\delta_1 \vec{\Phi}, \delta_2 \vec{\Phi}) = \int_0^1 (\delta_1 X^i \delta_2 \eta_i - \delta_2 X^i \delta_1 \eta_i), \tag{6.6.1}$$

where $\vec{\Phi} = (X^1, \ldots, X^n, \eta_1, \ldots, \eta_n)$ with $X^i : I \to \mathbb{R}$ and $\eta_i = \eta_i^\mu du^\mu$ with $\eta_i^\mu : I \to \mathbb{R}$ for all $i = 1, \ldots, n$. The constrained manifold is given by the EL equation

$$dX(u) + \alpha(X(u))\eta(u) = 0. \tag{6.6.2}$$

We will denote by C the space of solutions of (6.6.2). Let $\vec{\Phi} = (X, \eta)$ be a vector bundle morphism $TI \to T^* M$ and let β be a smooth function $I \to T^* M$ with $\beta(u) \in T_{X(u)}^* M$ for all $u \in I$ and $\beta(0) = \beta(1) = 0$. Define then

$$H_\beta = \int_0^1 \langle dX + \alpha\eta, \beta \rangle.$$

If we vary $\vec{\Phi}$ and let β depend on $\vec{\Phi}$, then H_β will define a Hamiltonian vector field ξ_β. The vector field ξ_β is the infinitesimal gauge transformation with gauge parameter β. It is define by

$$\iota_{\xi_\beta} \omega = dH_\beta.$$

Proposition 6.6.1 (Cattaneo–Felder[CF01d]) *Let $\vec{\Phi} = (X, \eta)$ be in C. Then the subspace of $T_{\vec{\Phi}} PM$ spanned by ξ_β for $\beta \in \Omega^0(I, X^* T^* M)$ is a closed subspace of codimension $2 \dim M$.*

Definition 6.6.2 (Koszul Lie Bracket) We define the Lie bracket on T^*PM as

$$[\beta, \gamma]_K := \mathrm{d}\langle \beta, \alpha\gamma \rangle - \iota_{\alpha\beta}\mathrm{d}\gamma + \iota_{\alpha\gamma}\mathrm{d}\beta, \quad \beta, \gamma \in \Omega^1(M).$$

We call $[\ ,\ ]_K$ the *Koszul Lie bracket*.

Remark 6.6.3 Locally, the Koszul Lie bracket is given as follows:

$$[\beta, \gamma]_K = (\partial_i\alpha^{jk}\beta_j\gamma_k + \alpha^{jk}\partial_j\beta_i\gamma_k + \alpha^{jk}\beta_j\partial_k\gamma_i)\mathrm{d}x^i.$$

Moreover, define $P_0\Omega^1(M)$ to be the Lie algebra of smooth maps $I \to \Omega^1(M)$ such that $\beta(0) = \beta(1) = 0$ endowed with the Koszul Lie bracket $[\beta, \gamma]_K(u) = [\beta(u), \gamma(u)]_K$. We consider the Hamiltonian

$$H_\beta(X, \eta) = \int_I \langle \mathrm{d}X(u) + \alpha(X(u))\eta(u), \beta(X(u), u) \rangle.$$

Note that the existence of a Hamiltonian vector field ξ with $\iota_\xi\omega = \mathrm{d}H$ is not guaranteed in the infinite-dimensional setting. Nevertheless, we get the following proposition:

Proposition 6.6.4 (Cattaneo–Felder[CF01d]) *The following holds:*

(1) For each $\beta \in P_0\Omega^1(M)$, there is a Hamiltonian vector field ξ_β generated by H_β.
(2) The Lie algebra $P_0\Omega^1(M)$ acts on T^*PM by the Hamiltonian vector fields ξ_β, i.e. the map $\beta \mapsto \xi_\beta$ is a Lie algebra homomorphism.
(3) The map $\mu\colon T^*PM \to P_0\Omega^1(M)^*$ satisfying $\langle \mu(X, \eta), \beta \rangle = H_\beta(X, \eta)$ is an equivariant moment map for this action.

6.6.2   The Phase Space and Its Symplectic Groupoid Structure

Note that the set $\mu^{-1}(0)$, i.e. the zeros of the moment map μ, is given by the constrained manifold $\mathcal{C}$. One is tempted to define the phase space by the Marsden–Weinstein quotient

$$T^*PM//H := \mathcal{C}/H,$$

where H is the group of symplectomorphisms generated by the flows of the Hamiltonian vector field ξ_β. The problem with this definition is that the manifold is infinite-dimensional and that the action of the group is very difficult to handle, thus the quotient will be very singular. However, one can still show that the distribution of tangent spaces of $\mathcal{C}$ spanned by ξ_β is integrable and that its integral manifolds are smooth of codimension $2 \dim M$ and are the orbits of H.

The algebraic groupoid structure of $\mathcal{G} = \mathcal{C}/H$ can be naturally defined in terms of composition of paths. In particular, we have an inlcusion $j : M \hookrightarrow \mathcal{G}$ sending a point $x \in M$ to the class of constant solution $X(u) = x$ and $\eta(u) = 0$. Let us define the maps $\ell, r : T^*PM \to M$ by

$$\ell(X, \eta) = X(0), \qquad r(X, \eta) = X(1).$$

It is not difficult to see that these maps are both invariant under the H-action since the symmetries preserve the endpoints and hence they descend to maps $\ell, r : \mathcal{G} \to M$. The composition law is encoded in the following lemma:

Lemma 6.6.5 *In each equivalence class* $[(X, \eta)] \in \mathcal{G} = \mathcal{C}/H$ *there is a representative with* $\eta(0) = \eta(1) = 0$. *Moreover, any representatives with this property can be related by an element of* H_0.

The composition law $[(X, \eta)] = [(X_1, \eta_1)] \bullet [(X_2, \eta_2)]$ can then be defined by choosing a representative as in Lemma 6.6.5 and setting

$$X(u) = \begin{cases} X_1(2u), & 0 \leq u \leq \frac{1}{2}, \\ X_2(2u - 1), & \frac{1}{2} \leq u \leq 1 \end{cases} \tag{6.6.3}$$

$$\eta(u) = \begin{cases} \eta_1^u(2u)\mathrm{d}u, & 0 \leq u \leq \frac{1}{2}, \\ 2\eta_2^u(2u - 1)\mathrm{d}u, & \frac{1}{2} \leq u \leq 1 \end{cases}, \tag{6.6.4}$$

whenever $X_1(1) = X_2(0)$. Here, we have written $\eta_i(u) = \eta_i^u(u)\mathrm{d}u$ for $i = 1, 2$.

Exercise 6.6.6 Check that $[(X, \eta)]$ satisfies the constrained equation (6.6.2) if $[(X_1, \eta_1)]$ and $[(X_2, \eta_2)]$ do.

Using the second part of Lemma 6.6.5, we can see that the class of (X, η) is independent of the choice of representative. The symplectic structure $\omega_\mathcal{G}$ on $\mathcal{G}$ is constructed by restricting the symplectic form ω on T^*PM to an H-invariant closed 2-form on $\mathcal{C}$ whose kernel is given by the tangent spaces to the orbits. Thus, we get that

$$\omega_\mathcal{G}(x)(\xi, \zeta) = \omega(\vec{\Phi})(\widehat{\xi}, \widehat{\zeta}), \quad \xi, \zeta \in T_x\mathcal{G},$$

is independent of the choice of $\vec{\Phi} \in \mathcal{C}$ such that $\pi(\vec{\Phi}) = x$ or of $\widehat{\xi}, \widehat{\zeta} \in T_{\vec{\Phi}}\mathcal{C}$ projecting to ξ, ζ and hence defines a symplectic form on $\mathcal{G}$.

The main theorem is then given as follows:

Theorem 6.6.7 (Cattaneo–Felder[CF01d]) *Assume that there exists a smooth manifold* $\mathcal{G}$ *and a smooth submersion* $\pi : \mathcal{C} \to \mathcal{G}$ *whose fibers are the* H-orbits. *Then* $\mathcal{G} \rightrightarrows M$ *is a symplectic groupoid over* M.

6.7 Deformation Quantization for Affine Poisson Structures

An affine Poisson structure on $\mathbb{R}^n$ is given by a Poisson bivector field α which is at most linear. The linear part α induces on the dual of $\mathbb{R}^n$ a Lie algebra structure while the constant part is a 2-cocycle in the Lie algebra cohomology with trivial coefficients. Let us denote this Lie algebra by $\mathfrak{h}$. The fields of the Poisson sigma model for an affine Poisson structure are then a map $X : \Sigma \to \mathfrak{h}^*$ and a 1-form η on Σ with values in $\mathfrak{h}$. Since $\mathfrak{h}$ is a Lie algebra, we can regard η as a connection 1-form on the trivial principal bundle P over Σ (with gauge group any Lie group whose Lie algebra is $\mathfrak{h}$). For definiteness, we will fix Σ to be $\mathbb{H}^2$ and we will require η to vanish at ∞ and on the boundary. The action is then given by

$$S = \int_\Sigma \left(\eta_i \, dX^i + \frac{1}{2} \alpha^{ij}(X) \eta_i \eta_j \right),$$

where

$$\alpha^{ij}(x) = \chi^{ij} + x^k f_k^{ij}$$

is the given affine Poisson structure on $\mathfrak{h}^*$. Using integration by parts, we can also rewrite it as

$$S = \int_\Sigma \left(\langle X, F_\eta \rangle + \frac{1}{2} \chi(\eta, \eta) \right),$$

where $\langle \ , \ \rangle$ denotes the canonical pairing between $\mathfrak{h}$ and $\mathfrak{h}^*$ while

$$(F_\eta)_i = d\eta_i + \frac{1}{2} f_i^{jk} \eta_j \eta_k$$

is the curvature 2-form of the connection 1-form η. Note that there is a Lie algebra $\mathfrak{g}$, which as a vector space consists of functions $\Sigma \to \mathfrak{h}$ vanishing at ∞ and on the boundary, that acts on the space of fields $\mathcal{M}$ and leaves the action invariant. The BRST operator on $\mathcal{M} \times \Pi\mathfrak{g}$ has the form as in Theorem 6.5.5 with $\epsilon = 1$. Geometrically, we can regard $\mathfrak{g}$ as the Lie algebra of infinitesimal gauge transformations of the principal bundle P; the field η actually transforms as a connection 1-form, while X transforms as a section of the coadjoint bundle in case $\chi = 0$. For $\chi \neq 0$, we can regard $X \oplus 1$ as a section of the coadjoint bundle for the Lie algebra $\widehat{\mathfrak{h}} \cong \mathfrak{h} \oplus \mathbb{R}$ obtained by central extension of $\mathfrak{h}$ through χ. The BRST operator on $\Pi\mathfrak{g}^* \times \mathfrak{g}^*$ has the usual form (6.4.6).

6.7.1 Gauge-Fixing and Feynman Diagrams

Choose a metric on Σ and define the gauge-fixing function $F(X, \eta) = \mathrm{d}^*\eta$. The gauge-fixing fermion is given by

$$\Psi_F = \int_\Sigma \langle \bar{c}, \mathrm{d}^*\eta \rangle$$

and the gauge-fixed action is given by

$$S_F = \int_\Sigma \left(\eta_i\, \mathrm{d}X^i + \frac{1}{2}\alpha^{ij}(X)\eta_i\eta_j + \lambda^i\mathrm{d}^*\eta_i - \bar{c}^k\mathrm{d}^*\left(\mathrm{d}c_k + \partial_k\alpha^{ij}(X)\eta_ic_j\right) \right).$$

Fix the value of X at ∞ to be given by x. We write $X = x + \xi$, where the field ξ has to vanish at infinity. We can observe that S_F has the same form as (6.3.13) (with y given by the collection of $\xi, \bar{c}, \lambda$ and z given by the collection of η, c). Hence, we can write $S_F = S_0 + S_1$ with

$$S_0 = \int_\Sigma \left(\eta_i\, \mathrm{d}X^i + \lambda^i\mathrm{d}^*\eta_i - \bar{c}^k\mathrm{d}^*\mathrm{d}c_k \right),$$

$$S_1 = \int_\Sigma \left(\frac{1}{2}\alpha^{ij}(x+\xi)\eta_i\eta_j - \bar{c}^k\mathrm{d}^*\left(\partial_k\alpha^{ij}(x+\xi)\eta_ic_j\right) \right),$$

and regard S_1 as a perturbation of S_0. If we consider superfields $\boldsymbol{\xi}$ and $\boldsymbol{\eta}$ as in (6.4.10), we can write

$$S_1 = \int_\Sigma \frac{1}{2}\alpha^{ij}(x+\boldsymbol{\xi})\boldsymbol{\eta}_i\boldsymbol{\eta}_j,$$

where integration on Σ is understood to select the 2-form component.

Remark 6.7.1 This shows that as long as the considered observables can be written as functions of the superfields, expectation values are then only computed in terms of the superpropagators (6.4.12). If we denote the superpropagator graphically as an arrow from $\boldsymbol{\eta}$ to $\boldsymbol{\xi}$, the perturbation S_1 is represented by the two vertices as in Fig. 6.10 with the bivalent vertex corresponding to $\alpha^{ij}(x) = \chi^{ij} + x^k f_k^{ij}$ and the trivial vertex corresponding to the structure constants f_k^{ij}.

Let us now consider the observable $\mathcal{O}_{f_1,\ldots,f_k;u_1,\ldots,u_k}$ as in Sect. 6.5.2. Since the evaluation at some point, which is defined as integration along a 0-cycle, is defined to select the 0-form component of a form, we can write

$$\mathcal{O}_{f_1,\ldots,f_k;u_1,\ldots,u_k} := f_1(x + \boldsymbol{\xi}(u_1)) \cdots f_k(x + \boldsymbol{\xi}(u_k)),$$

$$f_1, \ldots, f_k \in C^\infty(\mathbb{R}^n), \quad u_1, \ldots, u_k \in \partial\Sigma \cong \mathbb{R}, \quad u_1 < \cdots < u_k.$$

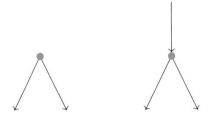

Fig. 6.10 The two vertices

Fig. 6.11 Example of an allowed graph and a non-allowed graph

Computing then the expectation value $\langle \mathcal{O}_{f_1,\ldots,f_k;u_1,\ldots,u_k}\rangle(x)$, we only need the superpropagator. The Feynman diagrams then have three types of vertices:

(1) bivalent vertices in the upper half-plane corresponding to $\alpha^{ij}(x)$,
(2) trivalent vertices in the upper half-plane corresponding to f_k^{ij},
(3) ℓ-valent vertices with $\ell \geq 0$ having only incoming arrows at one of the boundary points u_i corresponding to the ℓ-th derivative of f_i.

Recall that the normal ordering excludes all graphs containing a tadpole (i.e. an edge starting and ending at the same vertex). The combinatorics prevents automatically any vacuum subgraphs (see Fig. 6.11 for examples).

For the case when $k = 2$ and considering the gauge-fixing $d^*\eta = 0$ with respect to the Euclidean metric on $\mathbb{H}^2$, i.e. with the superpropagator determined by the 1-form ϑ as in Lemma 6.4.15, we get

$$\langle \mathcal{O}_{f,g;0,1}\rangle(x) = f \star g(x), \tag{6.7.1}$$

where $\star$ denotes Kontsevich's star product for the given affine Poisson structure.

Exercise 6.7.2 Prove (6.7.1).

6.7.2 Independence of the Evaluation Point

We want to give a formal proof of the independence of the expectation values of $\mathcal{O}_{f_1,\ldots,f_k;u_1,\ldots,u_k}$ from the evaluation points $u_1,\ldots,u_k$. Note first that

$$f(X(v)) - f(X(u)) = \int_u^v \mathrm{d}X^i \partial_i f(X) = \int_u^v \left(\mathrm{d}X^i + \mathrm{d}^*\lambda^i\right)\partial_i f(X) - \delta\Phi,$$

where

$$\Phi := \int_u^v \mathrm{d}^*\bar{c}\partial_i f(X).$$

Let (ω_r) be a sequence of 1-forms on Σ vanishing on the boundary and at ∞ that converges to the measure defined only on the interval $(u, v) \in \Sigma$. If we denote by (a, b), with $b \geq 0$, the coordinates on $\Sigma = \mathbb{H}^2$, a possible choice for this sequence is

$$\omega_r(a, b) = rb\exp(-rb^2/2)\chi_r(a)\mathrm{d}a,$$

where (χ_r) is a sequence of smooth, compactly supported functions converging almost everywhere to the characteristic function of the interval (u, v). Let $Y_{f,r}$ be the local vector field on $\widetilde{\mathcal{M}}$ corresponding to the infinitesimal displacement of η_i by $\omega_r \partial_i f(X)$. Then

$$f(X(v)) - f(X(u)) = \lim_{r \to \infty} Y_{f,r}(S_F) - \delta\Phi.$$

If $\mathcal{O}$ is a BRST-observable depending on the fields outside the closed interval $[u, v]$, we get

$$\langle(f(X(v)) - f(X(u)))\mathcal{O}\rangle = i\hbar \lim_{r \to \infty} \langle Y_{f,r}(\mathcal{O})\rangle - \langle\delta(\Phi\mathcal{O})\rangle = 0.$$

6.7.3 Associativity

The independence of the evaluation points gives us

$$\lim_{v \to u^+} \langle\mathcal{O}_{f,g,h;u,v,w}\rangle_0(x) = \lim_{v \to w^-} \langle\mathcal{O}_{f,g,h;u,v,w}\rangle_0(x).$$

The left-hand-side corresponds intuitively to evaluating first the expectation value of $\mathcal{O}_{f,g;u,v}$, then replacing the result at u and finally computing the expectation value of $\mathcal{O}_{\langle\mathcal{O}_{f,g;u,v}\rangle_0,h;w}$. The result is then given by $(f \star g) \star h$. Repeating the computation on the right-hand-side, we get $f \star (g \star h)$. This formal argument should explain why we expect the star product, defined through the general Poisson sigma model, to be indeed associative.

6.8 The Cattaneo–Felder Construction

We want to state the main theorem which relates Kontsevich's star product to the path integral quantization of the Poisson sigma model for general Poisson structures on $\mathbb{R}^d$. As it turns out, the general construction is based on a symplectic cohomological gauge theory formalism in the sense of Kijowski–Tulczyjew[KT79] provided by Batalin and Vilkovisky [BV77, BV81, BV83] applied to the Poisson sigma model. Interestingly, since the Poisson structure is general, this gauge formalism is actually needed. In fact, shortly after Kontsevich introduced his deformation quantization construction, Cattaneo and Felder have realized that one can use the path integral quantization of the Poisson sigma model in order to describe Kontsevich's star product as an asymptotic expansion of it. They proved that the graphs that appear in Kontsevich's formula coincide with the according Feynman diagrams. In order to describe the precise mathematical formulation of the Batalin–Vilkovisky formalism, we will need some notions from the theory of *supergeometry*.

6.8.1 Supermanifolds and Graded Manifolds

6.8.1.1 Sheaves and Presheaves

Definition 6.8.1 (Presheaf) A *presheaf* $\mathcal{F}$ of *rings* associates to some topological space M the following data:

(1) To any open set $U \subset M$, a ring $\mathcal{F}(U)$,
(2) To any inclusion of open sets $U \hookrightarrow V$ a map $\mathrm{res}_{U,V} : \mathcal{F}(V) \to \mathcal{F}(U)$ which is called the *restriction map* from $\mathcal{F}(V)$ to $\mathcal{F}(U)$. These maps satisfy the properties $\mathrm{res}_{U,U} = \mathrm{id}_{\mathcal{F}(U)}$ and $\mathrm{res}_{V,U} \circ \mathrm{res}_{W,V} = \mathrm{res}_{W,U}$ for inclusions of open sets $U \subset V \subset W$.

Moreover, an element $s \in \mathcal{F}(U)$ is called a *section* of $\mathcal{F}$ over U.

Remark 6.8.2 The objects that are assigned do not need to be rings. They can be also other algebraic objects such as e.g. groups or modules. However, (pre)sheaves of rings will be of particular interest for us.

Example 6.8.3 (C^∞-Functions) If M is a manifold, we can consider the application which assigns to each open set $U \subset M$ the ring of smooth functions $C^\infty(U)$. It is easy to see that this defines a presheaf on M where the restriction maps correspond to restrictions of smooth functions. This also holds more generally for C^k-functions on C^k-manifolds.

Example 6.8.4 (Sections of Vector Bundle) If E is some vector bundle on a topological space M, we can consider the application that assigns to an open subset $U \subset M$ the space of smooth sections $\Gamma(U, E)$. It is easy to show that this defines a presheaf of modules with the obvious restriction maps.

Definition 6.8.5 (Sheaf) A *sheaf* $\mathcal{F}$ of *rings* on some topological space M is a presheaf of rings satisfying the following additional properties for any open cover $\{U_i\}_{i \in I}$ of an open set $U \subset M$:

(1) Suppose that $f_i \in \mathcal{F}(U_i)$ are a collection of sections agreeing on overlaps, i.e. $\mathrm{res}_{U_i, U_i \cap U_j} f_i = \mathrm{res}_{U_j, U_i \cap U_j} f_j$ whenever the intersection exists. Then they lift to some $f \in \mathcal{F}(U)$ which has the property that $\mathrm{res}_{U, U_i} f = f_i$ for all $i \in I$,
(2) Suppose that $f, f' \in \mathcal{F}(U)$ and that $\mathrm{res}_{U, U_i} f = \mathrm{res}_{U, U_i} f'$ for all $i \in I$. Then we have $f = f'$.

Remark 6.8.6 In particular, point (1) says that if we have a collection of sections which agree on overlaps, we can glue them together to get sections over some larger set, and point (2) says that the sections are actually determined through their restrictions. Point (1) is usually called the *gluing axiom* and point (2) is usually called the *identity axiom*.

Definition 6.8.7 (Morphism of sheaves) Let M be a topological space and let $\mathcal{F}$ and $\mathcal{G}$ be sheaves of rings on M. A *morphism of sheaves* $\eta: \mathcal{F} \to \mathcal{G}$ is defined through the following properties:

(1) For each open set $U \subset M$, there is a morphism of rings $\eta_U : \mathcal{F}(U) \to \mathcal{G}(U)$,
(2) For open sets $U \subset V \subset M$, the following diagram commutes:

$$
\begin{array}{ccc}
\mathcal{F}(V) & \xrightarrow{\eta_V} & \mathcal{G}(V) \\
{\scriptstyle \mathrm{res}_{V,U}} \downarrow & & \downarrow {\scriptstyle \mathrm{res}_{V,U}} \\
\mathcal{F}(U) & \xrightarrow{\eta_U} & \mathcal{G}(U)
\end{array}
$$

Example 6.8.8 Consider a morphism $\eta: \mathcal{F} \to \mathcal{G}$ of sheaves of abelian groups. Then the presheaves $\mathrm{im}(\eta)$ and $\mathrm{coker}(\eta)$ defined by $\mathrm{im}(\eta)(U) := \mathrm{im}(\eta_U)$ and $\mathrm{coker}(\eta)(U) := \mathrm{coker}(\eta_U)$ will not be sheaves in general since the gluing axiom is not satisfied on the nose.

Example 6.8.9 (Holomorphic Functions) Let $M = \mathbb{C}$ together with the presheave of holomorphic functions $\mathcal{O}_M$. A morphism of presheaves is then given by locally taking the exponential $\exp: \mathcal{O}_M \to \mathcal{O}_M^\times$. Again, in this setting one can show that $\mathcal{O}_M$ fails to be a sheaf due to the gluing axiom.

Example 6.8.10 (Constant Functions) Let $M = \mathbb{R}^2$ and for an open subset $U \subseteq M$ define $\mathcal{P}(U) := \{f: U \to \mathbb{R} \mid f \text{ is constant}\}$. For $U \subseteq V$ denote by $\mathrm{res}_{V,U}: \mathcal{P}(V) \to \mathcal{P}(U)$ the restriction map. Now consider $U := U_1 \cup U_2$, where U_1 and U_2 are disjoint and non-empty open sets. Define then $\sigma_1 \in \mathcal{P}(U_1)$ by $\sigma_1(u_1) = 0$ for all $u_1 \in U_1$ and $\sigma_2 \in \mathcal{P}(U_2)$ by $\sigma_2(u_2) = c \in \mathbb{R}$ for all $u_2 \in U_2$. The overlap condition is vacuous since $U_1 \cap U_2 = \varnothing$, but there is no constant function $\sigma \in \mathcal{P}(U)$ such that $\sigma|_{U_i} = \sigma_i$ for $i = 1, 2$, i.e. the gluing axiom is not satisfied. Hence, $\mathcal{P}$ is not a sheaf.

Although there are presheaves which fail to be a sheaf, there is a procedure of how to construct a sheaf out of a presheaf. In order to understand this construction, we need to define the following notion:

Definition 6.8.11 (Stalk) Let $\mathcal{F}$ be a sheaf on a topological space M and let $p \in M$. The *stalk* $\mathcal{F}_p$ at p is the ring obtained by the following construction:

(1) The underlying set of $\mathcal{F}_p$ is given by

$$\mathcal{F}_p := \{(f, U) \mid U \subset M \text{ open containing } p, f \in \mathcal{F}(U)\}/ \sim,$$

where we identify $(f, U) \sim (g, V)$ if $\mathrm{res}_{U, U \cap V} f = \mathrm{res}_{V, U \cap V} g$,

(2) The ring operations are defined via $(f, U) + (g, V) = (f + g, U \cap V)$ and $(f, U) \cdot (g, V) = (fg, U \cap V)$. It is easy to check that this is well-defined.

Remark 6.8.12 It is actually easy to see that a morphism $\eta \colon \mathcal{F} \to \mathcal{G}$ of sheaves induces a morphism on the stalks $\mathcal{F}_p \to \mathcal{G}_p$. Indeed, using the universal property, since $\mathcal{F}_p$ represents morphisms from the space of sections $\mathcal{F}(U)$ which lie over p satisfying constraints by the restrictions maps, the map η induces such a family of morphisms into $\mathcal{G}_p$ after post-composing with $\mathcal{G}(U) \to \mathcal{G}_p$. This means we can take $(f, U) \in \mathcal{F}_p$ and map it to $(\eta_U(f), U) \in \mathcal{G}_p$, which is then well-defined.

Remark 6.8.13 (étale Space) Sheaves on a space M can be obtained by considering certain bundles over M and it can be shown that one can always consider sheaves as being extracted from a certain kind of bundles. In order to see this, one can consider the *étale space* of a sheaf $\mathcal{F}$ given by the disjoint union of all stalks $\bigsqcup_{p \in M} \mathcal{F}_p$. We can construct a bundle map $\pi \colon \bigsqcup_{p \in M} \mathcal{F}_p \to M$ by sending $f_p \in \mathcal{F}_p \mapsto p$, i.e. sending a germ at a point p to the point $p \in M$. The stalk $\mathcal{F}_p$ can be then regarded as the fiber of π over p in the usual sense of a fiber bundle.

Definition 6.8.14 (Sheafification I) Let $\mathcal{F}$ be a presheaf over a topological space M. The *sheafification* of $\mathcal{F}$ is a sheaf $\mathcal{F}^{\mathrm{sh}}$ and a map of presheaves $\alpha \colon \mathcal{F} \to \mathcal{F}^{\mathrm{sh}}$ such that for any other sheaf $\mathcal{G}$ and a map of presheaves $\eta \colon \mathcal{F} \to \mathcal{G}$, there is a unique map $\beta \colon \mathcal{F}^{\mathrm{sh}} \to \mathcal{G}$ such that the following diagram commutes:

There is an alternative definition given as follows:

Definition 6.8.15 (Sheafification II) Consider the étale space $\bigsqcup_{p \in M} \mathcal{F}_p$ and consider on it a topology as follows: For each $f \in \mathcal{F}(U)$, the set $\{f_p \mid p \in U\}$, where

f_p denotes the germ of f at p, is open in $\bigsqcup_{p \in M} \mathcal{F}_p$. We define the topology to be the one generated by these open sets. We then define

$$\mathcal{F}^{\mathrm{sh}}(U) := \left\{ s : U \to \bigsqcup_{p \in M} \mathcal{F}_p \,\middle|\, s \text{ continuous} \right\}$$

The map $\alpha : \mathcal{F} \to \mathcal{F}^{\mathrm{sh}}$ is then defined by sending $f \in \mathcal{F}(U)$ to the map $(p \mapsto f_p) \in \mathcal{F}^{\mathrm{sh}}(U)$.

Definition 6.8.16 (Ringed Space) A pair $(M, \mathcal{O}_M)$ is called a *ringed space* if M is a topological space and $\mathcal{O}_M$ is a sheaf of rings on M.

Definition 6.8.17 (Locally Ringed Space) A ringed space $(M, \mathcal{O}_M)$ is called *locally ringed space* if each stalk of $\mathcal{O}_M$ is a local ring, i.e. it has a unique maximal ideal.

Definition 6.8.18 (Morphism of Ringed Spaces) A *morphism of ringed spaces* $\pi : (M, \mathcal{O}_M) \to (N, \mathcal{O}_N)$ consists of a map $\pi : M \to N$ of topological spaces and a morphism of sheaves $\pi^{\sharp} : \mathcal{O}_N \to \pi_* \mathcal{O}_M$, where $\pi_* \mathcal{O}_M$ is the pushforward sheaf on N defined by $(\pi_* \mathcal{O}_N)(U) := \mathcal{O}_N(\pi^{-1}(U))$. A morphism of locally ringed spaces is a morphism of ringed spaces with the additional condition that, whenever $q \in \pi^{-1}(p)$, the induced map $(\mathcal{O}_N)_p \to (\mathcal{O}_M)_q$ is a local ring homomorphism, i.e. it maps the unique maximal ideal of $(\mathcal{O}_N)_p$ into the maximal ideal of $(\mathcal{O}_M)_q$.

6.8.1.2 Supermanifolds

Definition 6.8.19 (Supermanifold) A *supermanifold* is a locally ringed space $\mathcal{M} = (M, \mathcal{O}_M)$, which is locally isomorphic to

$$\left(U, C^{\infty}(U) \otimes \bigwedge V^* \right), \tag{6.8.1}$$

where $U \subset \mathbb{R}^d$ is open and V is some finite-dimensional real vector space.

Remark 6.8.20 The topological space M is usually called the *body* of the supermanifold $\mathcal{M}$.

Remark 6.8.21 Locally, we should think of a supermanifold to be given by *even* coordinates (x^i) on some open subset $U \subset \mathbb{R}^d$ together with smooth maps $C^{\infty}(U)$ which algebraically describes the commuting coordinates, i.e. we have $x^i x^j = x^j x^i$ for all i, j, and *odd* coordinates (θ_μ) with the property $\theta_\mu x^i = x^i \theta_\mu$ and $\theta_\mu \theta_\nu = -\theta_\nu \theta_\mu$ (anti-commutative) for all i, μ, ν. Then the algebra is described as $\widetilde{\mathbb{R}[x, \theta]}/ \sim$ where $\sim$ is given by the commutative relation of the even coordinates and the anti-commutative relation of the odd coordinates.

Remark 6.8.22 The isomorphism mentioned in Definition 6.8.19 is in the category of $\mathbb{Z}_2$-graded algebra, which is defined by the parity operator:

$$| \; |: \bigoplus_{k \geq 0} C^\infty(U) \otimes \bigwedge^k V^* \to \mathbb{Z}_2,$$

$$f \otimes \theta \mapsto |f \otimes \theta| := |\theta| = k \bmod 2.$$

In particular, this isomorphism induces that globally $C^\infty(\mathcal{M})$ is a graded commutative algebra. Namely, for two homogeneous elements $f, g \in C^\infty(\mathcal{M})$, we have $fg = (-1)^{|f||g|}gf$.

Example 6.8.23 If we consider $U = \mathbb{R}^q$ and $V = \mathbb{R}^p$, we can consider the standard local supermanifold $\mathbb{R}^{q|p}$ which is the supermanifold with body $\mathbb{R}^q$, i.e. we have q even coordinates, and structure sheaf

$$C^\infty(\mathbb{R}^q) \otimes \bigwedge \mathbb{R}^p,$$

i.e. we have p odd coordinates.

Example 6.8.24 (Odd Vector Bundle) Let $E \to M$ be a vector bundle. Then, similarly as we have seen in Sect. 5.2.7, we can associate to it the *odd vector bundle* ΠE, which is a supermanifold. Moreover, functions on ΠE are then given by

$$C^\infty(\Pi E) \cong \Gamma\left(\bigwedge E^*\right).$$

Example 6.8.25 (Odd (Co)tangent Bundle) Let M be a manifold. Then we can consider its *odd* tangent and cotangent bundle, given by ΠTM and ΠT^*M, respectively. They are determined by their ring of functions:

$$C^\infty(\Pi TM) \cong \Gamma\left(\bigwedge T^*M\right) \cong \bigoplus_{0 \leq i \leq \dim M} \Omega^i(M) =: \Omega^\bullet(M),$$

$$C^\infty(\Pi T^*M) \cong \Gamma\left(\bigwedge TM\right) \cong \mathcal{V}(M).$$

Definition 6.8.26 (Graded Manifold) A *graded manifold* is a locally ringed space $\mathcal{M} = (M, \mathcal{O}_M)$ which is locally isomorphic to

$$\left(U, C^\infty(U) \otimes \mathrm{Sym}(V^*)\right),$$

where $U \subset \mathbb{R}^d$ is open and V is some finite-dimensional real vector space.

Remark 6.8.27 When we talk about graded manifolds, we will always mean $\mathbb{Z}$-graded.

6.8.1.3 Berezinian Integration

Let $\pi : E \to M$ be a vector bundle of rank m over a manifold M with dim $M = d$. Let $\mathcal{M} = \Pi E$. We can then define the Berezinian line bundle of $\mathcal{M}$ as the real line bundle

$$\mathrm{Ber}(\mathcal{M}) = \overset{d}{\bigwedge} T^* M \otimes \overset{m}{\bigwedge} E.$$

A section in $\Gamma(\mathrm{Ber}(\mathcal{M}))$ is called a *Berezinian*. For any Berezianian $\sigma \in \Gamma(\mathrm{Ber}(\mathcal{M}))$, there is an $\mathbb{R}$-linear integration map

$$\int_{\mathcal{M}} \sigma : C^\infty(\mathcal{M}) \to \mathbb{R},$$

$$f \mapsto \int_{\mathcal{M}} \sigma f := \int_M \langle \sigma, (f)_m \rangle,$$

where $\langle\ ,\ \rangle$ denotes the fiberwise pairing between line bundle $\bigwedge^m E$ and $\bigwedge^m E^*$ and $(f)_m$ denotes the component of $f \in C^\infty(\Pi E) \cong \Gamma(\bigwedge E^*)$ in the top exterior power of E^*. In particular, $\langle \sigma, (f)_m \rangle \in \Gamma(\bigwedge^d T^* M)$. Sections of $\mathrm{Ber}(\mathcal{M})$ over M will correspond to Berezinians that are constant along the fiber direction of the bundle $\Pi E \to M$. We can consider the super vector bundle $\mathbf{Ber}(\mathcal{M}) := \mathrm{Ber}(\mathcal{M}) \otimes \bigwedge E^*$ over M such that $\Gamma(\mathbf{Ber}(\mathcal{M})) = \Gamma(\mathrm{Ber}(\mathcal{M})) \otimes_{C^\infty(M)} C^\infty(\mathcal{M})$, which is a module over $C^\infty(\mathcal{M})$. One can alternatively think of $\Gamma(\mathbf{Ber}(\mathcal{M}))$ as the space of sections of the pullback bundle $p^* \mathrm{Ber}(\mathcal{M})$ over $\mathcal{M}$ rather than M. Here, we have $p : \Pi E \to M$. In fact, it is given by the integration map

$$\int_{\mathcal{M}} : \Gamma(\mathbf{Ber}(\mathcal{M})) \to \mathbb{R}.$$

The odd tangent bundle $\mathcal{M} = \Pi T M$ is endowed with a distinguished Berezinian $\sigma_{\Pi T M}$ which is uniquely characterized as follows: For $f \in C^\infty(\Pi T M) \cong \Omega^\bullet(M)$, denote by $\tilde{f}$ the corresponding differential form on M. Then, $\sigma_{\Pi T M}$ satisfies

$$\int_{\Pi T M} \sigma_{\Pi T M} f = \int_M \tilde{f}.$$

In local coordinates we have $\sigma_{\Pi T M} = \prod_i \mathrm{d}x^i \mathrm{d}\theta_i \in \Gamma(\mathbf{Ber}(\Pi T M))$.

6.8.1.4   Change of Variables

Consider now a supermanifold $\mathcal{M}$ and let $J \in \mathrm{End}(\mathbb{R}^{q|p}) \otimes C^{\infty}(\mathcal{M})$ be an $\mathcal{M}$-dependent endomorphism of $\mathbb{R}^{q|p}$ of block form

$$J = \begin{pmatrix} A & B \\ C & D \end{pmatrix}$$

with blocks given by

$$A \in (\mathrm{End}(\mathbb{R}^q) \otimes C^{\infty}(\mathcal{M}))_{\mathrm{even}},$$
$$B \in (\mathrm{Hom}(\mathbb{R}^p, \mathbb{R}^q) \otimes C^{\infty}(\mathcal{M}))_{\mathrm{odd}},$$
$$C \in (\mathrm{Hom}(\mathbb{R}^q, \mathbb{R}^p) \otimes C^{\infty}(\mathcal{M}))_{\mathrm{even}},$$
$$D \in (\mathrm{End}(\mathbb{R}^p) \otimes C^{\infty}(\mathcal{M}))_{\mathrm{even}}.$$

Moreover, we assume that D is invertibel.

Definition 6.8.28 (Superdeterminant) The *superdeterminant* of J is given by

$$\mathrm{sdet}(J) = \det(A - BD^{-1}C)(\det D)^{-1} \in C^{\infty}(\mathcal{M}). \tag{6.8.2}$$

Remark 6.8.29 We can characterize the superdeterminant by the following two properties:

- (Multiplicativity) For $J, K \in \mathrm{End}(\mathbb{R}^{q|p}) \otimes C^{\infty}(\mathcal{M})$, we have

$$\mathrm{sdet}(JK) = \mathrm{sdet}(J)\mathrm{sdet}(K),$$

- For $j = \begin{pmatrix} a & b \\ c & d \end{pmatrix}$ an $\mathcal{M}$-dependent endomorphism of $\mathbb{R}^{q|p}$, we have

$$\mathrm{sdet}(\mathrm{id} + \varepsilon j) = 1 + \varepsilon \mathrm{sTr}\, j + O(\varepsilon^2).$$

We have denoted by $\mathrm{sTr}\, j = \mathrm{Tr}\, a - \mathrm{Tr}\, d$ the *supertrace* of j.

Using these two properties, it is easy to see that

$$\mathrm{sdet}\exp(j) = \exp(\mathrm{sTr}\, j).$$

Theorem 6.8.30 (Change of Variables) *Let $\mathbb{R}_{\mathrm{I}}^{q|p}$ and $\mathbb{R}_{\mathrm{II}}^{q|p}$ be two copies of the $(q \mid p)$-dimensional superspace endowed with coordinates (x^i, θ_μ) on the first copy and coordinates (y^i, ψ_μ) on the second copy. Let $\phi \colon \mathbb{R}_{\mathrm{I}}^{q|p} \to \mathbb{R}_{\mathrm{II}}^{q|p}$ be a smooth map of supermanifolds and $f(y, \psi) \in C_c^{\infty}(\mathbb{R}_{\mathrm{II}}^{q|p})$ a compactly supported function.*

Then the integral of f over $\mathbb{R}^{q|p}_{\mathrm{II}}$ with respect to to the standard Berezinian can be expressed as an integral of the pullback of f by ϕ as follows:

$$\int_{\mathbb{R}^{q|p}_{\mathrm{II}}} \mathrm{d}^q y \, \mathrm{d}^p \psi \, f(y, \psi) = \int_{\mathbb{R}^{q|p}_{\mathrm{I}}} \mathrm{d}^q x \, \mathrm{d}^p \theta \, \mathrm{sign \, det} \left(\frac{\partial y^i(x, 0)}{\partial x^j} \right) \mathrm{sdet} \frac{\partial(y, \psi)}{\partial(x, \theta)} f(y(x, \theta), \psi(x, \theta)).$$

(6.8.3)

6.8.1.5 Divergence of Vector Fields on Supermanifolds

Definition 6.8.31 (Divergence) Let $v \in \mathfrak{X}(\mathcal{M})$ be a vector field on a supermanifold $\mathcal{M}$ and let $\sigma \in \Gamma(\mathbf{Ber}(\mathcal{M}))$ be a Berezinian. The *divergence* $\mathrm{div}_\sigma(v) \in C^\infty(\mathcal{M})$ of v with respect to σ is defined by the property

$$\int_{\mathcal{M}} \sigma v(f) = - \int_{\mathcal{M}} \sigma \, \mathrm{div}_\sigma(v) f, \qquad \forall f \in C^\infty_c(\mathcal{M}).$$

Example 6.8.32 Let $\mathcal{M} = M$ be an ordinary manifold and let σ be a volume form on M. Then, by Stokes' theorem, we get

$$0 = \int_M L_v(\sigma f) = \int_M \sigma v(f) + \underbrace{(L_v \sigma)}_{=\sigma \, \mathrm{div}_\sigma(v)} f.$$

This shows that the definition of the divergence in this setting is compatible with the usual geometric construction where we can define $\mathrm{div}_\sigma(v) = \frac{1}{\sigma} L_v \sigma$. In fact, the divergence measures how the flow by v changes volumes of subsets of M as measured by σ.

Lemma 6.8.33 *Let σ, σ_0 be two Berezinians on $\mathcal{M}$ with $\sigma = \rho \sigma_0$ where $\rho \in C^\infty(\mathcal{M})$ is a non-vanishing function. Then, for a vector field $v \in \mathfrak{X}(\mathcal{M})$, we can relate the divergences with respect to σ and σ_0 as follows:*

$$\mathrm{div}_\sigma(v) = \mathrm{div}_{\sigma_0}(v) + \underbrace{\frac{1}{\rho} v(\rho)}_{=v(\log \rho)}.$$

(6.8.4)

Remark 6.8.34 Choosing coordinates (x^i, θ_μ) on a $(q \mid p)$-dimensional supermanifold $\mathcal{M}$, we can consider the standard Berezinian $\sigma_{\mathrm{std}} = \mathrm{d}^q x^i \wedge \mathrm{d}^p \theta$. We can then express a vector field locally as

$$v = \sum_i v^i(x, \theta) \frac{\partial}{\partial x^i} + \sum_\mu v^\mu(x, \theta) \frac{\partial}{\partial \theta_\mu}.$$

The divergence is then locally given by

$$\text{div}_{\sigma_{\text{std}}}(v) = \sum_i \frac{\partial}{\partial x^i} v^i - (-1)^{|v|} \sum_\mu \frac{\partial}{\partial \theta_\mu} v^\mu.$$

This can be written more compactly as

$$\text{div}_{\sigma_{\text{std}}}(v) = \sum_i v^i \frac{\overleftarrow{\partial}}{\partial x^i} - \sum_\mu v^\mu \frac{\overleftarrow{\partial}}{\partial \theta_\mu}.$$

6.8.1.6 Odd-symplectic Supermanifolds

Definition 6.8.35 (Odd-symplectic Supermanifold) Let $\mathcal{M}$ be a supermanifold. An odd-symplectic structure on $\mathcal{M}$ is a 2-form ω on $\mathcal{M}$ such that ω is closed, odd, i.e. in local coordinates (x^i, θ_μ) on $\mathcal{M}$ it has the form $\sum_{i,\mu} \omega_{i\mu}(x, \theta) dx^i \wedge d\theta_\mu$ with $(\omega_{i\mu}(x, \theta))$ a matrix of local functions on $\mathcal{M}$, and non-degenerate, i.e. the matrix of coefficients $(\omega_{i\mu}(x, \theta))$ is invertible. A supermanifold $\mathcal{M}$ endowed with an odd-symplectic form is called *odd-symplectic supermanifold*.

Theorem 6.8.36 (Schwarz[Sch93]) *Let $(\mathcal{M}, \omega)$ be an odd-symplectic supermanifold with body M.*

(1) In the neighborhood of any point of M, one can find local coordinates (x^i, ξ_i) on $\mathcal{M}$ such that $\omega = \sum_i dx^i \wedge d\xi_i$.

(2) There is a (global) symplectomorphism $\phi \colon (\mathcal{M}, \omega) \to (\Pi T^* M, \omega_{\text{std}})$ with ω_{std} the standard symplectic structure on the odd cotangent bundle, locally given by $\omega_{\text{std}} = \sum_i dx^i \wedge d\xi_i$.

Remark 6.8.37 Note that point (1) of Theorem 6.8.36 is an analogue odd version of the Darboux theorem in ordinary symplectic geometry, whereas point (2) is unlike the analogue of the ordinary case, since it says that, up to symplectomorphism, all odd-symplectic supermanifolds are (odd) cotangent bundles.

6.8.2 The Batalin–Vilkovisky Formalism

The main difference between the Batalin–Vilkovisky (BV) formalism and the BRST formalism is the usage of a symplectic structure within the cohomological setting which allows to use the techniques of symplectic geometry and, in fact, interpret gauge-fixing as a choice of Lagrangian submanifold of some extended space of fields. We will start with the definition of a BV manifold:

Definition 6.8.38 (BV Manifold) A *BV manifold* is a triple $(\mathcal{M}_{\text{BV}}, \omega_{\text{BV}}, S_{\text{BV}})$ such that $\mathcal{M}_{\text{BV}}$ is a $\mathbb{Z}$-graded supermanifold, ω_{BV} an odd-symplectic form of degree

-1 and S_{BV} an even function on $\mathcal{M}_{\mathrm{BV}}$ of degree 0. Moreover, if we denote by $(\ ,\ )$ the Poisson bracket induced by ω_{BV}, we require that

$$(S_{\mathrm{BV}}, S_{\mathrm{BV}}) = 0. \tag{6.8.5}$$

Equation (6.8.5) is called the *classical master equation (CME)*.

Remark 6.8.39 We will usually call $\mathcal{M}_{\mathrm{BV}}$ the *BV space of fields*, ω_{BV} the *BV symplectic form* and S_{BV} the *BV action (functional)*. In the physics literature, the odd Poisson bracket $(\ ,\ )$ of degree $+1$ is often denoted by round brackets which is the notation due to Batalin and Vilkovisky [BV77] and is in the physics literature usually called the *BV bracket* or the *anti-bracket*.

Remark 6.8.40 (Ghost Number) The ghost number in this setting corresponds to the $\mathbb{Z}$-grading on $\mathcal{M}_{\mathrm{BV}}$. The $\mathbb{Z}_2$-grading (odd and even) of the supermanifold structure corresponds in the physics literature to the commuting and anti-commuting coordinates representing *bosons* (even) and *fermions* (odd), respectively. For some function f on $\mathcal{M}_{\mathrm{BV}}$, we will denote by $\mathrm{gh}(f)$ its degree with respect to the $\mathbb{Z}$-grading.

Remark 6.8.41 Using the symplectic formalism, we can consider the symplectic cohomological vector field Q given by the Hamiltonian vector field of S_{BV} with respect to the symplectic form ω_{BV}, i.e. Q is uniquely defined through the equation

$$\iota_Q \omega_{\mathrm{BV}} = \mathrm{d}S_{\mathrm{BV}},$$

where d here denotes the de Rham differential on the BV space of fields $\mathcal{M}_{\mathrm{BV}}$. It is easy to check that Q is then indeed *symplectc*, i.e. $L_Q \omega_{\mathrm{BV}} = 0$, and that it is *cohomological*, i.e. $Q^2 = 0$. Moreover, the reason why we have chosen the degree of ω_{BV} to be -1 is to make sure that Q is of degree $+1$ such that it really defines a differential and thus the Q-cohomology is well-defined. Note also that, by definition, we can write $Q = (S_{\mathrm{BV}},\ )$ and thus Eq. (6.8.5) is then translated into $Q(S_{\mathrm{BV}}) = 0$. This will encode the Euler–Lagrange equations similarly as in the BRST formalism, where we impose $\delta S = 0$ with δ the BRST operator.

Definition 6.8.42 (BV Theory) A *BV theory* is the assignment of a BV manifold to any source manifold Σ.

Remark 6.8.43 Note that we are interested in the case where our theory is local, i.e. the objects of interest, such as e.g. ω_{BV} or S_{BV}, are given by integrals over certain densities. Hence, when talking about a BV theory as in Definition 6.8.42, the source manifold serves as a generalized space-time manifold over which we integrate.

Definition 6.8.44 (s-Density on Supermanifold) A density ρ of weight $s \in \mathbb{R}$ (or s-density) on a supermanifold $\mathcal{M}$, covered by an atlas of coordinate charts U_α with local coordinates $(x^i_{(\alpha)}, \theta^{(\alpha)}_\mu)$, is a collection of locally defined functions $\rho_{(\alpha)}(x_{(\alpha)}, \theta^{(\alpha)})$ satisfying the following transformation rule on the intersections

$U_\alpha \cap U_\beta$:

$$\rho_{(\alpha)}(x_{(\alpha)}, \theta^{(\alpha)}) = \rho_{(\beta)}(x_{(\beta)}, \theta^{(\beta)}) \left| \mathrm{sdet} \frac{\partial(x_{(\alpha)}, \theta^{(\alpha)})}{\partial(x_{(\beta)}, \theta^{(\beta)})} \right|^s,$$

where sdet denotes the superdeterminant as in (6.8.2).

Remark 6.8.45 The space of smooth s-densities on $\mathcal{M}$ will be denoted by $\mathrm{Dens}^s(\mathcal{M})$.

6.8.2.1   The BV Laplacian

Let now $(\mathcal{M}, \omega)$ be an odd symplectic supermanifold, i.e. a supermanifold $\mathcal{M}$ endowed with an odd symplectic form ω. We can introduce a second order differential operator $\Delta \colon C^\infty(\mathcal{M}) \to C^\infty(\mathcal{M})$, called the *BV Laplacian*, which is locally defined on Darboux charts as

$$\Delta := \sum_i \frac{\partial^2}{\partial x^i \partial \theta_i}.$$

Exercise 6.8.46   Show that Δ squares to zero.

If σ denotes any Berezinian on $\mathcal{M}$, we can define the operator $\Delta_\sigma \colon C^\infty(\mathcal{M}) \to C^\infty(\mathcal{M})$ by setting:

$$\Delta_\sigma(f) := \frac{1}{2} \mathrm{div}_\sigma X_f,$$

where X_f denotes the Hamiltonian vector field generated by the Hamiltonian f. We can then define an anti-bracket on $\mathcal{M}$ by

$$(f, g) := (-1)^{|f|} X_f(g).$$

Locally, on a Darboux chart (x^i, θ_i) on $\mathcal{M}$, when assuming that the Berezinian has the form $\sigma = \rho(x, \theta) \mathrm{d}^d x \mathrm{d}^d \theta$ with ρ a local density function, we get

$$\Delta_\sigma(f) = \sum_i \frac{\partial^2}{\partial x^i \partial \theta_i} f + \frac{1}{2}(\log \rho, f).$$

Moreover, the local form of the anti-bracket is given by

$$(f, g) = \sum_i f \left(\frac{\overleftarrow{\partial}}{\partial x^i} \frac{\overrightarrow{\partial}}{\partial \theta_i} - \frac{\overleftarrow{\partial}}{\partial \theta_i} \frac{\overrightarrow{\partial}}{\partial x^i} \right) g$$

Exercise 6.8.47 Show that Δ_σ satisfies

$$\Delta_\sigma(fg) = \Delta(f)g + (-1)^{|f|}f\Delta(g) + (-1)^{|f|}(f, g).$$

Remark 6.8.48 One can see that the BV Laplacian Δ_σ does not automatically square to zero. In fact, it will only square to zero whenever the Berezinian σ is compatible[1] with the odd symplectic supermanifold $(\mathcal{M}, \omega)$ and in this case $\Delta_\sigma = \Delta$.

As it was shown in [Khu04], one can define canonically an operator $\Delta \colon \mathrm{Dens}^{\frac{1}{2}}(\mathcal{M}) \to \mathrm{Dens}^{\frac{1}{2}}(\mathcal{M})$ locally given by

$$\Delta \colon \rho(x, \theta)\mathrm{d}^{\frac{1}{2}}x\mathrm{d}^{\frac{1}{2}}\theta \mapsto \left(\sum_i \frac{\partial^2}{\partial x^i \partial \theta_i}\rho(x, \theta)\right)\mathrm{d}^{\frac{1}{2}}x\mathrm{d}^{\frac{1}{2}}\theta.$$

The operator Δ is called *canonical BV Laplacian*.

6.8.2.2 BV Integrals and the BV Theorem

We first note that, for a d-manifold M, we can make the following identification:

$$\mathrm{Ber}(\Pi T^*M) \cong \left(\bigwedge^d T^*M\right)^{\otimes 2}.$$

Similarly, for a supermanifold $\mathcal{N}$, we have

$$\mathrm{Ber}(\Pi T^*\mathcal{N})|_{\mathcal{N}} \cong \mathrm{Ber}(\mathcal{N})^{\otimes 2}.$$

This implies that there is a canonical map which sends Berezinians σ on $\Pi T^*\mathcal{N}$ to Berezinians $\sqrt{\sigma|_\mathcal{N}}$ on $\mathcal{N}$. Let X^i be coordinates on $\mathcal{N}$ containing even and odd parity and denote by (X^i, Ξ_i) the corresponding Darboux coordinates on $\Pi T^*\mathcal{N}$. Then, a Berezianina $\sigma = \rho(X, \Xi)\mathrm{d}X\mathrm{d}\Xi$ on $\Pi T^*\mathcal{N}$ is mapped to a Berezinian $\sqrt{\sigma|_\mathcal{N}} := \sqrt{\rho(X, 0)}\mathrm{d}X$ on $\mathcal{N}$. Now let $(\mathcal{M}, \omega, \sigma)$ be an odd-symplectic supermanifold with a compatible Berezinian.

Definition 6.8.49 (BV Integral) A *BV integral* on $(\mathcal{M}, \omega, \sigma)$ is defined by

$$\int_{\mathcal{L} \subset \mathcal{M}} f\sqrt{\sigma|_\mathcal{L}},$$

---

[1] A Berezinian σ on an odd-symplectic $(q \mid q)$-dimensional supermanifold $(\mathcal{M}, \omega)$ is *compatible* with ω if there is an atlas of Darboux charts (x^i, ξ_i) on $\mathcal{M}$ such that locally $\sigma = \mathrm{d}^q x \mathrm{d}^q \xi$ is the standard coordinate Berezinian in all charts of the atlas.

where $\mathcal{L}$ is a Lagrangian submanifold of $\mathcal{M}$ and $f \in C^\infty(\mathcal{M})$ is a function satisfying $\Delta_\sigma f = 0$.

Theorem 6.8.50 (Batalin–Vilkovisky–Schwarz[BV77, Sch93]) *Let $(\mathcal{M}, \omega, \sigma)$ be an odd-symplectic supermanifold with compact body endowed with a compatible Berezinian.*

(1) For any function $g \in C^\infty(\mathcal{M})$ and Lagrangian submanifold $\mathcal{L} \subset \mathcal{M}$, we have

$$\int_{\mathcal{L}} \Delta_\sigma g \sqrt{\sigma}|_{\mathcal{L}} = 0. \tag{6.8.6}$$

(2) Let $\mathcal{L}$ and $\mathcal{L}'$ be two Lagrangian submanifolds whose bodies are homologous cycles in the body of $\mathcal{M}$ and let $f \in C^\infty(\mathcal{M})$ be a function satisfying $\Delta_\sigma f = 0$. Then we get

$$\int_{\mathcal{L}} f \sqrt{\sigma}|_{\mathcal{L}} = \int_{\mathcal{L}'} f \sqrt{\sigma}|_{\mathcal{L}'}. \tag{6.8.7}$$

Proof Using point (2) of Theorem 6.8.36, we can assume without loss of generality that $\mathcal{M} = \Pi T^* M$ for some d-manifold M. Define then the *odd Fourier transform*

$$\text{OFT} \colon C^\infty(\Pi T^* M) \to C^\infty(\Pi T M)$$

which in local coordinates (x^i, ξ_i) on $\Pi T^* M$ and local coordinates (x^i, θ_i) on $\Pi T M$, assuming the Berezinian is of the form $\sigma = \rho(x, \xi) \mathrm{d}^d x \mathrm{d}^d \xi$, is defined by

$$f(x, \xi) \mapsto \tilde{f}(x, \theta) := \int_{\Pi T_x^* M} \sqrt{\rho(x, \xi)} \mathrm{d}^d \xi \exp(\langle \theta, \xi \rangle) f(x, \xi).$$

The odd Fourier transform maps the BV Laplacian Δ_σ on $C^\infty(\Pi T^* M)$ to the de Rham differential on $\Omega^\bullet(M) \cong C^\infty(\Pi T M)$, i.e. $\text{OFT} \circ \Delta_\sigma = \mathrm{d} \circ \text{OFT}$. Consider now the *conormal Lagrangian submanifold* defined as the $(k \mid d - k)$-dimensional Lagrangian submanifold $\mathcal{L}_C \subset \Pi T^* M$ which is constructed as the odd conormal bundle of a closed k-dimensional submanifold $C \subset M$, i.e. $\mathcal{L}_C = \Pi N^* C$. Then the BV integral

$$\int_{\mathcal{L}_C} f \sqrt{\sigma}|_{\mathcal{L}_C} = \int_C \tilde{f},$$

where on the right-hand-side we have an integral of a differential form $\tilde{f} = \text{OFT}(f)$ on M. If we restrict everything to Lagrangians of the form $\mathcal{L}_C$, we get by Stokes' theorem on M

$$\int_{\mathcal{L}_C} \Delta_\sigma g \sqrt{\sigma}|_{\mathcal{L}_C} = \int_C \mathrm{d}\tilde{g} = 0,$$

and

$$\int_{\mathcal{L}'_C} f\sqrt{\sigma|_{\mathcal{L}'_C}} - \int_{\mathcal{L}_C} f\sqrt{\sigma|_{\mathcal{L}_C}} = \left(\int_{C'} - \int_C\right)\tilde{f} = \int_D \tilde{f},$$

where $D \subset M$ is a submanifold with boundary $\partial D = C' - C$. Note that we need to use that $\tilde{f}$ is a closed form on M in order to use Stokes' theorem which follows from the assumption $\Delta_\sigma f = 0$. As it was shown in [Sch93], given a Lagrangian $\mathcal{L}$ in an odd-symplectic manifold $\mathcal{M}$ with body M, there is a submanifold $C \subset M$ and a symplectomorphism $\phi\colon \mathcal{M} \to \Pi T^*M$ such that ϕ maps $\mathcal{L} \subset \mathcal{M}$ to $\mathcal{L}_C = \Pi N^*C \subset \Pi T^*M$. Using this, when starting with a general Lagrangian submanifold $\mathcal{L} \subset \Pi T^*M$ in (6.8.6), we can always reduce to a Lagrangian of the form $\mathcal{L}_C$ for some $C \subset M$. Another result of [Sch93] is that any Lagrangian $\mathcal{L} \subset \Pi T^*M$ can be obtained from a Lagrangian of the form $\mathcal{L}_C = \Pi N^*C$, for some $C \subset M$, as a graph of $d\Psi$ where $\Psi \in C^\infty(\mathcal{L}_C)_{\mathrm{odd}}$ (note that we can use Weinstein's tubular neighborhood theorem (Theorem 3.6.5) to identify ΠT^*M in the neighborhood of $\mathcal{L}_C$ with $\Pi T^*\mathcal{L}_C$.). Using this, when starting with a general Lagrangian submanifold $\mathcal{L} \subset \Pi T^*M$ in (6.8.7), we can reduce to a Lagrangian submanifold of the form $\mathcal{L}_C$ by the following computation: Let $\mathcal{L}_t$ be a smooth family of Lagrangian submanifolds of $\mathcal{M}$ with $t \in [0, 1]$ such that $\mathcal{L}_{t+\varepsilon} = \mathrm{graph}(\varepsilon d\Psi_t + O(\varepsilon^2))$ with $\Psi_t \in C^\infty(\mathcal{L}_t)$. Then, for $f \in C^\infty(\mathcal{M})$ such that $\Delta_\sigma f = 0$, we get

$$\frac{d}{dt}\int_{\mathcal{L}_t} f\sqrt{\sigma|_{\mathcal{L}_t}} = \int_{\mathcal{L}_t} \Delta(f\Psi_t)\sqrt{\sigma|_{\mathcal{L}_t}} = 0.$$

Note that the integral on the right-hand-side vanishes because of (6.8.6). Therefore, we can choose a family $\mathcal{L}_t$ which connects a given Lagrangian submanifold $\mathcal{L} \subset \Pi T^*M$ to a Lagrangian submanifold of the form $\mathcal{L}_C$. Such a family exists by the arguments before and moreover, the BV integral is constant along such families by the computation above.   □

Remark 6.8.51 In the setting of quantum field theory we are mainly interested in the case where $f = \exp(iS/\hbar)$. Thus the condition $\Delta_\sigma f = 0$ translates then to the equation $\Delta \exp(iS/\hbar) = 0$. The latter equation is called *quantum master equation (QME)*.

Proposition 6.8.52 *The QME $\Delta \exp(iS/\hbar) = 0$ is equivalent to*

$$\frac{1}{2}(S, S) - i\hbar\Delta S = 0.$$

Proof Note that for any function X we have $\Delta X^n = nX^{n-1}\Delta X + \frac{n(n-1)}{2}X^{n-2}(X, X)$ which can be proven by using induction on n and using

Exercise 6.8.47. Thus, we get

$$\Delta \exp(X) = \Delta \left(\sum_{n=0}^{\infty} \frac{X^n}{n!} \right) = \left(\Delta X + \frac{1}{2}(X, X) \right) \exp(X).$$

If we set $X = \frac{i}{\hbar} S$, we get

$$\Delta \exp(iS/\hbar) = (i\hbar)^{-2} \left(\frac{1}{2}(S, S) - i\hbar \Delta S \right) \exp(iS/\hbar)$$

and thus $\Delta \exp(iS/\hbar) = 0$ if and only if $\frac{1}{2}(S, S) - i\hbar \Delta S = 0$. $\square$

Remark 6.8.53 Note that in the classical limit $\hbar \to 0$, the QME reduces to the CME $(S, S) = 0$.

6.8.2.3 BRST via BV

Let us consider a BRST theory $(\mathcal{M}_{BRST}, S_{BRST}, \delta_{BRST})$. Then we can construct the BV data out of it in the following way: the BV space of fields is given by the (-1)-shifted cotangent bundle of the BRST space of fields, i.e. we have $\mathcal{M}_{BV} = T^*[-1]\mathcal{M}_{BRST}$, and the BV symplectic form ω_{BV} is given by the canonical symplectic form on the cotangent bundle. The BV action is defined by $S_{BV} = p^* S_{BRST} + \widetilde{\delta_{BRST}}$, where $p \colon \mathcal{M}_{BV} \to \mathcal{M}_{BRST}$ denotes the projection to the base of the cotangent bundle and $\widetilde{\delta_{BRST}}$ denotes the lift of δ_{BRST} to a function on the total space which is linear in the fibers. The cohomological vector field is then given by $Q = X_{p^* S_{BRST}} + \delta_{BRST}^{cot.lift}$, where $X_{p^* S_{BRST}}$ denotes the Hamiltonian vector field of $p^* S_{BRST}$ and $\delta_{BRST}^{cot.lift}$ denotes the cotangent lift of the vector field δ_{BRST} on the base to the total space of the cotangent bundle. If we choose local coordinates (Φ^α) on $\mathcal{M}_{BRST}$, we have the corresponding Darboux coordinates $(\Phi^\alpha, \Phi_\alpha^+)$ on $\mathcal{M}_{BV}$. The base coordinates are usually called *fields* and the fiber coordinates Φ_α^+ are called *anti-fields*. Then we can express our objects locally as:

$$\omega_{BV} = \sum_{\alpha} d\Phi^\alpha \wedge d\Phi_\alpha^+, \tag{6.8.8}$$

$$S_{BV}(\Phi, \Phi^+) = S_{BRST}(\Phi) + \sum_{\alpha} \Phi_\alpha^+ L_{\delta_{BRST}} \Phi^\alpha, \tag{6.8.9}$$

and

$$Q = \sum_{\alpha} \left(\frac{\partial}{\partial \Phi^{\alpha}} S_{\text{BRST}}(\Phi) \right) \frac{\partial}{\partial \Phi_{\alpha}^{+}} +$$

$$+ \sum_{\alpha} L_{\delta_{\text{BRST}}} \Phi^{\alpha} \frac{\partial}{\partial \Phi^{\alpha}} + \sum_{\alpha,\beta} (-1)^{(|\Phi^{\alpha}|+1)|\Phi^{\beta}|} \Phi_{\alpha}^{+} \left(\frac{\partial}{\partial \Phi^{\beta}} L_{\delta_{\text{BRST}}} \Phi^{\alpha} \right) \frac{\partial}{\partial \Phi_{\alpha}^{+}}$$

$$(6.8.10)$$

For the quantum picture, we use the gauge-fixing Lagrangian submanifold $\mathcal{L}_{\Psi} =$ graph$(\mathrm{d}\Psi) \subset T^{*}[-1]\mathcal{M}_{\text{BRST}}$ for a gauge-fixing fermion $\Psi(\Phi)$ of degree -1. If we denote by $\mathcal{L}_{0}$ the zero section of $\mathcal{M}_{\text{BV}} = T^{*}[-1]\mathcal{M}_{\text{BRST}}$, we can replace the integral

$$\int_{\mathcal{L}_{0}} \sqrt{\sigma_{\text{BV}}} \exp(\mathrm{i} S_{\text{BV}}(\Phi, \Phi^{+})/\hbar) \qquad (6.8.11)$$

by the integral

$$\int_{\mathcal{L}_{\Psi}} \sqrt{\sigma_{\text{BV}}} \exp(\mathrm{i} S_{\text{BV}}(\Phi, \Phi^{+})/\hbar), \qquad (6.8.12)$$

where $\sigma_{\text{BV}} = (\sigma_{\text{BRST}})^{\otimes 2}$. Note that on the zero section, S_{BV} reduces to S_{BRST} and thus (6.8.11) reduces to $\int_{\mathcal{M}_{\text{BRST}}} \exp(\mathrm{i} S_{\text{BRST}}(\Phi)/\hbar)\mathscr{D}[\Phi]$. In order to evaluate (6.8.12), we shall note first that $S_{\text{BV}}|_{\mathcal{L}_{\Psi}}$ is given by

$$S_{\text{BV}}\left(\Phi^{\alpha}, \Phi^{+} = \pm \frac{\partial \Psi}{\partial \Phi^{\alpha}} \right) = S_{\text{BRST}} + \delta_{\text{BRST}}\Psi.$$

Hence, (6.8.12) is given by $\int_{\mathcal{M}_{\text{BRST}}} \exp(\mathrm{i}/\hbar(S_{\text{BRST}} + \delta_{\text{BRST}}\Psi))$. Note that the BV gauge-fixing which corresponds to the Lagrangian homotopy $\mathcal{L}_{0} \mapsto \mathcal{L}_{\Psi}$ is exactly given by the BRST gauge-fixing procedure, which is shifting the BRST-action by a δ_{BRST}-exact term.

6.8.2.4   Faddeev–Popov via BV

Recall that for the Faddeev–Popov construction we start with some d-dimensional manifold $\mathcal{M}$ together with a symmetry Lie group G which acts on it. Moreover, we consider some G-invariant action functional $S \in C^{\infty}(\mathcal{M})^{G}$ and an invariant measure $\mu_{\mathcal{M}} \in \Omega^{d}(\mathcal{M})^{G}$. The procedure to link the Faddeed–Popov construction to the BV construction is by considering at first the BRST method out of the Faddeev–Popov method and then construct the BV data by considering the (-1)-shifted cotangent bundle of the BRST space of fields. The minimal BV space of fields constructed out of the Faddeev–Popov data is given by

$$\widetilde{\mathcal{M}}_{BV} = T^*[-1](\mathcal{M} \times \mathfrak{g}[1]) = T^*[-1]\mathcal{M} \times \mathfrak{g}[1] \times \mathfrak{g}^*[-2].$$

Denote by x^i the coordinates (classical fields) on $\mathcal{M}$ of degree 0, by c^a the coordinates (ghosts) on $\mathfrak{g}[1]$ of degree $+1$, by x_i^+ the coordinates (anti-fields for classical fields) on the fibers of $T^*[-1]\mathcal{M}$ and by c_a^+ the coordinates (anti-fields for ghosts) on $\mathfrak{g}^*[-2]$. We can then formulate the minimal BV symplectic form as the canonical odd-symplectic form on the shifted cotangent bundle

$$\widetilde{\omega}_{BV} = \sum_i dx^i \wedge dx_i^+ + \sum_a dc^a \wedge dc_a^+.$$

The minimal BV action is given by

$$\widetilde{S}_{BV} = S(x) + \sum_{i,a} c^a v_a^i(x) x_i^+ + \frac{1}{2} \sum_{a,b,c} f_{ab}^c c^b c^c c_a^+.$$

We can consider the minimal volume form $\widetilde{\mu}_{\mathcal{M}}$ locally to be given by $\widetilde{\mu}_{\mathcal{M}} = \rho(x)d^d x$ where ρ is a local density function. The Berezinian on $\widetilde{\mathcal{M}}_{BV}$ is then locally given by

$$\widetilde{\mu}_{BV} = \rho(x)^2 d^d x d^d x^+ d^m c d^m c^+.$$

The non-minimal BV data is given by adding auxiliary fields

$$\mathcal{M}_{BV} = \widetilde{\mathcal{M}}_{BV} \times T^*[-1](\mathfrak{g}^*[-1] \oplus \mathfrak{g}^*).$$

The auxiliary fields are then given by λ_a (Lagrange multiplier) the coordinates on $\mathfrak{g}^*$ of degree 0, by $\bar{c}_a$ (anti-ghosts) the coordinates on $\mathfrak{g}^*[-1]$ of degree -1, by λ^{+a} (anti-fields for Lagrange multiplier) the coordinates on the fiber of $T^*[-1]\mathfrak{g}^*$ of degree -1 and by $\bar{c}^{+a}$ (anit-fields for anti-ghosts) the coordinates on the fiber of $T^*[-1]\mathfrak{g}^*[-1]$ of degree 0. Then the BV symplectic form is given by

$$\omega_{BV} = \widetilde{\omega}_{BV} + \sum_a d\lambda_a \wedge d\lambda^{+a} + \sum_a d\bar{c}_a \wedge d\bar{c}^{+a}.$$

The BV action is given by

$$S_{BV} = \widetilde{S}_{BV} + \sum_a \lambda_a \bar{c}^{+a}.$$

The Berezinian is given by

$$\mu_{BV} = \widetilde{\mu}_{BV} d^m \lambda d^m \lambda^+ d^m \bar{c} d^m \bar{c}^+.$$

The gauge-fixing Lagrangian for the non-minimal BV construction is then given by

$$\mathcal{L}_{FP} = \mathrm{graph}(d\Psi_F) \subset \mathcal{M}_{BV},$$

where $\Psi_F := \langle \bar{c}, F(x) \rangle$, with $F \colon \mathcal{M} \to \mathfrak{g}$ being the gauge-fixing function. Locally, the Lagrangian $\mathcal{L}_{FP}$ is defined by

$$\mathcal{L}_{FP} := \left\{ x,\, c,\, \lambda,\, \bar{c} \text{ are free},\ x^+ = -(d_x F)^T \bar{c},\ c^+ = 0,\ \lambda^+ = 0,\ \bar{c}^+ = F(x) \right\}.$$

It is then easy to see that

$$S_{BV}|_{\mathcal{L}_{FP}} = S_F,$$

i.e. the BV action restricted to the Lagrangian $\mathcal{L}_{FP}$ gives the Faddeev–Popov gauge-fixed action as in (6.4.3).

6.8.3   Formulation of the Main Theorem

Theorem 6.8.54 (Cattaneo–Felder[CF00]) *Consider the Poisson manifold* $(\mathbb{R}^d, \pi)$ *with Poisson structure* π *together with the corresponding Poisson sigma model on a disk* D. *The fields are then given by a map* $X \colon D \to \mathbb{R}^d$ *and a 1-form* $\eta \in \Gamma(D, T^*D \otimes X^*T^*\mathbb{R}^d)$ *with boundary condition* $\eta|_{S^1} = 0$. *Moreover, let* $0, 1$ *and* ∞ *be cyclically ordered points on* $\partial D = S^1$, *i.e. if we start at* 0 *and move counterclockwise on* S^1 *we will first meet* 1 *and then* ∞ *(see Fig. 6.12), and impose the condition* $\eta(\infty) = 0$. *Then Kontsevich's star product between two functions* $f, g \in C^\infty(\mathbb{R}^d)$ *evaluated at the point* $x = X(\infty) \in \mathbb{R}^d$ *is given by*

$$f \star g(x) = \int_{X(\infty)=x} f(X(1))g(X(0)) \exp(iS(X, \eta)/\hbar)\, \mathscr{D}[X]\mathscr{D}[\eta], \qquad (6.8.13)$$

where the right-hand-side should be understood by perturbative expansion in terms of Feynman diagrams.

Fig. 6.12 Cyclically ordered points on $\partial D = S^1$

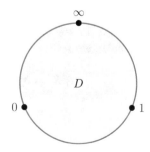

Remark 6.8.55 The global field theoretic approach was given in [CMW20] using cutting and gluing methods for manifolds with boundary by introducing a formal global action as more generally developed in [CMW19]. Moreover, also the case for manifolds with corners was covered. The main gauge formalism that was used is known as the *BV-BFV formalism* [CMR14, CMR17, CM20], which can bee seen as an extension of the Batalin–Vilkovisky formalism for manifolds with boundary. A first approach to a global formulation of the Poisson sigma model on closed manifolds using the Batalin–Vilkovisky formalism was given in [BCM12].

6.8.4   Proof Sketch of the Main Theorem

Let us consider the proof for any open subset $M \subset \mathbb{R}^d$. Note at first that for the path integral (6.8.13) we need to consider gauge-fixing and renormalization as usual. We can see that the Poisson sigma model is invariant with respect to the transformations with infinitesimal parameter $\beta_i \in \Gamma(X^*T^*M)$ such that $\beta_i|_{\partial D} = 0$ given by

$$\delta_\beta X^i = \alpha^{ij}(X)\beta_j, \tag{6.8.14}$$

$$\delta_\beta \eta_i = -\mathrm{d}\beta_i - \partial_i \alpha^{jk}(X)\eta_j\beta_k. \tag{6.8.15}$$

We note that the commutator of two gauge transformations is a gauge transformation only on shell (modulo equations of motion).

$$[\delta_\beta, \delta_{\beta'}]X^i = \delta_{\{\beta,\beta'\}}X^i,$$

$$[\delta_\beta, \delta_{\beta'}]\eta_i = \delta_{\{\beta,\beta'\}}\eta_i - \partial_i\partial_k\alpha^{rs}\beta_r\beta'_s(\mathrm{d}X^k + \alpha^{kj}(X)\eta_j).$$

We have denoted $\{\beta, \beta'\}_i = -\partial_i\alpha^{jk}(X)\beta_i\beta'_k$. The Euler–Lagrange equation for S is given by $\mathrm{d}X^k + \alpha^{kj}\eta_j = 0$. Since the BRST operator δ_0 acts then as follows

$$\delta_0 X^i = \alpha^{ij}(X)\beta_j, \tag{6.8.16}$$

$$\delta_0 \eta_i = -\mathrm{d}\beta_i - \partial_i\alpha^{kl}(X)\eta_k\beta_l, \tag{6.8.17}$$

$$\delta_0 \beta_i = \frac{1}{2}\alpha^{jk}(X)\beta_j\beta_k. \tag{6.8.18}$$

It is easy to see that δ_0 is only a differential modulo equations of motion. In particular, we have

$$\delta_0^2 X^i = \delta_0^2\beta_i = 0, \tag{6.8.19}$$

$$\delta_0^2\eta_i = -\frac{1}{2}\partial_i\partial_k\alpha^{rs}\beta_r\beta_s(\mathrm{d}X^k + \alpha^{kj}(X)\eta_j). \tag{6.8.20}$$

Table 6.1 Ghost number and form degree of the fields. The horizontal line denotes the form degree and the vertical line denotes the ghost degree

	0	1	2
−2			β^{+i}
−1		η^{+i}	X_i^+
0	X^i	η_i	
1	β_i		

The ghost numbers are given by $\mathrm{gh}(X^i) = \mathrm{gh}(\eta_i) = 0$ and $\mathrm{gh}(\eta_i) = +1$ and the form degrees of the fields, seen as differential forms on the disk, are given by $\deg(X^i) = \deg(\beta_i) = 0$ and $\deg(\eta_i) = +1$. In general, the BRST operator does not[2] square to zero (only modulo equations of motion) and thus we need to use the BV procedure. Consider the anti-fields X^+, η^+, β^+ with ghost number and form degree as in Table 6.1.

Then we can consider the BV action S_{BV} for the Poisson sigma model and the corresponding QME

$$(S_{BV}, S_{BV}) - 2i\hbar \Delta S_{BV} = 0.$$

The renormalization of the BV Laplacian Δ actually turns out to be rather trivial and additionally we have $\Delta S_{BV} = 0$ and the CME $(S_{BV}, S_{BV}) = 0$. The cohomological vector field is given by $Q = (S_{BV}, \ )$ and it acts on the fields Φ^α and anti-fields Φ_α^+ as follows

$$Q(\Phi^\alpha) = (-1)^{\mathrm{gh}(\Phi^\alpha)} \frac{\overrightarrow{\partial} S_{BV}}{\partial \Phi_\alpha^+}, \tag{6.8.21}$$

$$Q(\Phi_\alpha^+) = (-1)^{\mathrm{gh}(\Phi^\alpha)+\deg(\Phi^\alpha)} \frac{\overrightarrow{\partial} S_{BV}}{\partial \Phi^\alpha}. \tag{6.8.22}$$

In components, the BV action is given by

$$S_{BV} = S + \int_D \left(X_i^+ Q(X^i) + \eta^{+i} Q(\eta_i) - \beta^{+i} Q(\beta_i) \right) - \frac{1}{4} \int_D \eta^{+i}\eta^{+j} \partial_i \partial_j \alpha^{kl}(X) \beta_k \beta_l$$

$$= \int_D \left(\eta_i dX^i + \frac{1}{2}\alpha^{ij}(X)\eta_i \eta_j + X_i^+ \alpha^{ij}(X)\beta_j - \eta^{+i}(d\beta_i + \partial_i \alpha^{kl}(X)\eta_k \beta_l) \right.$$

$$\left. - \frac{1}{2}\beta^{+i}\partial_i \alpha^{jk}(X)\beta_j \beta_k - \frac{1}{4}\eta^{+i}\eta^{+j} \partial_i \partial_j \alpha^{kl}(X)\beta_k \beta_l \right). \tag{6.8.23}$$

Lemma 6.8.56 *If the regularization is appropriate, we get* $\Delta S_{BV} = 0$.

---

[2] In the case when $M = \mathfrak{g}^*$ with linear Poisson structures, the second derivatives of α vanish and thus the BRST operator does square to zero.

Proof of Lemma 6.8.56 Note that the only terms contributing to the BV Laplacian of the BV action contain both a field and an anti-field. Thus, we have

$$
\Delta S_{\mathrm{BV}} = \Delta \int_D \left(X_i^+ \alpha^{ij}(X)\beta_j - \eta^{+i}\partial_i \alpha^{kl}(X)\eta_k \beta_l - \frac{1}{2}\beta^{+i}\partial_i \alpha^{jk}(X)\beta_j \beta_k \right)
$$

$$
= (1 - 2 + 1)C \int_D \partial_i \alpha^{ij}(X)\beta_j \mathrm{dVol}
$$

$$
= 0,
$$

$$(6.8.24)$$

where C denotes some infinite constant.   □

Remark 6.8.57 The Hodge dual of the anti-fields have to have the same boundary conditions as their fields. In fact, since the boundary conditions we impose are that for $u \in \partial D$, we have $\beta_i(u) = 0$ and $\eta_i(u)|_{T_u \partial D} = 0$, we get that $\beta^{+i}(u) = 0$ and $\eta_i^+(u)|_{N_u \partial D} = 0$.

The superfields of the theory are given as

$$
\mathbf{X}^i = X^i + \eta_\mu^{+i}\theta^\mu - \frac{1}{2}\beta_{\mu\nu}^{+i}\theta^\mu \theta^\nu,
$$

$$(6.8.25)$$

$$
\boldsymbol{\eta}_i = \beta_i + \eta_{i,\mu}\theta^\mu + \frac{1}{2}X_{1,\mu\nu}^+ \theta^\mu \theta^\nu.
$$

$$(6.8.26)$$

Let $\mathbf{D} := \frac{\partial}{\partial u^\mu}\theta^\mu$ be the superdifferential. Then, the cohomological vector field Q acts on the superfields as a derivation of degree $+1$ as follows:

$$
Q(\mathbf{X}^i) = \mathbf{D}\mathbf{X}^i + \alpha^{ij}(\mathbf{X})\boldsymbol{\eta}_j,
$$

$$(6.8.27)$$

$$
Q(\boldsymbol{\eta}_i) = \mathbf{D}\boldsymbol{\eta}_i + \frac{1}{2}\partial_i \alpha^{jk}(\mathbf{X})\boldsymbol{\eta}_j \boldsymbol{\eta}_k.
$$

$$(6.8.28)$$

On the components we get

$$
Q(X^i) = \alpha^{ij}(X)\beta_j,
$$

$$(6.8.29)$$

$$
Q(\eta^{+i}) = -\mathrm{d}X^i - \alpha^{ij}(X)\eta_j - \partial_k \alpha^{ij}(X)\eta^{+k}\beta_j,
$$

$$(6.8.30)$$

$$
Q(\beta^{+i}) = -\mathrm{d}\eta^{+i} - \alpha^{ij}(X)X_j^+ + \frac{1}{2}\partial_k \partial_l \alpha^{ij}(X)\eta^{+k}\eta^{+l}\beta_j +
$$

$$
+ \partial_k \alpha^{ij}(X)\eta^{+k}\eta_j + \partial_k \alpha^{ij}(X)\beta^{+k}\beta_j.
$$

$$(6.8.31)$$

and

$$Q(\beta_i) = \frac{1}{2}\partial_i \alpha^{kl}(X)\beta_k \beta_l, \tag{6.8.32}$$

$$Q(\eta_i) = -\mathrm{d}\beta_i - \partial_i \alpha^{kl}(X)\eta_k \beta_l - \frac{1}{2}\partial_i \partial_j \alpha^{kl}(X)\eta^{+j}\beta_k \beta_l, \tag{6.8.33}$$

$$Q(X_i^+) = \mathrm{d}\eta_i + \partial_i \alpha^{kl}(X)X + k^+ \beta_l - \partial_i \partial_j \alpha^{kl}(X)\eta^{+j}\eta_k \beta_l + \frac{1}{2}\partial_i \alpha^{kl}(X)\eta_k \eta_l -$$

$$- \frac{1}{4}\partial_i \partial_j \partial_p \alpha^{kl}(X)\eta^{+j}\eta^{+p}\beta_k \beta_l - \frac{1}{2}\partial_i \partial_j \alpha^{kl}(X)\beta^{+j}\beta_k \beta_l. \tag{6.8.34}$$

The BV action of the Poisson sigma model is then given by

$$S_{\mathrm{BV}} = \int_D \left(\eta_i \mathbf{D}X^i + \frac{1}{2}\alpha^{ij}(\mathbf{X})\eta_i \eta_j \right). \tag{6.8.35}$$

Note that the integral (6.8.35) chooses the 2-form part on the disk of the expression $\eta_i \mathbf{D}X^i + \frac{1}{2}\alpha^{ij}(\mathbf{X})\eta_i \eta_j$ such that it is well-defined. Note also that

$$Q\left(\eta_i \mathbf{D}X^i + \frac{1}{2}\alpha^{ij}(\mathbf{X})\eta_i \eta_j \right) = \mathbf{D}\left(\eta_i \mathbf{D}X^i \right),$$

and thus it is Q-closed and hence Q applied to the 2-form part of $\eta_i \mathbf{D}X^i + \frac{1}{2}\alpha^{ij}(\mathbf{X})\eta_i \eta_j$ yields a 1-form vanishing on the boundary ∂D.

The gauge-fixing that we choose is the *Lorentz gauge* $\mathrm{d} * \eta_i = 0$. Let $\Omega := Q - i\hbar\Delta$ be the *quantum BV operator* and consider the Lagrangian

$$\mathcal{L} := \mathrm{graph}(\mathrm{d}\Psi) = \left\{ \left(\Phi^\alpha, \Phi^+_\alpha = \frac{\partial \Psi}{\partial \Phi^\alpha} \right) \right\},$$

for some gauge-fixing fermion Ψ of ghost number -1. Then we consider the path integral

$$\int_{\mathcal{L}} \exp(iS_{\mathrm{BV}}/\hbar)\mathcal{O}, \tag{6.8.36}$$

where $\mathcal{O}$ is an observable which is closed with respect to the quantum BV operator:

$$\Omega\mathcal{O} = \underbrace{Q(\mathcal{O})}_{(S_{\mathrm{BV}}, \mathcal{O})} - i\hbar\Delta\mathcal{O} = 0.$$

In order to make the integral (6.8.36) well-defined, we need to choose Ψ in an appropriate way. We introduce anti-commuting scalar fields (anti-ghosts) γ^i of ghost number -1 and scalar Lagrange multiplier fields λ^i of ghost number 0 together with their anti-fields γ_i^+ and λ_i^+. We impose the boundary conditions $\lambda^i|_{\partial D}$ and $\gamma^i|_{\partial D} = \text{const.}$ The fields are added to the BV action by adding the term $-\int_D \lambda^i \gamma_i^+$. Moreover, we have

$$Q(\lambda) = Q(\gamma^+) = 0, \tag{6.8.37}$$

$$Q(\lambda^+) = -\gamma^+, \tag{6.8.38}$$

$$Q(\gamma) = \lambda. \tag{6.8.39}$$

It is easy to see that $S_{\mathrm{BV}} - \int_D \lambda^i \gamma_i^+$ satisfies the CME

$$\left(S_{\mathrm{BV}} - \int_D \lambda^i \gamma_i^+, S_{\mathrm{BV}} - \int_D \lambda^i \gamma_i^+ \right) = 0.$$

We can encode the gauge-fixing $\mathrm{d}*\eta = 0$ in $\Psi = -\int_D \mathrm{d}\gamma^i * \eta_i$. The Lagrangian submanifold is then given by

$$\mathcal{L} = \mathrm{graph}(\mathrm{d}\Psi) = \{X^+ = \beta^+ = \lambda^+ = 0, \ \gamma_i^+ = \mathrm{d}*\eta_i + \Xi(\partial D), \ \eta^{+i} = *\mathrm{d}\gamma^i\}, \tag{6.8.40}$$

where $\Xi(\partial D)$ is some boundary term whose form degree will not matter. The gauge-fixed action is then given by

$$S_{\mathrm{gf}} = \int_D \left(\eta_i \mathrm{d}X^i + \frac{1}{2}\alpha^{ij}(X)\eta_i \eta_j - *\mathrm{d}\gamma^i(\mathrm{d}\beta_i + \partial_i \alpha^{kl}(X)\eta_k \beta_l) - \right.$$
$$\left. - \frac{1}{4}*\mathrm{d}\gamma^i * \mathrm{d}\gamma^j \partial_i \partial_j \alpha^{kl}(X)\beta_k \beta_l - \lambda^i \mathrm{d}*\eta_i \right) \tag{6.8.41}$$

Let us look at the Feynman rules for the perturbative expansion in $\hbar$ around the classical solution $X(u) = x$ and $\eta(u) = 0$. We write $X(u) = x + \xi(u)$, where $\xi(u)$ is some fluctuation field with $\xi(\infty) = 0$. Consider then the *kinetic term* of the gauge-fixed action (6.8.41)

$$S_{\mathrm{gf}}^0 := \int_D \left(\eta_i \mathrm{d}\xi^i - *\mathrm{d}\gamma^i \mathrm{d}\beta_i - \lambda^i \mathrm{d}*\eta_i \right)$$
$$= \int_D \left(\eta_i(\mathrm{d}\xi^i + *\mathrm{d}\lambda^i) + \beta_i \mathrm{d} * \mathrm{d}\gamma^i \right) \tag{6.8.42}$$

The second equality of (6.8.42) holds by the property of the Hodge dual and integration by parts (Stokes' theorem). In particular, by the property of the Hodge dual we have

$$\eta_i * d\lambda^i = d\lambda^i * \eta_i.$$

Then by integration by parts we get

$$\int_D d\lambda^i * \eta_i = \int_{\partial D} \lambda^i * \eta_i - \int_D \lambda^i d * \eta_i.$$

Since $*\eta_i|_{\partial D} = 0$, we get that the integral over the boundary of the disk vanishes and thus

$$\int_D d\lambda^i * \eta_i = -\int_D \lambda^i d * \eta_i,$$

which yields the claim. Similarly, using the Hodge dual property together with integration by parts, we have

$$\int_D d\beta_j * d\gamma^i = \int_{\partial D} \beta_j * d\gamma^i - \int_D \beta_j d * d\gamma^i.$$

Again, using that the integral over the boundary of the disk vanishes due to the boundary condition $\beta_j|_{\partial D} = 0$, we get

$$\int_D d\beta_j * d\gamma^i = -\int_D \beta_j d * d\gamma^i,$$

which yields the claim. We can write (6.8.42) slightly different as

$$S_{gf}^0 = \int_D \left(\eta_i (d \oplus *d)(\xi^i \oplus \lambda^i) + \beta_i d * d\gamma^i \right). \tag{6.8.43}$$

We denote the remaining (perturbation) part of S_{gf} by S_{gf}^1. In order to get the propagators, we need to invert the operators

$$d \oplus *d \colon \Omega^0(D) \oplus \Omega_0^0(D) \to \Omega^1(D), \tag{6.8.44}$$

$$d * d \colon \Omega^0(D) \to \Omega^2(D), \tag{6.8.45}$$

where Ω_0 denotes the space of forms λ^i with Dirichlet boundary conditions, i.e. $\lambda^i|_{\partial D} = 0$. Note that $d \oplus *d$ and $d * d$ are both surjective and

$$\dim \ker d \oplus *d = \dim \ker d * d = 1,$$

with $\ker d \oplus *d = \ker d * d = \Omega_{const}^0(D)$, where $\Omega_{const}^0(D)$ denotes the space of constant functions on D.

Next, we want to conformally map the disk onto the upper half-plane $\mathbb{H}^2$. Then we can describe the integral kernel of $(d * d)^{-1}$ by the Green's function $\frac{1}{2\pi}\psi(z, w)$ with

$$\psi(z, w) = \log \left| \frac{z - w}{z - \bar{w}} \right|, \quad z, w \in \mathbb{H}^2.$$

The integral kernel of $d \oplus *d$ is given by the Green's function

$$G(w, z) = \frac{1}{2\pi}(*d_z\psi(z, w) \oplus d_z\phi(z, w)),$$

where $d_z = dz\frac{\partial}{\partial z} + d\bar{z}\frac{\partial}{\partial \bar{z}}$ and

$$\phi(z, w) = \frac{1}{2i} \log \left(\frac{(z - w)(z - \bar{w})}{(\bar{z} - \bar{w})(\bar{z} - w)} \right), \quad z, w \in \mathbb{H}^2.$$

We can observe that

$$d_w * d_w\psi(z, w) = d_w * d_w\phi(z, w) = 2\pi\delta_z(w),$$

where $\delta_z(w)$ denotes the Dirac delta distribution 2-form and, for $w \in \mathbb{H}^2$, we have Dirichlet boundary conditions for ψ and Neumann boundary conditions for ϕ. Thus, we get the propagators:

$$\left\langle \gamma^k(w)\beta_j(z) \right\rangle = \frac{i\hbar}{2\pi}\delta^i_j\psi(z, w), \tag{6.8.46}$$

$$\left\langle \xi^k(w)\eta_j(z) \right\rangle = \frac{i\hbar}{2\pi}\delta^k_j d_z\phi(z, w), \tag{6.8.47}$$

$$\left\langle \lambda^k(w)\eta_j(z) \right\rangle = \frac{i\hbar}{2\pi}\delta^k_j * d_z\psi(z, w). \tag{6.8.48}$$

By observing that $*d_w\psi(z, w) = d_w\phi(z, w)$, we get that

$$\left\langle *d\gamma^k(w)\beta_j(z) \right\rangle = \frac{i\hbar}{2\pi}\delta^k_j d_w\phi(z, w),$$

and hence we can put everything into a *superpropagator*

$$\left\langle \xi^k(w)\eta_j(z) \right\rangle + \left\langle *d\gamma^k(w)\beta_j(z) \right\rangle = \frac{i\hbar}{2\pi}\delta^k_j d\phi(z, w),$$

with $d = d_z + d_w$. Let

$$\xi^k(w, \zeta) := \xi^k(w) + \eta_\mu^{+k}(w)\zeta^\mu, \tag{6.8.49}$$

$$\eta_j(z, \theta) := \beta_j(z) + \eta_{j,\mu}(z)\theta^\mu, \tag{6.8.50}$$

be the *superfields*, where $\eta^{+j} = *d\gamma^j$. Then the superpropagator can be written as

$$\langle\xi^k(w, \zeta)\eta_j(z, \theta)\rangle = \frac{i\hbar}{2\pi}\delta_j^k \mathbf{D}\phi(z, w), \tag{6.8.51}$$

where $\mathbf{D} := \theta^\mu \frac{\partial}{\partial z^\mu} + \zeta^\mu \frac{\partial}{\partial w^\mu}$ is the *superdifferential*. The perturbative expansion for the expectation value of an observable $\mathcal{O}$ is then given by

$$\int \exp(iS_{\mathrm{gf}}/\hbar)\mathcal{O} = \sum_{n\geq 0}\frac{i^n}{\hbar^n n!}\int \exp(iS_{\mathrm{gf}}^0/\hbar)(S_{\mathrm{gf}}^1)^n\mathcal{O}. \tag{6.8.52}$$

Using Theorem 6.2.5, we can compute (6.8.52) as

$$\int \exp(iS_{\mathrm{gf}}^0/\hbar)\xi^{k_1}(w_1, \zeta_1)\cdots\xi^{k_N}(w_N, \zeta_N)\eta_{j_1}(z_1, \theta_1)\cdots\eta_{j_N}(z_N, \theta_N)\delta_x(X(\infty))$$

$$= \sum_{\sigma\in S_N}\langle\xi^{k_{\sigma(1)}}(w_{\sigma(1)}, \zeta_{\sigma(1)})\eta_{j_1}(z_1, \theta_1)\rangle\cdots\langle\xi^{k_{\sigma(N)}}(w_{\sigma(N)}, \zeta_{\sigma(N)})\eta_{j_N}(z_N, \theta_N)\rangle.$$

$$\tag{6.8.53}$$

If we impose the normalization

$$\int \exp(iS_{\mathrm{gf}}^0/\hbar)\delta_x(X(\infty)) = 1,$$

we can observe that for the zero Poisson structure we will obtain back the ordinary product. Note that here we have denoted

$$\delta_x(X(t)) = \prod_{1\leq i\leq d}\delta(X^i(t) - x^i)\gamma^i(t),$$

which fixes the value of the zero modes, which are constant functions, of X. The γ^i are needed in order to get an integral different from zero due to the fact that we have also zero modes when integrating over γ. The perturbation term of the gauge-fixed action together with the given observable determine the vertices in the Feynman graph expansion. Expanding them in powers of the superfields yields

$$S_{\mathrm{gf}}^1 = \frac{1}{2}\int_D\int d^2\theta\sum_{k\geq 0}\frac{1}{k!}\partial_{j_1}\cdots\partial_{j_k}\alpha^{ij}(x)\xi^{j_1}\cdots\xi^{j_k}\eta_i\eta_j, \tag{6.8.54}$$

where the Berezin integration only selects the 2-form part. The observable of interest is chosen to be

$$\mathcal{O} = f(\mathbf{X}(1))g(\mathbf{X}(0))\delta_x(X(\infty)), \quad f, g \in C^{\infty}(M).$$

The expansion of f and g in powers of the superfields will induce the expansion in Feynman graphs. In fact, the graphs with n vertices as in (6.8.54) are given by the Kontsevich graphs Γ of order n, but possibly with short loops. These tadpole graphs can be renormalized by using appropriate counterterms and the method of *point-splitting regularization*.

Let us now consider the general case of multivector fields and multidifferential operators. Let α be a multivector field on M. Let $S = S_0 + S_\alpha$ with

$$S_0 = \int_D \int d^2\theta\, \boldsymbol{\eta}_j \mathbf{DX}^j - \int_D \lambda^i \gamma_i^+, \tag{6.8.55}$$

and

$$S_\alpha = \sum_{p=0}^{d-1} \int_D \int d^2\theta\, \frac{1}{(p+1)!} \alpha^{j_0 \cdots j_p}(\mathbf{X}(u, \theta)) \boldsymbol{\eta}_{j_0}(u, \theta) \cdots \boldsymbol{\eta}_{j_p}(u, \theta).$$

We can then look at the correlation functions of boundary fields which are associated to functions $f_0, \ldots, f_m \in C^{\infty}(M)$:

$$U(\alpha)(f_0 \otimes \cdots \otimes f_m)(x) = \int \exp(iS/\hbar)\mathcal{O}_x(f_0, \ldots, f_m), \tag{6.8.56}$$

where

$$\mathcal{O}_x(f_0, \ldots, f_m) = \int_{\mathcal{B}_m} [f_0(\mathbf{X}(t_0, \theta_0)) \cdots f_m(\mathbf{X}(t_m, \theta_m))]^{(m-1)} \delta_x(X(\infty)). \tag{6.8.57}$$

Note that the integral in (6.8.56) is over the Lagrangian submanifold (6.8.40). We have denoted by $\mathcal{B}_m$ the moduli space of $m + 1$ cyclically ordered points on the boundary of the disk modulo conformal transformations. Explicitly, we have

$$\mathcal{O}_x(f_0, \ldots, f_m) = \int_{1 > t_1 > \cdots > t_{m-1} > 0} f_0(X(1)) \prod_{k=1}^{m-1} \partial_{i_k} f_k(X(t_k)) \eta^{+i_k}(t_k) f_m(X(0))\delta_x(X(\infty)).$$

If we expand everything in powers of $\hbar$, we can observe that we end up with a map U which associates to each multivector field α, a formal power series with coefficients given by multidifferential operators. In particular, we get

$$U_n(\alpha_1, \ldots, \alpha; \hbar)(f_0 \otimes \cdots \otimes f_m)(x) = \int \exp(iS_0/\hbar) \frac{i}{\hbar} S_{\alpha_1} \cdots \frac{i}{\hbar} S_{\alpha_n} \mathcal{O}_x(f_0, \ldots, f_m).$$

If α_i is homogeneous of degree p_i, we get that S_{α_i} is the integral of the 2-form component of $\frac{1}{(p_i+1)!} \alpha_i^{j_0 \cdots j_{p_i}}(\mathbf{X}) \eta_{j_0} \cdots \eta_{j_{p_i}}$ and is of ghost number $p_i - 1$. A first consequence of this is that the integral over the t-variables will only be performed on the $(m-1)$-form component of $\prod_i f_i(\mathbf{X}(t_i))$, which has ghost number $1 - m$ and for the path integral to be different from zero, we get the ghost number condition

$$1 - m + \sum_{i=1}^{n} (p_i - 1) = 0,$$

which is equivalent to

$$m = 1 - n + \sum_{i=1}^{n} p_i.$$

This implies that U_n is indeed a map of degree $1 - n$ from the n-fold tensor product of multivector fields on M to multidifferential operators on M. The $\hbar$-dependence is obtained explicitly as

$$U_n(\alpha_1, \ldots, \alpha_n; \hbar)(f_0 \otimes \cdots \otimes f_m) = (i\hbar)^{n+m-1} U_n(\alpha_1, \ldots, \alpha_n)(f_0 \otimes \cdots \otimes f_m),$$

with $U_n(\alpha_1, \ldots, \alpha_n) = U_n(\alpha_1, \ldots, \alpha_n; \hbar = 1/i)$ being independent of $\hbar$. This is due to the fact that each vertex has a $\frac{1}{\hbar}$ and each propagator has an $\hbar$ contribution. A second consequence is that U_n is graded anti-symmetric, i.e. we have

$$U_n(\ldots, \alpha_i, \ldots, \alpha_j, \ldots) = (-1)^{(p_i-1)(p_j-1)} U_n(\ldots, \alpha_j, \ldots, \alpha_i, \ldots).$$

It remains to show that U is indeed an L_∞-morphism. Note that, with S_0 given as in (6.8.55) and α_j homogeneous multivector fields of degree p_j for $j = 1, \ldots, n$, we have

$$\int \Delta \left(\exp(iS_0/\hbar) \prod_{1 \le i \le n} S_{\alpha_i} \mathcal{O}_x(f_0, \ldots, f_m) \right) = 0. \tag{6.8.58}$$

This is due to condition (1) of Theorem 6.8.50 which says that the integral of any element in the image of the BV Laplacian vanishes. We want to evaluate the left-hand-side of (6.8.58). In order to do this, we use the property of the BV Laplacian as in Exercise 6.8.47 and that $\Delta S_0 = \Delta S_\alpha = 0$. Moreover, we use that

$$(S_0, S_\alpha) \sim \int_D \int \mathrm{d}^2\theta \, \mathbf{D} \left(\alpha^{j_0 \cdots j_p}(\mathbf{X}) \eta_{j_0} \cdots \eta_{j_p} \right),$$

where $\sim$ means proportionality, and that the right-hand-side vanishes due to the boundary conditions for η_j. Therefore, we get

$$(-1)^{m-1} \int \exp(\mathrm{i}S_0/\hbar) \prod_{1 \le i \le n} S_{\alpha_i} \frac{\mathrm{i}}{\hbar} \big(S_0, \mathcal{O}_x(f_0, \ldots, f_m) \big) +$$

$$+ \int \exp(\mathrm{i}S_0/\hbar) \sum_{1 \le j < k \le n} \varepsilon_{jk}(S_{\alpha_j}, S_{\alpha_k}) \prod_{i \ne j,k} S_{\alpha_i} \mathcal{O}_x(f_0, \ldots, f_m) = 0, \qquad (6.8.59)$$

where we have used the sign

$$\varepsilon_{ij} = (-1)^{(g_1 + \cdots + g_j)g_j + (g_1 + \cdots + g_{j-1} + g_{j+1} + \cdots + g_{k-1})g_k}, \qquad g_j := p_j - 1.$$

We can observe that the BV bracket $(S_{\alpha_j}, S_{\alpha_k})$ is again of the form S_α. In particular, we have

$$(S_{\alpha_j}, S_{\alpha_k}) = -S_{[\alpha_j, \alpha_k]_{\mathrm{SN}}},$$

where $[\ ,\ ]_{\mathrm{SN}}$ denotes the Schouten–Nijenhuis bracket. Moreover, note that

$$(S_0, f(\mathbf{X}(t), \theta)) = \mathbf{D}f(\mathbf{X}(t), \theta). \qquad (6.8.60)$$

In components we have

$$(S_0, f(X(t))) = 0, \qquad (6.8.61)$$

$$(S_0, \partial_i f(X(t))\eta^{+i}(t)) = -\mathrm{d}f(X(t)). \qquad (6.8.62)$$

Note also that the BV bracket satisfies a graded Leibniz rule

$$(f, gh) = (f, g)h + (-1)^{(\mathrm{gh}(f)-1)\,\mathrm{gh}(g)} g(f, h). \qquad (6.8.63)$$

Using (6.8.60) together with (6.8.63), we can see that the integral in (6.8.57) over $\mathcal{B}_m$ will reduce to an integral over the boundary of some suitable configuration space. The boundary of $\mathcal{B}_m$ will consist of disk bubbling (or pair of disks) which appear due to the collision of points on the boundary. The connected components are obtained by distributing the points on the two disks in every possible way which is compatible with the cyclic ordering (see Fig. 6.13). Now let $S_{\ell, n-\ell}$ be the subset of the symmetric group S_n consisting of shuffle permutations as in Definition 5.2.59.

Fig. 6.13 Example of a
component of $\partial \mathcal{B}_m$. Note that
t_{m+1} would denote the point
∞. The blue point denotes
the point of collapse

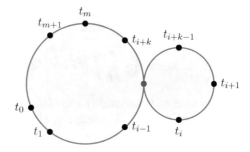

For $\sigma \in S_{\ell, n-\ell}$, we introduce the sign

$$\varepsilon(\sigma) := (-1)^{\sum_{r=1}^{\ell} g_{\sigma(r)} \left(\sum_{s=1}^{\sigma(r)-1} g_s - \sum_{s=1}^{r-1} g_{\sigma(s)} \right)}.$$

Then we will get

$$\sum_{\ell=0}^{n} \sum_{k=1}^{m-1} \sum_{i=0}^{m-k} \sum_{\sigma \in S_{\ell, n-\ell}} \varepsilon(\sigma)(-1)^{k(i+1)}(-1)^m U_\ell(\alpha_{\sigma(1)}, \ldots, \alpha_{\sigma(\ell)}) \Big(f_0 \otimes \cdots \otimes f_{i-1} \otimes$$

$$\otimes U_{n-\ell}(\alpha_{\sigma(\ell-1)}, \ldots, \alpha_{\sigma(n)})(f_i \otimes \cdots \otimes f_{i+k}) \otimes f_{i+k+1} \otimes \cdots \otimes f_m \Big)$$

$$= \sum_{i < j} \varepsilon_{ij} U_{n-1}([\alpha_i, \alpha_j], \alpha_1, \ldots, \widehat{\alpha}_i, \ldots, \widehat{\alpha}_j, \ldots, \alpha_n)(f_0 \otimes \cdots \otimes f_m). \qquad (6.8.64)$$

Thus, U is indeed an L_∞-morphism.

Remark 6.8.58 (Renormalization) A final remark concerns the regularization of
tadpoles. In the path integral approach to Kontsevich's star product we can have
tadpole diagrams which can be removed by considering a *point-split regularization*.
The problem arises when having the ill-defined factor $d\phi(z, z)$, i.e. when we
consider the superpropagator on the diagonal. To overcome this problem, we can
define $d\phi(z, z)$ as the limit

$$d\phi(z, z) = \kappa(z; \zeta) = \lim_{\epsilon \to 0} d\phi(z, z + \epsilon \zeta(z)),$$

where $\zeta(z)$ denotes a vector field on D which does not vanish in the interior of the
disk. It can be shown that this limit exists but, however, depends on the vector field
$\zeta(z)$. If we write $\zeta(z) = r(z) \exp(i\vartheta(z))$, we have

$$\kappa(z; \zeta) = d\vartheta(z).$$

The Feynman graphs will then have a finite-dimensional ambiguity. We can resolve this by adding the counterterm

$$S_{\text{c.t.}} = \frac{i\hbar}{2\pi} \int_D \int d^2\theta \, \partial_i \alpha^{ij}(\mathbf{X}) \eta_j \kappa, \quad \kappa := \theta^\mu \kappa_\mu. \tag{6.8.65}$$

For the case of the formality map $U(\alpha)$ we also have tadpole graphs which can be removed similarly. In particular, either by choosing a constant angle ϑ, or by replacing S_α by

$$\widetilde{S}_\alpha = S_\alpha - \frac{i\hbar}{2\pi} \frac{1}{p!} \int_D \int d^2\theta \, \kappa \, \partial_k \alpha^{kj_1 \cdots j_p} \eta_{j_1} \cdots \eta_{j_p}.$$

Finally, note that we can then still use the same arguments as before since

$$\Delta \widetilde{S}_\alpha = (S_0, \widetilde{S}_\alpha) = 0, \tag{6.8.66}$$

$$(\widetilde{S}_\alpha, \widetilde{S}_\beta) = -\widetilde{S}_{[\alpha,\beta]}. \tag{6.8.67}$$

6.8.5   Other Similar Constructions

The way of using quantum field theory to obtain certain mathematical constructions has been proven to be a very interesting and deep method relating seemingly purely mathematical constructions to physics. Other examples include Witten's famous construction to obtain the 4-manifold invariants described by Donaldson [Don83] using a certain deformation of the supersymmetric *Yang–Mills* action and a special type of observables [Wit88a]. Computing the expectation value of this observable with respect to the theory formulated by this special action, one can recover the Donaldson polynomials of the given 4-manifold. Another important and surprising result of Witten [Wit89] was that the expectation value of a *Wilson loop* observable representing a given knot with respect to the *Chern–Simons* action [CS74, AS91, AS94] will give the *Jones polynomial* [Jon85] of the knot, which is an important knot invariant. These results have lead to many mathematical conjectures, deeper insights in physics and created a whole new perspective towards the interplay between mathematics and quantum field theory.

6.9   A More General Approach: The AKSZ Construction

As it turns out, the Poisson sigma model can be actually regarded as a special type of a more general theory developed by Alexandrov, Kontsevich, Schwarz and Zaboronsky in [Ale+97]. More precisely, the AKSZ construction gives a method to obtain cohomological symplectic field theories compatible with the Batalin–Vilkovisky formalism, i.e. they form a subclass of Batalin–Vilkovisky theories.

6.9.1 Differential Graded Symplectic Hamiltonian Manifolds

In order to define AKSZ theories, we need the notion of a differential graded symplectic Hamiltonian manifold. Let therefore $\mathcal{X}$ be a graded manifold endowed with an exact symplectic form $\omega = \mathrm{d}\alpha$ of degree d and $\Theta \in C^\infty(\mathcal{X})$ a Hamiltonian function on $\mathcal{X}$ of degree $d + 1$ such that

$$\{\Theta, \Theta\}_\omega = 0,$$

where $\{\ ,\ \}_\omega$ denotes the Poisson bracket on $C^\infty(\mathcal{X})$ induced by ω. We will call the triple $(\mathcal{X}, \omega, \Theta)$ a differential graded (dg) symplectic Hamiltonian manifold of degree d.

6.9.2 AKSZ Sigma Models

Let Σ be a compact, connected source d-manifold, possibly with boundary, and consider its shifted tangent bundle $\Pi T\Sigma$. Moreover, let $(\mathcal{X}, \omega, \Theta)$ be a dg symplectic Hamiltonian manifold of degree $d - 1$. Then we define the *AKSZ space of fields* as

$$\mathcal{M}_{\mathrm{AKSZ}} := \mathrm{Map}(\Pi T\Sigma, \mathcal{X}).$$

On $\mathcal{M}_{\mathrm{AKSZ}}$, we can define a cohomological vector field Q_Σ by the lift of the de Rham differential on Σ, denoted by d_Σ and the Hamiltonian vector field $Q_\mathcal{X}$ defined by $\iota_{Q_\mathcal{X}}\omega = \mathrm{d}\Theta$. Namely, we can define $Q_\Sigma = \widehat{\mathrm{d}_\Sigma} + \widehat{Q_\mathcal{X}}$, where the hat denotes the lift to the mapping space. Consider now the diagram

$$\mathrm{Map}(\Pi T\Sigma, \mathcal{X}) \times \Pi T\Sigma \xrightarrow{\ \mathrm{ev}\ } \mathcal{X}$$
$$\mathrm{pr}_1 \downarrow$$
$$\mathrm{Map}(\Pi T\Sigma, \mathcal{X})$$

where ev denotes the *evaluation* map and pr_1 denotes the *projection* onto the first factor. We can then define a map $\mathscr{T} \colon \Omega^\bullet(\mathcal{X}) \to \Omega^\bullet(\mathrm{Map}(\Pi T\Sigma, \mathcal{X}))$ by setting $\mathscr{T} := (\mathrm{pr}_1)_* \mathrm{ev}^*$, where ev_* denotes the pullback of the evaluation map and pr_1 denotes integration along the fibers. The map $\mathscr{T}$ is usually called *transgression map*. Using the transgression map, we can extract a BV symplectic form ω_Σ on

$\mathcal{M}_{\text{AKSZ}}$ out of the symplectic form ω on $\mathcal{X}$ by setting

$$\omega_\Sigma = \mathcal{T}(\omega) = (\text{pr}_1)_* \text{ev}^* \omega = \int_{\Pi T \Sigma} \text{ev}^* \omega.$$

Indeed, we can see that the degree of ω_Σ is equal to $(d-1) - d = -1$ which is compatible with the degree of a BV symplectic form. Moreover, we can define an action S_{AKSZ} by using the transgression by

$$S_{\text{AKSZ}} = \iota_{\widehat{\text{d}_\Sigma}} \mathcal{T}(\alpha) + \mathcal{T}(\Theta). \tag{6.9.1}$$

If we denote the coordinates on $\mathcal{M}_{\text{AKSZ}}$ by Φ, we can rewrite the action as

$$S_{\text{AKSZ}} = \int_\Sigma \left(\alpha_i(\Phi) \text{d}_\Sigma \Phi^i + \Theta(\Phi) \right), \tag{6.9.2}$$

where α_i denote the components of the primitive 1-form α.

Theorem 6.9.1 (AKSZ[Ale+97]) *The field theory* $(\mathcal{M}_{\text{AKSZ}}, S_{\text{AKSZ}}, \omega_\Sigma, Q_\Sigma)$ *defines a BV theory, i.e. the CME is satisfied:*

$$(S_{\text{AKSZ}}, S_{\text{AKSZ}}) = 0,$$

where $(\ ,\ )$ *is the odd Poisson bracket (BV anti bracket) induced by* ω_Σ.

Theorem 6.9.2 (Cattaneo–Felder[CF01a]) *The Poisson sigma model is an AKSZ theory.*

Remark 6.9.3 Many important field theories do actually appear as AKSZ theories, such as e.g. Chern–Simons theory [AS91, AS94], Donaldson–Witten theory [Wit88a], the A- and B-twisted sigma model [Wit88b, Ale+97], $2D$ Yang–Mills theory [IM19], the Courant sigma model [Roy07, CQZ10] or (abelian) BF theory. Since they form a subclass of Batalin–Vilkovisky theories, they can be naturally extended to the case where the source manifold has boundary, similarly as in the case of deformation quantization where the Poisson sigma model is considerd on the disk. In particular, this concept can be generalized to cutting and gluing through the BV-BFV formalism [CMR14, CMR17, CM20]. Obtaining the Moyal product through gluing has been tested in [CMW17] and is expected to hold for the general case of Kontsevich's star product.

Bibliography

[AB84]   M.F. Atiyah, R. Bott, The moment map and equivariant cohomology. Topology **23**, 1–28 (1984)

[Ale+16]   A. Alekseev, C.A. Rossi, C. Torossian, T. Willwacher, Logarithms and deformation quantization. Invent. Math. **206**, 1–28 (2016)

[Ale+97]   M. Alexandrov, M. Kontsevich, A. Schwarz, O. Zaboronsky, The geometry of the master equation and topological quantum field theory. Int. J. Modern Phys. A **12**(7), 1405–1429 (1997)

[AN01]   V.I. Arnold, S.P. Novikov, *Dynamical Systems IV*. Encyclopedia of Mathematical Sciences (Springer, Berlin, 2001)

[Arn78]   V.I. Arnold, *Mathematical Methods of Classical Mechanics*. Graduate Texts in Mathematics, vol. 60 (Springer, New York, 1978)

[AS91]   S. Axelrod, I.M. Singer, Chern-Simons perturbation theory, in *Differential Geometric Methods in Theoretical Physics, Proceedings, New York*, vol. 1 (1991), pp. 3–45

[AS94]   S. Axelrod, I.M. Singer, Chern-Simons perturbation theory II. J. Differ. Geom. **39**(1), 173–213 (1994)

[Ati88]   M.F. Atiyah, Topological quantum field theories. Publ. Math. IHÉS **68**(1), 175–186 (1988)

[Bay+78a]   F. Bayen, M. Flato, C. Fronsdal, A. Lichnerowicz, D. Sternheimer, Deformation theory and quantization. I. Deformations of symplectic structures. Ann. Phys. **111**(1), 61–110 (1978)

[Bay+78b]   F. Bayen, M. Flato, C. Fronsdal, A. Lichnerowicz, D. Sternheimer, Deformation theory and quantization. II. Physical applications. Ann. Phys. **111**(1), 111–151 (1978)

[BC05]   H. Bursztyn, M. Crainic, Dirac structures, momentum maps, and quasi-Poisson manifolds, in *The Breadth of Symplectic and Poisson Geometry, Progress in Mathematics*, vol. 232 (Birkhäuser, Boston, 2005), pp. 1–40

[BCM12]   F. Bonechi, A.S. Cattaneo, P. Mnev, The Poisson sigma model on closed surfaces. J. High Energy Phys. **1**,099, 26 (2012)

[BD95]   J.C. Baez, J. Dolan, Higher dimensional algebra and topological quantum field theory. J. Math. Phys. **36**, 6073–6105 (1995)

[Ber01]   R. Berndt. *An introduction to Symplectic Geometry*. Graduate Studies in Mathematics (American Mathematical Society Providence, 2001)

[BGV92]   N. Berline, E. Getzler, M. Vergne. *Heat Kernels and Dirac Operators*. Grundlehren der Mathematischen Wissenschaften, vol. 298 (Springer, Berlin, 1992)

[Bot10]   R. Bott, Some aspects of invariant theory in differential geometry, in *Differential Operators on Manifolds* (Springer, Berlin, 2010), pp. 49–145

© The Author(s), under exclusive license to Springer Nature Switzerland AG 2022   321
N. Moshayedi, *Kontsevich's Deformation Quantization and Quantum Field Theory*,
Lecture Notes in Mathematics 2311, https://doi.org/10.1007/978-3-031-05122-7

[BRS74]   C. Becchi, A. Rouet, R. Stora, The abelian Higgs Kibble model, unitarity of the S-operator. Phys. Lett. B **52**(3), 344–346 (1974)

[BRS75]   C. Becchi, A. Rouet, R. Stora, Renormalization of the abelian Higgs-Kibble model. Commun. Math. Phys. **42**(2), 127–162 (1975)

[BRS76]   C. Becchi, A. Rouet, R. Stora, Renormalization of gauge theories. Ann. Phys. **98**(2), 287–321 (1976)

[BT82]   R. Bott, L.W. Tu, *Differential Forms in Algebraic Topology*. Springer Graduate Texts in Mathematics (Springer, Berlin, 1982)

[BV77]   I.A. Batalin, G.A. Vilkovisky, Relativistic S-matrix of dynamical systems with boson and fermion constraints. Phys. Lett. B **69**(3), 309–312 (1977)

[BV81]   I.A. Batalin, G.A. Vilkovisky, Gauge algebra and quantization. Phys. Lett. B **102**(1), 27–31 (1981)

[BV83]   I.A. Batalin, G.A. Vilkovisky, Quantization of gauge theories with linearly dependent generators. Phys. Rev. D **28**(10), 2567–2582 (1983)

[BW04]   H. Bursztyn, A. Weinstein, *Poisson geometry and Morita equivalence* (2004). arXiv: math/0402347

[BW12]   S. Bates, A. Weinstein, *Lectures on the Geometry of Quantization*. University Reprints (American Mathematical Society, Providence, 2012)

[Cal+17] D. Calaque, T. Pantev, B. Toën, M. Vaquié, G. Vezzosi, Shifted Poisson structures and deformation quantization. J. Topol. **10**(2), 483–584 (2017)

[Can08]   A. Cannas da Silva, *Lectures on Symplectic Geometry*. Lecture Notes in Mathematics (Springer, Berlin, 2001). Corrected 2nd Printing (2008)

[Cat+05] A.S. Cattaneo, B. Keller, C. Torossian, A. Bruguières, *Déformation, Quantification, Théorie de Lie, Panoramas et Syntheses 20* (Société Mathématique de France, 2005)

[Cat04]   A.S. Cattaneo, On the Integration of Poisson Manifolds, Lie Algebroids, and Coisotropic Submanifolds. Lett. Math. Phys. **67**, 33–48 (2004)

[Cat18]   A.S. Cattaneo, *Notes on Manifolds* (2018). Available here 2018

[CF00]   A.S. Cattaneo, G. Felder, A path integral approach to the Kontsevich quantization formula. Commun. Math. Phys. **212**, 591–611 (2000)

[CF01a]   A.S. Cattaneo, G. Felder, On the AKSZ formulation of the Poisson sigma model. Lett. Math. Phys. **56**(2), 163–179 (2001)

[CF01b]   A.S. Cattaneo, G. Felder, On the globalization of Kontsevich's star product and the perturbative Poisson sigma model. Progre. Theor. Phys. Suppl. **144**, 38–53 (2001)

[CF01c]   A.S. Cattaneo, G. Felder, Poisson sigma models and deformation quantization. Mod. Phys. Lett. A **16**, 179–190 (2001)

[CF01d]   A.S. Cattaneo, G. Felder, Poisson sigma models and symplectic groupoids, in *Quantization of Singular Symplectic Quotients* (Birkhäuser Basel, 2001), pp. 61–93

[CF03]   M. Craininc, R.L. Fernandes, Integrability of lie brackets. Ann. Math. **157**(2), 575–620 (2003)

[CF04]   M. Crainic, R.L. Fernandes, Integrability of Poisson brackets. J. Differ. Geom. **66**, 71–137 (2004)

[CF10]   A.S. Cattaneo, G. Felder, Effective Batalin–Vilkovisky theories, equivariant configuration spaces and cyclic chains. Higher Struct. Geomet. Phys. **287**, 111–137 (2010)

[CFS92]   A. Connes, M. Flato, D. Sternheimer, Closed star products and cyclic cohomology. Lett. Math. Phys. **24**(1), 1–12 (1992)

[CFT02a] A.S. Cattaneo, G. Felder, L. Tomassini, Fedosov connections on jet bundles and deformation quantization, in *Halbout G. Deformation Quantization* (2002), pp. 191–202

[CFT02b] A.S. Cattaneo, G. Felder, L. Tomassini, From local to global deformation quantization of Poisson manifolds. Duke Math. J. **115**(2), 329–352 (2002)

[CI05]   A.S. Cattaneo, D. Indelicato, Formality and star products, in *Poisson Geometry, Deformation Quantisation and Group Representations*, ed. by S. Gutt, J. Rawnsley, D. Sternheimer. London Mathematical Society Lecture Note Series, vol. 323 (Cambridge University Press, Cambridge, 2005), pp. 79–144

[CM20]   A.S. Cattaneo, N. Moshayedi, Introduction to the BV-BFV formalism. Rev. Math. Phys. **32**, 67 (2020)

[CMR14]   A.S. Cattaneo, P. Mnev, N. Reshetikhin, Classical BV theories on manifolds with boundary. Commun. Math. Phys. **332**(2), 535–603 (2014)

[CMR17]   A.S. Cattaneo, P. Mnev, N. Reshetikhin, Perturbative quantum gauge theories on manifolds with boundary. Commun. Math. Phys. **357**(2), 631–730 (2017)

[CMW17]   A.S. Cattaneo, N. Moshayedi, K. Wernli, Relational symplectic groupoid quantization for constant Poisson structures. Lett. Math. Phys. **107**(9), 1649–1688 (2017)

[CMW19]   A.S. Cattaneo, N. Moshayedi, K. Wernli, Globalization for perturbative quantization of nonlinear split AKSZ sigma models on manifolds with boundary. Commun. Math. Phys. **372**(1), 213–260 (2019)

[CMW20]   A.S. Cattaneo, N. Moshayedi, K. Wernli, On the globalization of the Poisson sigma model in the BV-BFV formalism. Commun. Math. Phys. **375**(1), 41–103 (2020)

[Con85]   A. Connes, Non-commutative differential geometry. Publ. Math. IHÉS **62**, 41–144 (1985)

[Con95]   J. Conn, Normal forms for smooth Poisson structures. Ann. of Math. **121**, 565–593 (1995)

[Cou90a]   T. Courant, Dirac manifolds. Trans. Amer. Math. Soc. **319**(2), 631–661 (1990)

[Cou90b]   T. Courant, Tangent Dirac structures. J. Phys. A. **23**(22), 5153–5168 (1990)

[CQZ10]   A.S. Cattaneo, J. Qiu, M. Zabzine, 2D and 3D topological field theories for generalized complex geometry. Adv. Theor. Math. Phys. **14**(2), 695–725 (2010)

[CS74]   S.-S. Chern, J. Simons, Characteristic forms and geometric invariants. Ann. Math. **99**(1), 48–69 (1974)

[Dar82]   G. Darboux, Sur le problème de Pfaff. Bull. Sci. Math. **6**, 14–36, 49–68 (1882)

[DeD30]   T. DeDonder, *Théorie invariantive du calcul des variations* (Gauthier-Villars, Paris, 1930)

[Del95]   P. Deligne, Déformations de l'algèbre des fonctions d'une variété symplectique: comparaison entre Fedosov et De Wilde, Lecomte. Sel. Math. New Ser. **1**(4), 667–697 (1995)

[Dir30]   P.A.M. Dirac. *The Principles of Quantum Mechanics* ((Oxford University Press, Oxford, 1930)

[DK00]   J.J. Duistermaat, J.A.C. Kolk. *Lie Groups*. Universitext (Springer, Berlin, 2000)

[DK90]   S.K. Donaldson, P. Kronheimer, *The Geometry of Four-Manifolds*. Oxford Mathematical Monographs (The Clarendon Press, Oxford University Press, New York, 1990)

[DL83]   M. DeWilde, P.B.A. Lecomte, Existence of star-products and of formal deformations of the Poisson Lie algebra of arbitrary symplectic manifolds. Lett. Math. Phys. **7**(6), 487–496 (1983)

[Dol05]   V. Dolgushev, Covariant and equivariant formality theorems. Adv. Math. **191**(1), 147–177 (2005)

[Don83]   S.K. Donaldson, An application of gauge theory to four dimensional topology. J. Differ. Geom. **18**(2), 279–315 (1983)

[Don96]   S.K. Donaldson, Symplectic submanifolds and almost-complex geometry. J. Differ. Geom. **44**, 666–705 (1996)

[DZ05]   J.-P. Dufour, N.T. Zung, *Poisson Structures and Their Normal Forms*. Progress in Mathematics, vol. 242 (Birkhäuser, Basel, 2005)

[EG98]   Y. Eliashberg, M. Gromov, Lagrangian intersection theory: finite-dimensional approach. Amer. Math. Soc. Transl. **186**(2), 27–118 (1998)

[Fed94]   B.V. Fedosov, A simple geometrical construction of deformation quantization. J. Differ. Geom. **40**(2), 213–238 (1994)

[Fed96]   B.V. Fedosov, *Deformation Quantization and Index Theory*. Mathematical Topics, vol. 9 (Akademie Verlag, Berlin, 1996), p. 325

[Fer00]   R.L. Fernandes, Connections in Poisson geometry I: holonomy and invariants. J. Differ. Geom. **54**, 303–366 (2000)

[Fer02]   R.L. Fernandes, Lie algebroids, holonomy and characteristic classes. Adv. Math. **170**, 119–179 (2002)

[Fey42]   R.P. Feynman, The principle of least action in quantum mechanics. Thesis (Ph.D.), Department of Physics, Princeton University, Princeton, NJ, 1942

[Fey49]   R.P. Feynman, Space-time approach to quantum electrodynamics. Phys. Rev. **76**(6), 769–789 (1949)

[Fey50]   R.P. Feynman, Mathematical formulation of the quantum theory of electro-magnetic interaction. Phys. Rev. **80**(3), 440–457 (1950)

[FH65]   R.P. Feynman, A.R. Hibbs, *Quantum Mechanics and Path Integrals*. International Series in Pure and Applied Physics (McGraw-Hill, New York, 1965)

[FM94]   W. Fulton, R. MacPherson, A compactification of configuration spaces. Ann. Math. **139**(1), 183–225 (1994)

[FP67]   L.D. Faddeev, V.N. Popov, Feynman diagrams for the Yang-Mills field. Phys. Lett. B **25**(1), 29–30 (1967)

[Fro77]   G. Frobenius, Über das Pfaffsche problem. J. für Reine und Angew. Math. **8**, 230–315 (1877)

[Ger63]   M. Gerstenhaber, The cohomology structure of an associative ring. Ann. Math. **78**(2), 267–288 (1963)

[GF69]   I.M. Gelfand, D.B. Fuks, The cohomology of the Lie algebra of vector fields on a smooth manifold. J. Funct. Analy. **33**, 194–210 (1969)

[GF70]   I.M. Gelfand, D.B. Fuks, The cohomology of the Lie algebra of formal vector fields. Izv. AN SSR **34**, 110–116 (1970)

[GG01]   V. Ginzburg, A. Golubev, Holonomy on Poisson manifolds and the modular class. Israel J. Math. **122**, 221–242 (2001)

[GK71]   I.M. Gelfand, D.A. Kazhdan, Some problems of the differential geometry and the calculation of cohomologies of Lie algebras of vector fields. Dokl. Akad. Nauk Ser. Fiz. **200**, 269–272 (1971)

[GL92]   V.L. Ginzburg, J.-H. Lu, Poisson cohomology of Morita-equivalent Poisson manifolds. Internat. Math. Res. Notices **10**, 199–205 (1992)

[Got82]   M. Gotay, On coisotropic imbeddings of presymplectic manifolds. Proc. Amer. Math. Soc. **84**, 111–114 (1982)

[GR99]   S. Gutt, J. Rawnsley, Equivalence of star products on a symplectic manifold; an introduction to Deligne's Čech cohomology classes. J. Geom. Phys. **29**(4), 347–392 (1999)

[Gro46]   H.J. Groenewold, On the principles of elementary quantum mechanics. Physics **12**, 405–460 (1946)

[Gro68]   A. Grothendieck, Crystals and the de Rham cohomology of schemes. In: *Dix exposes sur la cohomologie des schemas* (1968), pp. 306–358

[GRS05]   S. Gutt, J. Rawnsley, D. Sternheimer, *Poisson Geometry, Deformation Quantisation and Group Representations*. London Mathemaical Society Lecture Notes Series, vol. 323 (Cambridge University Press, Cambridge, 2005)

[GS77]   V. Guillemin, S. Sternberg. *Geometric Asymptotics*. Mathematical Surveys and Monographs, vol. 14 (American Mathematical Society, Providence, 1977)

[Gut05]   S. Gutt, Deformation Quantization: an introduction, in *3rd Cycle, Monastir (Tunisie)* (2005), p. 60

[GW92]   V. Ginzburg, A. Weinstein, Lie-Poisson structures on some Poisson Lie groups. J. Amer. Math. Soc. **5**, 445–453 (1992)

[Her61]   H. Hermann, *Zur allgemeinen Theorie der Bewegung der Flüssigkeiten* Göttingen, Dieterische Universität Buchdruckerei (1861)

[HKR62]   G. Hochschild, B. Kostant, A. Rosenberg, Differential forms on regular affine algebras. Trans. Amer. Math. Soc. **2**, 383–408 (1962)

[Ike94]   N. Ikeda, Two-dimensional gravity and nonlinear Gauge theory. Ann. Phys. **235**(2), 435–464 (1994)

[IM19]   R. Iraso, P. Mnev, Two-dimensional Yang-Mills theory on surfaces with corners in Batalin-Vilkovisky formalism. Commun. Math. Phys. **370**, 637–702 (2019)

[Jon85]   V.F.R. Jones, A polynomial invariant for knots via von Neumann algebras. Bull. Amer. Math. Soc. **12**(1), 103–112 (1985)

[JSW02]   B. Jurco, P. Schupp, J. Wess, Noncommutative line bundle and Morita equivalence. Lett. Math. Phys. **61**, 171–186 (2002)

[Kac77]   V.G. Kac, Lie superalgebras. Adv. Math. **26**(1), 8–96 (1977)

[Kat70]   N. Katz, Nilpotent connections and the monodromy theorem: applications of a result of Turrittin. Publ. Mathématiques de l'IHÉS **39**, 175–232 (1970)

[Kat79]   V.J. Katz, The history of Stokes' theorem. Math. Mag. **52**(3), 146–156 (1979)

[Khu04]   H.M. Khudaverdian, Semidensities on odd symplectic supermanifolds. Commun. Math. Phys. **247**(2), 353–390 (2004)

[Kir85]   A. Kirillov, Geometric quantization. Dyn. Syst. **4**(4), 141–176 (1985)

[KN63]   S. Kobayashi, K. Nomizu, *Foundations of Differential Geometry, Vol 2*. Interscience Tracts in Pure and Applied Mathematics, vol. 1 (Interscience Publishers, New York, 1963)

[KN69]   S. Kobayashi, K. Nomizu, *Foundations of Differential Geometry, vol 2.*. Interscience Tracts in Pure and Applied Mathematics, vol. 1 (Interscience Publishers, New York, 1969)

[Kon03]   M. Kontsevich, Deformation quantization of Poisson manifolds. Lett. Math. Phys. **66**(3), 157–216 (2003)

[Kon94]   M. Kontsevich, Feynman diagrams and low-dimensional topology. English. in *First European Congress of Mathematics Paris, July 6–10, 1992*. ed. by A. Joseph, F. Mignot, F. Murat, B. Prum, R. Rentschler. Progress in Mathematics, vol. 120 (Birkhäuser, Basel, 1994), pp. 97–121

[Kon99]   M. Kontsevich, Operads and motives in deformation quantization. Lett. Math. Phys. **48**, 35–72 (1999)

[KS00]   M. Kontsevich, Y. Soibelman, Deformations of algebras over operads and Deligne's conjecture, in *Proceedings, Conference Moshe Flato : Quantization, Deformations, and Symmetries* I/II (2000), pp. 255–308

[KSS05]   A. Kotov, P Schaller, T. Strobl, Dirac sigma models. Commun. Math. Phys. **260**, 455–480 (2005)

[KT79]   J. Kijowski, W.M. Tulczyjew, *A Symplectic Framework for Field Theories*. Springer Lecture Notes in Physics (Springer, Berlin, 1979)

[Lee02]   J.M. Lee, *Introduction to Smooth Manifolds*. Springer Graduate Texts in Mathematics (Springer, Berlin, 2002)

[Lic77]   A. Lichnerowicz, Les variétés de Poisson et leurs algèbres de Lie associées". J. Differ. Geom. **12**, 253–300 (1977)

[Lio55]   J. Liouville, Note sur l'intégration des équations différentielles de la Dynamique. J. de Mathématiques Pures et Appliquées **20**, 137–138 (1855)

[LM95]   T. Lada, M. Markl, Strongly homotopy Lie algebras. Comm. Algebra **23**(6), 2147–2161 (1995)

[LS93]   T. Lada, J. Stasheff, Introduction to SH Lie algebras for physicists. Int. J. Theo. Phys. **32**, 1087–1103 (1993)

[Lur09]   J. Lurie, On the classification of topological field theories. Curr. Dev. Math. **2008**, 129–280 (2009)

[Lur17]   J. Lurie. *Higher Algebra* (2017). Available at https://www.mathiasedu/~lurie/

[Mac71]   S. Mac Lane. *Categories for the Working Mathematician*. Graduate Text in Mathematics. (Springer Science+Business Media, Berlin, 1971)

[Man58]   Y.I. Manin, Algebraic curves over fields with differentiation. Izv. Akad. Nauk SSSR Ser. Mat. **22**(6), 737–756 (1958)

[Max73]   J.C. Maxwell, *A Treatise on Electricity and Magnetism*, vol. 1 (The Clarendon Press, Oxford University Press, Oxford, 1873)

[MF78]   A.S. Mishchenko, A.T. Fomenko, Generalized Liouville method of integration of Hamiltonian systems. Funct. Anal. Appl. **12**, 113–121 (1978)

[Mne19]   P. Mnev, *Quantum Field Theory: Batalin–Vilkovisky Formalism and Its Applications*. University Lecture Series, vol. 72 (American Mathematical Society (AMS), Providence, 2019), p. 192

[Mor58]   K. Morita, Duality for modules and its applications to the theory of rings with minimum condition. Sci. Rep. Tokyo Kyoiku Daigaku Sect. A **6**, 83–142 (1958)

[Mos19]   N. Moshayedi, On globalized traces for the Poisson sigma model. Commun. Math. Phys. **393**, 583–629 (2022)

[Mos65]   J. Moser, On the volume elements on a manifold. Trans. Amer. Math. Soc. **120**, 286–294 (1965)

[Moy49]   J.E. Moyal, Quantum mechanics as a statistical theory. Math. Proc. Camb. Philos. Soc. **45**(01), 99 (1949)

[MS95]   D. McDuff, D. Salamon, *Introduction to Symplectic Topology*. Oxford Mathematical Monographs (Oxford University Press, New York, 1995)

[MS99]   J.E. McClure, J.H. Smith, *A Solution of Deligne's conjecture* (1999). arXiv: math/9910126

[MW74]   J. Marsden, A. Weinstein, Reduction of symplectic manifolds with symmetry. Rep. Math. Phys. **5**, 121–130 (1974)

[MX00]   K. Mackenzie, P. Xu, Integration of Lie bialgebroids. Toplogy **39**, 445–467 (2000)

[Noe18]   E. Noether, Invarianten beliebiger Differentialausdrücke. Gött. Nachr. **1918**, 37–44 (1918)

[NT95]   R. Nest, B. Tsygan, Algebraic index theorem. Commun. Math. Phys. **172**(2), 223–262 (1995)

[Pan+13]   T. Pantev, B. Toën, M. Vaquié, G. Vezzosi, Shifted symplectic structures. Publ. Mathématiques de l'IHÉS **117**, 271–328 (2013)

[Pol05]   M. Polyak, Feynman diagrams for pedestrians and mathematicians. Proc. Symp. Pure Math. **73**, 15–42 (2005)

[Rha31]   G. de Rham, Sur l'analysis situs des variétés à n dimensions. J. de Mathématiques Pures et Appliquées **10**, 115–200 (1931)

[Rie74]   M.A. Rieffel, Morita equivalence for C^*-algebras and W^*-algebras. J. Pure Appl. Algebra **5**(1), 51–96 (1974)

[Roy07]   D. Roytenberg, AKSZ–BV formalism and courant algebroid-induced topological field theories. Lett. Math. Phys. **79**, 143–159 (2007)

[RT91]   N.Y. Reshetikhin, V.G. Turaev, Invariants of 3-manifolds via link polynomials and quantum groups. Invent. Math. **103**(1), 547–597 (1991)

[Sch93]   A. Schwarz, Geometry of Batalin-Vilkovisky quantization. Commun. Math. Phys. **155**(2), 249–260 (1993)

[Sch98]   A. Schwarz, Morita equivalence and duality. Nuclear Phys. B **534**, 720–738 (1998)

[Seg88]   G.B. Segal, The definition of conformal field theory. Diff. Geom. Methods Theor. Phys. **250**, 165–171 (1988)

[Šev05]   P. Ševera, *On Deformation Quantization of Dirac Structures* (2005). arXiv: math/0511403

[SS94]   P. Schaller, T. Strobl, Poisson structure induced (topological) field theories. Mod. Phys. Lett. A **09**(33), 3129–3136 (1994)

[SS95]   P. Schaller, T. Strobl, Introduction to Poisson sigma models, in *Low-Dimensional Models in Statistical Physics and Quantum Field Theory*, ed. by H. Grosse, L. Pittner (Springer, Berlin, 1995), pp. 321–333

[Sta92]   J. Stasheff, Differential graded Lie algebras, quasi-Hopf algebras and higher homotopy algebras, in *Quantum Groups*. Lecture Notes in Mathematics, vol. 1510 (Springer, Berlin, 1992)

[Tam03]   D.E. Tamarkin, Formality of chain operad of little discs. Lett. Math. Phys. **66**, 65–72 (2003)

[Tam98]   D.E. Tamarkin, *Another Proof of M. Kontsevich Formality Theorem* (1998). arXiv: math/9803025

[Tu17]   L.W. Tu, *Differential Geometry*. Graduate Text in Mathematics (Springer, Berlin, 2017)

[Tyu76]   I.V. Tyutin, Gauge invariance in field theory and statistical physics in operator formalism. *Preprints of P.N. Lebedev Physical Institute, No. 39* (1976)

[Var04]   V.S. Varadarajan, *Supersymmetry for Mathematicians: An Introduction*. Courant Lecture Notes, vol. 11 (American Mathematical Society, Providence, 2004)

[Wal07]   S. Waldmann, *Poisson-Geometrie und Deformierungsquantisierung* (Springer, Berlin, 2007)

[War83]   F.W. Warner, *Foundations of Differentiable Manifolds and Lie Groups*. Graduate Texts in Mathematics (Springer, New York, 1983)

[Wei71]   A. Weinstein, Symplectic manifolds and their Lagrangian submanifolds. Adv. Math. **6**, 329–346 pp

[Wei77]   A. Weinstein, *Lectures on Symplectic Manifolds*. Regional Conference Series in Mathematics, vol. 29 (American Mathematical Society, Providence, 1977)

[Wei81]   A. Weinstein, Neighborhood classification of isotropic embeddings. J. Differ. Geom. **16**, 125–128 (1981)

[Wei83]   A. Weinstein, The local structure of Poisson manifolds. J. Differ. Geom. **18**, 523–557 (1983)

[Wey31]   H. Weyl, *The Theory of Groups and Quantum Mechanics. Dover, New York, translated from* Quantenmechanik und Gruppentheorie *Z. Physik (1927)*, vol. 46 (1931), pp. 1–46

[Wey35]   H. Weyl, Geodesic fields in the calculus of variation for multiple integrals. Ann. Math. **36**(3), 607–629 (1935)

[Whi34]   H. Whitney, Analytic extensions of functions defined in closed sets. Trans. Amer. Math. Soc. **36**(1), 63–89 (1934)

[Wig32]   E.P. Wigner, Quantum corrections for thermodynamic equilibrium. Phys. Rev. **40**, 749–759 (1932)

[Wit88a] E. Witten, Topological quantum field theory. Commun. Math. Phys. **117**(3), 353–386 (1988)

[Wit88b] E. Witten, Topological sigma models. Commun. Math. Phys. **118**(3), 411–449 (1988)

[Wit89]   E. Witten, Quantum field theory and the Jones polynomial. Commun. Math. Phys. **121**(3), 351–399 (1989)

[Woo97]   N. Woodhouse, *Geometric Quantization* (Oxford University Press (OUP), Oxford, 1997)

[Xu04]   P. Xu, Momentum maps and morita equivalence. J. Differ. Geom. **67**, 289–333 (2004)

[Xu91]   P. Xu, Morita equivalence and poisson manifolds. Commun. Math. Phys. **142**, 493–509 (1991)

[Zin94]   J. Zinn-Justin, *Quantum Field Theory and Critical Phenomena*. International Series of Monographs on Physics, vol. 85 (Oxford Science Publications, The Clarendon, Oxford University Press, New York, 1994)

Index

LECTURE NOTES IN MATHEMATICS

Editors in Chief: J.-M. Morel, B. Teissier;

Editorial Policy

1. Lecture Notes aim to report new developments in all areas of mathematics and their applications – quickly, informally and at a high level. Mathematical texts analysing new developments in modelling and numerical simulation are welcome.

   Manuscripts should be reasonably self-contained and rounded off. Thus they may, and often will, present not only results of the author but also related work by other people. They may be based on specialised lecture courses. Furthermore, the manuscripts should provide sufficient motivation, examples and applications. This clearly distinguishes Lecture Notes from journal articles or technical reports which normally are very concise. Articles intended for a journal but too long to be accepted by most journals, usually do not have this "lecture notes" character. For similar reasons it is unusual for doctoral theses to be accepted for the Lecture Notes series, though habilitation theses may be appropriate.

2. Besides monographs, multi-author manuscripts resulting from SUMMER SCHOOLS or similar INTENSIVE COURSES are welcome, provided their objective was held to present an active mathematical topic to an audience at the beginning or intermediate graduate level (a list of participants should be provided).

   The resulting manuscript should not be just a collection of course notes, but should require advance planning and coordination among the main lecturers. The subject matter should dictate the structure of the book. This structure should be motivated and explained in a scientific introduction, and the notation, references, index and formulation of results should be, if possible, unified by the editors. Each contribution should have an abstract and an introduction referring to the other contributions. In other words, more preparatory work must go into a multi-authored volume than simply assembling a disparate collection of papers, communicated at the event.

3. Manuscripts should be submitted either online at www.editorialmanager.com/lnm to Springer's mathematics editorial in Heidelberg, or electronically to one of the series editors. Authors should be aware that incomplete or insufficiently close-to-final manuscripts almost always result in longer refereeing times and nevertheless unclear referees' recommendations, making further refereeing of a final draft necessary. The strict minimum amount of material that will be considered should include a detailed outline describing the planned contents of each chapter, a bibliography and several sample chapters. Parallel submission of a manuscript to another publisher while under consideration for LNM is not acceptable and can lead to rejection.

4. In general, **monographs** will be sent out to at least 2 external referees for evaluation.

   A final decision to publish can be made only on the basis of the complete manuscript, however a refereeing process leading to a preliminary decision can be based on a pre-final or incomplete manuscript.

   Volume Editors of **multi-author works** are expected to arrange for the refereeing, to the usual scientific standards, of the individual contributions. If the resulting reports can be forwarded to the LNM Editorial Board, this is very helpful. If no reports are forwarded or if other questions remain unclear in respect of homogeneity etc, the series editors may wish to consult external referees for an overall evaluation of the volume.

5. Manuscripts should in general be submitted in English. Final manuscripts should contain at least 100 pages of mathematical text and should always include

   – a table of contents;
   – an informative introduction, with adequate motivation and perhaps some historical remarks: it should be accessible to a reader not intimately familiar with the topic treated;
   – a subject index: as a rule this is genuinely helpful for the reader.
   – For evaluation purposes, manuscripts should be submitted as pdf files.

6. Careful preparation of the manuscripts will help keep production time short besides ensuring satisfactory appearance of the finished book in print and online. After acceptance of the manuscript authors will be asked to prepare the final LaTeX source files (see LaTeX templates online: https://www.springer.com/gb/authors-editors/book-authors-editors/manuscriptpreparation/5636) plus the corresponding pdf- or zipped ps-file. The LaTeX source files are essential for producing the full-text online version of the book, see http://link.springer.com/bookseries/304 for the existing online volumes of LNM). The technical production of a Lecture Notes volume takes approximately 12 weeks. Additional instructions, if necessary, are available on request from lnm@springer.com.

7. Authors receive a total of 30 free copies of their volume and free access to their book on SpringerLink, but no royalties. They are entitled to a discount of 33.3 % on the price of Springer books purchased for their personal use, if ordering directly from Springer.

8. Commitment to publish is made by a *Publishing Agreement*; contributing authors of multiauthor books are requested to sign a *Consent to Publish form*. Springer-Verlag registers the copyright for each volume. Authors are free to reuse material contained in their LNM volumes in later publications: a brief written (or e-mail) request for formal permission is sufficient.

Addresses:
Professor Jean-Michel Morel, CMLA, École Normale Supérieure de Cachan, France
E-mail: moreljeanmichel@gmail.com

Professor Bernard Teissier, Equipe Géométrie et Dynamique,
Institut de Mathématiques de Jussieu – Paris Rive Gauche, Paris, France
E-mail: bernard.teissier@imj-prg.fr

Springer: Ute McCrory, Mathematics, Heidelberg, Germany,
E-mail: lnm@springer.com

Printed in the United States
by Baker & Taylor Publisher Services